Planning and Installing Sustainable Onsite Wastewater Systems

About the Author

S. M. Parten is a licensed professional engineer with over 25 years of experience researching and designing decentralized and centralized wastewater systems. Since 1992, Parten has owned and operated Community Environmental Services, Inc., a civil engineering firm specializing in the design and management of decentralized wastewater systems. In addition to authoring a number of technical articles, Parten was a member of the Water Environment Federation's Technical Committee Task Force for Natural Systems for Wastewater Treatment, developed Fact Sheets for onsite wastewater treatment systems currently referenced in the U.S. EPA's *Onsite Wastewater Treatment Systems Manual*, and was principal investigator for a nationwide Water Environment Research Foundation (WERF) study published in 2008 on the performance of large/community-sized decentralized wastewater systems in the United States. Recently, Parten has served as a consultant to onsite wastewater regulatory authorities in several U.S. states and territories, assisting with the development of improved standards and management practices.

Planning and Installing Sustainable Onsite Wastewater Systems

S. M. Parten, P.E.

New York Chicago San Francisco
Lisbon London Madrid Mexico City
Milan New Delhi San Juan
Seoul Singapore Sydney Toronto

The McGraw·Hill Companies

Cataloging-in-Publication Data is on file with the Library of Congress

Copyright © 2010 by S. M. Parten. All rights reserved. Printed in the United States of America. Except as permitted under the United States Copyright Act of 1976, no part of this publication may be reproduced or distributed in any form or by any means, or stored in a database or retrieval system, without the prior written permission of the publisher.

1 2 3 4 5 6 7 8 9 0   DOC/DOC   0 1 5 4 3 2 1 0 9

ISBN 978-0-07-162463-3
MHID 0-07-162463-5

Sponsoring Editor
 Joy Bramble

Editing Supervisor
 Stephen M. Smith

Production Supervisor
 Richard C. Ruzycka

Acquisitions Coordinator
 Michael Mulcahy

Project Manager
 Somya Rustagi,
 Glyph International

Copy Editor
 Surendra Nath Shivam,
 Glyph International

Proofreader
 Anju Panthari,
 Glyph International

Art Director, Cover
 Jeff Weeks

Composition
 Glyph International

Printed and bound by RR Donnelley.

McGraw-Hill books are available at special quantity discounts to use as premiums and sales promotions, or for use in corporate training programs. To contact a representative, please e-mail us at bulksales@mcgraw-hill.com.

This book is printed on acid-free paper.

Information contained in this work has been obtained by The McGraw-Hill Companies, Inc. ("McGraw-Hill") from sources believed to be reliable. However, neither McGraw-Hill nor its authors guarantee the accuracy or completeness of any information published herein, and neither McGraw-Hill nor its authors shall be responsible for any errors, omissions, or damages arising out of use of this information. This work is published with the understanding that McGraw-Hill and its authors are supplying information but are not attempting to render engineering or other professional services. If such services are required, the assistance of an appropriate professional should be sought.

To those teachers and practitioners who've
committed themselves to making information as
accessible and understandable as possible to their
students and the public, and to the memory of
Sherwood (Woody) Reed, P.E., who exemplified
that spirit of generosity in his distinguished
career as a natural treatment expert.

Contents

Forewords

Water management scenarios have been dominated by either a centralized collection, treatment, and discharge model or an individual septic tank as a temporary solution model. This is changing. Receiving waters accepting treated sewage are compromised, and soil-based systems are becoming increasingly adapted to the use of a limiting constituent model in permitting and design.

The land limiting or LDP approach described in this work provides practitioners at all levels with sound tools to better site, size, design, install, operate, manage, and maintain onsite and decentralized wastewater systems as the permanent and essential element of the wastewater infrastructure they represent. These land-based systems have been utilized for over 100 years, mostly in rural areas. That model is changing. Today, these systems are utilized increasingly in suburban and urban fringe areas in part because of limitations associated with receiving waters.

Do we know all there is to know regarding this decentralized infrastructure, all we need to know about soil loadings and retention times to achieve pollutant removals in the plant soil system? Probably not. Do we know about assimilative capacities in receiving streams? Probably not, and that has not stopped installations of centralized infrastructure with associated stream discharge.

This book provides a rational model for assessing assimilative capacities of soil systems, and this model should serve as a foundation for effective rules and regulations to enable the better use of decentralized systems.

Robert Rubin, Professor Emeritus
Biological and Agricultural Engineering
North Carolina State University, and
Senior Environmental Scientist, McKim & Creed

These days project managers have to be something no other team member is: the "inspired generalist." Whether an architect, engineer, or construction manager, the inspired generalist is the team member who must know a good deal about almost everything being built or installed into a successful construction project. In these days of specialists for everything from low-voltage wiring systems to waste management, it is imperative that there be a symphony conductor who can bring it all together so that the result is a system of complimenting parts. As architects and engineers move into project management, positions, and situations where coordination of various elements of small and large projects is needed, it's important that there be an understanding of some of the concepts that are fundamental to sustainable wastewater systems planning and implementation.

This book is a "must have" for persons wanting a practical reference guide for designing and constructing decentralized systems, including the inspired generalist. I know of no other book on this subject that so effectively links essential scientific principles and practical design and construction details needed for sound and sustainable systems. The author breaks down a very complicated and complex subject into bite-sized chunks of useful information that is logically laid out and easy to absorb—whether it be for the construction sciences student, design professional, permit reviewer, or construction project manager. Unlike what we often end up finding when we do Internet searches for information on a particular subject, what's presented here is not "naïve source knowledge." It is real, accurate, and vetted thanks to the author's important combination of deep subject-matter knowledge, education, and many years of relevant practical field experience.

The book presents many complex scientific principles integral to sustainable wastewater planning in a way that's accessible for practical application. The many photo illustrations of methods and materials for construction help greatly to inform the reader. Most college texts and professional books don't adequately link academically based technical information with step-by-step practical information needed to implement those concepts in a sound manner. The book is also written so that the basic concepts are understandable to persons not necessarily having significant technical training in these areas of science and engineering.

While it's clearly impossible to anticipate and evaluate all of the constantly evolving decentralized wastewater systems, the book walks readers through a logical progression of explanations for key factors to consider relative to their long-term sustainability. That same basic approach can be used for comparing new technologies and processes as they appear on the market and become available for consideration. The comparative tables presented in several of the chapters are very helpful as quick reference guides. The consumer-style rating of wastewater treatment processes commonly employed for decentralized systems in Chap. 6 serves as a very good tool for making basic comparisons of system types in a wide range of settings and conditions.

Last, but not least, the inspired generalist has to have the ability to be creative when attacking the inevitable project challenges. To that end, the examples included in the final chapter of the book present a diverse array of projects and site conditions that promote creative thinking about the use of natural treatment processes and recycled materials that compliment the very important notion of "sustainable" architecture and development.

Peter L. Pfeiffer, FAIA, LEED AP
Principal, Barley & Pfeiffer Architects

Preface

Engineers and project planners dealing with decentralized, or "onsite," wastewater systems are routinely asked by property owners how to choose the most suitable and cost-effective systems. Complete answers to those questions are often fairly complex, and based on a wide range of engineering principles and science spanning several technical disciplines. Most engineers leave their formal training in engineering colleges without having had the diverse training needed for designing sound and sustainable decentralized systems of varying sizes and in differing geophysical conditions. As the body of knowledge within each technical discipline increases, we see more and more specialization among scientists and engineers, making it harder for practitioners to obtain the training and experience needed to keep up with multiple specialized areas within each major discipline. Even within a single major technical discipline such as civil engineering, persons having a thorough understanding of and high level of competence with larger centralized wastewater systems typically don't keep up with methods, materials, and state-of-the-art industry practices applicable to small- to midsized decentralized wastewater systems.

Questions from property owners and developers often focus on the merits and cost-effectiveness of obtaining centralized service, if available, versus implementing an appropriate decentralized wastewater solution. While many of the basic scientific principles routinely used for municipal-scale wastewater systems are applicable to small-scale systems, methods and materials found to work most effectively for smaller-scale systems are often quite different from those used successfully for large-scale systems. It usually takes many years of practice for engineers and planners to be prepared to help guide project owners with implementing the most sound and cost-effective wastewater service approaches.

Onsite, or decentralized, wastewater treatment systems industry practices around the world have tended to evolve based on a mixture of local geographic, socioeconomic, educational, and cultural conditions. In geopolitical areas where rules are in place for the design and construction of those systems, such rules have often developed around empirical observations and local influences, including design practitioner familiarity with and bias toward certain methods and materials. Rules and practices are also often greatly influenced by vendors of certain products lobbying for a significant share in local markets.

Larger municipal-scale wastewater systems in the United States and many other countries have historically been much more strongly influenced by nationally and internationally vetted sources of information for their design, construction, and management. University/college engineering programs and other wastewater industry educational curricula guiding the implementation of larger centralized systems tend to draw upon a fairly well-organized body of scientific research and literature on a myriad of wastewater treatment systems issues. It is not surprising to find more "centralized" bodies of technical information for larger-scale wastewater systems since they serve population centers around the world. In that most centralized wastewater systems discharge to streams/rivers and surface water bodies, with potentially direct impacts on downstream users of those water supplies, they are somewhat analogous to other large inter-connected utility grids. National and international organizations have evolved over time that identify research and funding needs for these systems, direct those research efforts, and help federal, state, and local governments with education and information dissemination on industry design and construction practices.

Potential impacts to the surrounding environment from decentralized systems applying effluent below the ground's surface tend to be much less visible and are harder to track than those from systems using direct surface discharge. For systems relying on final land/soil disposition of wastewater effluent, consideration of the character of local conditions is obviously critical to effecting sound design and construction practices and accompanying rules. However, due to the more localized focus and distributed character of onsite wastewater systems, they have tended to be disadvantaged with fewer organized sources of funding for research and education. Relatively few college engineering curricula devote appreciable time and resources specifically to decentralized wastewater systems coursework. Absent that formal specialized training, engineers/designers dealing with those systems tend to (1) design smaller-scale systems in ways consistent with their formal training on municipal-scale wastewater systems, (2) rely primarily on state or local rules to guide their practices, which in most cases are essentially minimally applicable standards, (3) rely largely on manufacturers' technical representatives dealing with certain product lines, or (4) strive to educate themselves on methods and materials more suited to decentralized systems. In practice, onsite systems designs are usually the product of a combination of two or more of those influences.

Even though the best engineers and designers seem to be those who spend appreciable time at construction sites for their systems, very few owners of small- to midsized decentralized wastewater systems can or are willing to pay for engineers/designers to spend much time at the site during construction. The result is that most onsite systems designers tend to spend a fairly limited amount of time at those construction sites, and often only the minimum number of visits to satisfy regulatory permitting requirements. With most onsite systems components located below the ground's surface, and piping, valves, and other critical elements often already buried when the system designer goes to the site to make any obligatory checks, there are limited opportunities for most designers to observe the "devil" that creeps into many of the details. Photo illustrations of actual systems installations included in this book will hopefully be of help to systems owners and others for participation in the oversight of their projects during construction.

Much of the body of science and engineering practices associated with large municipal-scale land treatment systems is directly applicable to small onsite/decentralized wastewater systems. In recognition of the important role that natural land treatment processes play in the implementation of more cost-effective and well-performing domestic wastewater systems, the U.S. EPA, the Water Environment Federation, and other entities have researched and compiled a number of excellent technical resources on that subject and on natural treatment processes in general. Those books, design manuals, and many technical guidance documents and articles offer a very well-founded and logical design approach, and are used in many college/university courses covering larger-scale land treatment systems.

This book is, in large part, intended to help bring together those two worlds: the more centrally organized body of technical information developed for larger municipal-scale domestic wastewater systems, and the more diffuse and widely distributed world of small- to midsized decentralized systems. The land-limiting constituent (LLC) or limiting design parameter (LDP) concept was presented several decades ago in the context of larger systems using land/soil disposition of wastewater effluent. In graduate engineering school at the University of Texas—Austin, I had the good fortune to study land treatment processes under Dr. Raymond Loehr, one of the leading experts on applying LLC concepts to the design of sound and sustainable land treatment systems. In later years, through both research and design work, I gained increasing appreciation for the relevance and importance of those basic land treatment engineering concepts.

Each of us tends to build upon and benefit from the experience of others, and time in the field with good contractors has offered many opportunities to learn details that could not possibly have all been covered in even the best of installer training programs. Work on a number of decentralized systems demonstration projects has also offered invaluable insights into the pros and cons associated with the use of certain methods and materials. Several chapters in this book cover details of construction for certain types of systems that may or may not be included in training programs for installers, and would almost certainly not be covered in educational curricula for engineers/designers. While frequent visits to construction sites should, in my opinion, be an essential element of all engineers'/designers' professional practices, it is hoped that imparting some of that field experience here can help prevent some lessons from having to be learned the hard way.

The chapters of this book have been organized to guide readers through a logical progression of concepts that build upon each other. The early chapters present sustainability considerations and planning concepts that apply to essentially all domestic decentralized wastewater systems. From there, chapters cover methods and materials starting basically from the building's sewer stub-out, progressing to treatment processes, and then to final effluent land/soil disposition. Summarizing tables are presented in several chapters that may be helpful for later referencing once the reader is familiar with the context of information presented. Because of the abundance of technical support available to industry practitioners for most proprietary treatment and dispersal methods, detailed design and construction information presented in several chapters is devoted to certain nonproprietary methods that are considered some of the most sustainable in use today. The last chapter presents several project examples intended to put previously discussed planning concepts together as they relate to specific geophysical and project conditions.

Additional tables, resources, and helpful links are available at www.mhprofessional.com/paisows.

S. M. Parten, P.E.
Principal Engineer
Community Environmental Services, Inc.
SMParten@ces-txvi.com

Acknowledgments

Project experience leading up to this book would not have occurred without the dedication to quality and sustainable building demonstrated by architects, project managers, and owners with whom I've had the privilege to work during my career. Over the years with the design and construction of small- to mid-sized decentralized systems, I've had the opportunity to work with some exceptional systems installers from whom I've learned and continue to learn much, including Mr. Mike Clark.

Reviewers from different sectors of industry practice contributed generously with detailed review of and comments on this book, taking considerable time from their lives to do so. Many thanks go to Dr. Bob Rubin, Dr. Howard Liljestrand, Peter Pfeiffer, Peter Reynolds, Dave Deane, Eric Schweizer, and Seyed Miri for their time and energy in that regard. Special thanks go to Ms. Cheryl Rae, without whose enthusiasm, encouragement, and technical support this book would not have happened.

Finally, many thanks go to the McGraw-Hill Professional editorial and production teams for their hard work and dedication, with special thanks going to Michael Mulcahy and Stephen Smith.

CHAPTER 1

Introduction

1.1 Introduction

Onsite or decentralized wastewater systems make up a large and critical portion of the world's wastewater service. The U.S. EPA estimates that approximately a quarter of U.S. households rely on onsite or decentralized systems for their wastewater service. Those include both new installations and existing homes that may at some point need repairs or replacement. Although, historically these systems have often been considered temporary wastewater service solutions, they are increasingly recognized as a necessary element of long-term sustainable development and infrastructure. Decentralized systems are often needed to serve single residences or businesses, developments, and resorts located along sensitive and pristine watersheds. Implementing sound and sustainable decentralized wastewater treatment practices is essential to protecting the world's critical water resources.

The terms "onsite" and "decentralized" are sometimes used interchangeably for wastewater systems not connected to centralized collection and treatment "grids" (e.g., municipal systems). In a strict sense "onsite" refers to decentralized systems for which the final disposition of the wastewater effluent occurs on the property where the wastewater is generated. "Decentralized wastewater systems" constitutes a broader category of systems serving either individual or multiple properties (e.g., "cluster" or "collective" systems). For clustered systems, wastewater from multiple properties is collected and conveyed to a common location(s) used for final treatment and effluent disposition. *Onsite* systems are, therefore, a subset of *decentralized* systems.

The principal goals for this book are to

- Frame and provide explanations for sustainability considerations for planning and implementing decentralized systems
- Dispel certain myths surrounding onsite/decentralized wastewater systems that have tended to persist within either the industry or the public
- Provide basic technical explanations and guidance for evaluating project sites

1

- Provide basic descriptions of technologies to assist persons with making sound and sustainable choices in the selection of overall approaches, methods, and materials for their decentralized wastewater projects

- Direct engineers/designers of systems to other technical resources for more detailed scientific information and design guidance on systems

- Provide helpful information to project owners/managers, contractors, and engineers/designers on methods and materials used for constructing sustainable systems

- Provide more detailed construction information on several nonproprietary treatment and dispersal methods considered to be very sustainable where appropriate for use, and for which less installation guidance is available as compared with proprietary systems.

- Provide information on operation, maintenance, and management practices essential to reliable performance and long useful service lives for systems

Onsite or decentralized wastewater systems play a critical role in sustainable water services infrastructure. In areas outlying communities or larger cities having centralized wastewater service grids, it is often much less cost-effective and may be very environmentally disruptive to continue further extending larger centralized sewer lines. Those larger lines must also be built prior to the construction of development that it serves, in contrast to decentralized systems that can be staged in simultaneously with residential and/or commercial building projects. Due to the higher initial capital costs often associated with expansion of centralized wastewater service grids, there are, in many cases, pressures to overdevelop lands in sensitive environmental settings to increase numbers of users paying for that centralized system. Decentralized wastewater systems are very compatible with less densely populated areas having lower impervious cover, and low impact development (LID) concepts in general.

Implementation of sound decentralized wastewater systems often confronts some of the same economic challenges as off-the-grid solar/photovoltaic electric systems serving single or limited numbers of dwellings or businesses. Due to economies of scale, initial costs of installing the system tend to be higher as compared with systems serving larger numbers of users, given the ability to spread costs out among more persons. However, for such systems serving individual homes or business, there are no monthly user charges unless the system is managed by an outside entity, unlike centralized wastewater systems for which users are typically charged monthly service fees.

Because of the on-lot investment for decentralized systems components on properties served, it's especially important to select the most sustainable overall approaches. Methods and materials need to be used that will result in the least long-term costs with the longest useful service lives, while meeting water quality and public health goals and any applicable regulatory requirements. Recurring costs that can add up quickly over the life of a system include routine operation and maintenance expenses and periodic replacement or repair of components. As well as reducing long-term costs, systems using less electric power to operate and that generate less waste sludge contribute to much more sustainable wastewater service approaches. Information presented throughout this book on different types of decentralized wastewater technologies and components is intended to provide a basis for comparing, selecting, and implementing more sustainable systems.

There are currently many technical publications and texts available to help engineers and planners with the conceptual planning, theory, and detailed calculations associated with a wide variety of decentralized treatment systems. Examples include the books *Small and Decentralized Wastewater Management Systems* (Tchobanoglous and Crites, McGraw-Hill, 1998), and *Advanced Onsite Wastewater Systems Technologies* (Jantrania and Gross, Taylor & Francis, 2006). Another excellent planning resource is the U.S. EPA's *Onsite Wastewater Systems Treatment Manual* (EPA/625/R-00/008, February 2002). There are, however, very few technical resources that guide planners on the selection and planning of the most sustainable onsite treatment options. Information presented in this book is intended help guide engineers and systems planners with the selection of the most sustainable and cost-effective options, and to provide practical detailed guidance on the planning and installation of those options.

Increasing numbers of property owners and developers are interested in making more "green" or sustainable choices for the methods and materials used for their residences and development projects. Those interests are based on a variety of concerns and factors such as energy usage and water quality protection. The focus throughout this book is in presenting detailed practical information on options that have proven to be the most sustainable and cost-effective on a long-term basis, and to have long useful service lives when designed and built properly. Basic considerations for sustainable wastewater planning are presented and discussed in Chap. 2.

Regulators sometimes comment that engineers/designers of small- and large-scale decentralized wastewater systems seem to design systems with which they are familiar, rather than those necessarily best suited for site conditions.[1] Achieving the most sustainable wastewater service solutions necessitates selecting a method of treatment and effluent disposition most appropriate to the site, as well as considering other factors related to sustainability. Regulators tend to leave it to designers of systems and property owners to make choices relative to sustainability and cost-effectiveness, and simply require that systems meet applicable standards at a minimum.

Some designers with knowledge on a variety of system types may assume that property owners prefer the option that is least costly up front, rather than explore owners' interests in systems that are the most sustainable and cost-effective in the long term. For these reasons it's important for property owners to take an active interest in the planning process to ensure that the methods and materials chosen are those best serving their interests.

Many engineers, scientists, and planners who may be very knowledgeable on the subject of sustainable wastewater systems may find themselves unable to spend appreciable time "in the field" during the construction of these systems. Even in cases where engineers may have that experience, smaller-scale systems property owners are often not willing or able to pay for their engineers to spend enough time overseeing the construction of an onsite system to ensure its proper installation. County or local inspectors may perform only a limited number of inspections of those installations, and oftentimes just one.

A number of U.S. states have some type of basic training or certification program for onsite wastewater systems installers, but most of these training programs are held over a period of just a few days, and not able to go into the kind of detail needed to impart enough information needed to install various types of systems properly. Training is often only in classrooms for 2 to 3 days. It may take 10 to 15 years constructing some types of systems for contractors to really learn the "tricks of the trade" needed to avoid certain problems. Without experienced engineers on the job site to verify that contractors are installing systems properly, property owners have to trust the quality of work performed by contractors who themselves may not understand the reason for doing things in a certain way when their intentions are to do installations properly. This book is intended to serve as an aid to both designers and installers of systems, and help through the many illustrations and explanations to serve as an important resource for project owners in observing and controlling the quality of their own projects.

Many manufacturers of proprietary treatment and effluent dispersal systems have developed and distribute detailed documentation, including videos and printed installation and operating instructions for their products. Most engineers, and the installers of systems they design, routinely refer to manufacturers' technical support and guidance documentation for the proprietary systems they deal with. That may also explain why those systems may be used more, short of detailed guidance information on nonproprietary methods.

Several types of nonproprietary systems considered sustainable and appropriate for use in various geographic settings are discussed in detail, with key construction steps illustrated and explained.

The technologies covered in Chaps. 4, 5, 7, 8, and 10 are nonproprietary and for which there may not be such detailed manufacturers' instructions and technical support. Chapter 5 covers primary treatment components in detail, which is applicable to all onsite wastewater systems, whether proprietary or not. Chapters 7 and 8 cover two types of natural treatment systems in detail, with much of the information also applicable to other types of treatment systems. Chapter 10 covers in detail a method of final subsurface effluent dispersal that is considered one of the most sustainable land/soil dispersal methods for a variety of reasons explained in Chaps. 9 and 10. That method of subsurface effluent dispersal may be used with essentially all levels and methods of treatment, including just primary or septic tank pretreatment where soil and site conditions are acceptable for that.

By being provided with detailed, photo-illustrated steps in the construction of several basic types of systems capable of reliably providing high-quality onsite wastewater treatment, engineers, contractors, and project owners can hopefully benefit in different ways.

1.2 Current State of the Onsite Wastewater Industry in the United States

In recent decades, and following passage of the U.S. Clean Water Act in 1972, greater funding tended to be directed toward wastewater treatment plant upgrades and scientific investigations associated with larger centralized systems, as compared with smaller decentralized systems. In recent years, however, the U.S. EPA has focused greater attention on decentralized and smaller-scale systems, recognizing their role and importance for achieving sustainable and cost-effective wastewater service. In a January 2005 EPA document (EPA 832-R-05-002) entitled "Decentralized Wastewater Treatment Systems: A Program Strategy," EPA stated that their vision was that "decentralized wastewater treatment systems are appropriately managed, perform effectively, protect human health and the environment, and are a key component of our nation's wastewater infrastructure." EPA followed by stating it was their mission to "provide national direction and support to improve the performance of decentralized systems by promoting the concept of continuous management and facilitating upgraded professional standards of practice."

As with the history of so many technical industries, the development of practices, materials, and processes used for decentralized and small-scale onsite systems has gone through many changes during the past few decades. Prior to the second half of the 20th century, the vast majority of systems consisted of very basic approaches that relied less on science and more on successes or failures with "out of sight out of mind" practices. Many such systems and practices still exist today, and especially in developing countries. As more research and development has occurred with onsite systems' methods, materials, and technologies since the 1970s, accepted practices have changed dramatically in most U.S. states. As more data and observations have been reported on the performance of systems using those various technologies, the industry has continued to change.

An example of one of the earlier technologies that were adapted to and began to be used for residential scale wastewater systems is aerated tank units (ATUs). These units, sometimes also referred to as "aerobic treatment units," use the suspended growth treatment process also used for most municipal wastewater treatment plants. It is however (as with municipal plants) necessary to effectively control the treatment process in these units, though without the constant presence and benefit of treatment plant operators. Various product modifications have therefore occurred for those units over time, with many other types of proprietary manufactured units emerging on the market. However, problems persist with ATUs due to operational problems and service neglect, and there continues to be a need for more regulatory monitoring and auditing programs to ensure adequate performance is occurring.

Other technologies continue to be developed and used to overcome some of the challenges inherent to small-scale wastewater treatment systems. The benefits and limitations associated with different types of predispersal treatment technologies are discussed in Chap. 6. As EPA and states have continued to focus more attention and resources on developments in the onsite industry, the increasing availability of better performing and more reliable systems continues.

Essentially all of the information presented in this book pertains to *domestic* wastewaters, and not to those waste streams considered to be industrial or hazardous.

Industrial wastewaters are those contaminated in some way by industrial or commercial activities prior to their subsequent (and required) treatment and release into the environment, or the reuse of that treated water. Examples of industrial waste flows would include waters contaminated from manufacturing processes (e.g., semiconductors), commercial food processing (e.g., meat packing facility), and other processes and activities producing nonhuman wastes. Hazardous wastes make up a very diverse range of materials, and can be solid, liquid, sludges, or contained gases. Examples of hazardous wastes include oil-based paints and thinners, pesticides, and many cleaning fluids.

Domestic wastewaters are human-generated sewage from homes and businesses. They are wastes produced from sanitary facilities serving residences, cities, mobile home parks, subdivisions, restaurants, rest homes, resorts, and so on. Systems handling that category of waste, and which use some method of soil/land-based final disposition of effluent, are the focus of this book.

Reference

1. S. M. Parten, *Analysis of Existing Community-Sized Decentralized Wastewater Treatment Systems*, Water Environment Research Foundation, Alexandria, VA, July 2008.

CHAPTER 2

Sustainability Considerations for Decentralized Wastewater Systems

Some of the basic concepts associated with wastewater treatment processes integral to sustainable wastewater planning are presented in this chapter. While it's beyond the scope or intention of this book to thoroughly cover those concepts, some understanding is useful for comparisons of technologies and processes covered in later chapters. It is impossible to anticipate and provide examples of all of the potential situations and geophysical settings where decentralized systems are relied upon. An increased understanding of many of the "whys" and "hows" can hopefully assist systems designers with better evaluating options best suited to and more sustainable for specific conditions. As new technologies develop over time, most of the same treatment process considerations will likely continue to apply.

Criteria used for selecting the most cost-effective and appropriate wastewater system to serve one or more residences or businesses have long included such factors as initial capital costs and water quality protection. However, there is increasing attention placed on long-term impacts and more *hidden* costs as related to sustainability. Energy consumption and reliability of power supplies is an increasingly important factor today in assessing the long-term costs associated with each type of system. "Residuals management" or sludge and septage pumping and hauling for decentralized systems is another major cost and environmental factor. Presented below are a variety of factors that might be considered by project owners and planners in determining the most appropriate, cost-effective, and sustainable decentralized wastewater service approach for a given project.

Those considerations listed here are certainly not a comprehensive set of sustainability factors. They are very dependent on the specific geographic setting. Our understanding of sustainability issues will change over time as more information is gathered on the interconnectedness of the world's resources, their use, and associated impacts. The list below should, however, provide examples of the types of considerations needed for planning sustainable wastewater treatment systems.

- Ability of the system to reliably meet treatment levels that will not result in short- or long-term degradation of ground or surface water resources ("receiving waters"), and will protect public health.
- Local climatic and seasonal conditions, as they impact operating conditions and potential for problems or poor performance from the system.
- Initial materials costs for the system.
- Initial system installation costs.
- Land area requirements for the system.
- Energy consumption needed for meeting target treatment levels during normal/design operating conditions.
- Ability to recycle and/or reuse all or a portion of the treated waste stream.
- Sludge/septage production associated with the treatment process(es); expected time interval between pumping and hauling based on primary tank sizing, performance data and operating histories; proximity and availability of permitted facilities able to receive waste sludge/septage pumped from the system, and the costs associated with that.
- Are there treatment by-products discharged from the system that will immediately or over time result in certain adverse environmental impacts?
- Expected useful service life of major and critical components of the system, including durability of materials used, resistance to corrosion, and other local environmental factors.
- Local availability of materials for use in constructing the system.
- Weight and transport costs associated with components needing to be shipped to the project site.

- Energy consumption needed to operate the system, and maintain "ready for use" treatment conditions during and following periods of lower or no usage (such as seasonal homes and businesses or vacations from permanent dwellings).

- Expected repairs or components likely to need replacement over time, and their costs.

- Labor and material costs associated with operating the system.

- Labor and material costs associated with routine maintenance of the system.

- Aesthetic and cultural considerations, including potential for odors, or unattractive features of the system.

A discussion of each of these considerations is provided later in this chapter. The Leadership in Energy and Environmental Design (LEED) Green Building Rating System is also described as it relates to wastewater systems planning and implementation.

Some general understanding of basic wastewater treatment processes and physical phenomena are necessary for understanding some of the factors relating to the sustainability of systems. A general discussion of that is presented in the following section.

2.1   Basic Concepts and Science Related to Onsite/Decentralized Wastewater Systems

Onsite wastewater treatment and dispersal systems discussed in this book consist of some type of pretreatment, followed by final soil/land treatment and disposition of the effluent. While some decentralized wastewater systems use some method of treatment followed by direct surface water discharge of the treated effluent, there are certain concerns related to water quality and sustainability with that approach. Small to midsized decentralized systems typically don't have enough management and monitoring needed to reliably prevent certain adverse short- or long-term water quality impacts from direct surface discharge of effluent. Surface drinking water supplies have been found to be increasingly laden with prescription and nonprescription drugs, at least in small concentrations, along with many other organic chemicals that make their way through wastewater treatment systems and are then directly discharged back to surface waters.[1,2] The very high numbers of bacteria naturally present in soils are much better able to break down those organic chemicals than most mechanized wastewater treatment plants.

As discussed further in this chapter, natural land and unsaturated soil conditions tend to offer opportunities for treatment of a wide range of domestic wastewater pollutants before the wastewater effluent may reach any receiving waters. Domestic waste streams typically contain a wide range of human pollutants in varying concentrations, including pharmaceuticals and personal care products (PPCPs) and other chemicals not easily removed or decomposed through most smaller-scale treatment methods. The billions of natural bacteria present in relatively small volumes of soil are capable of decomposing many such wastewater constituents given enough time. Sufficient hydraulic residence or retention time along with other favorable natural conditions in the soil are needed for those processes to occur. That relates to both soil type and rate of application of the effluent to a soil, as discussed in Chaps. 3 and 9.

Some land types have less soil treatment capabilities than others, and require more pretreatment of the wastewater prior to soil dispersal to prevent adverse environmental or public health impacts. For example, some geographic areas are characterized by rock outcrops and have little to no topsoil or overall depth of soil for treatment to occur. That land type therefore has substantially lower treatment capabilities for several important wastewater constituents as compared with areas with deeper soils such as sandy loams.

There are several principal wastewater constituents around which successful designs should be based. These include

- Wastewater flow (or volume of wastewater produced daily)
- Total suspended solids or TSS (removed partially through primary treatment/settling)
- Biochemical oxygen demand (BOD)*
- Chemical oxygen demand (COD)
- Nitrogen in several forms including ammonia/ammonium (NH_3/NH_4^+), nitrate (NO_3), nitrite (NO_2) (nitrite is a highly unstable or "transient" form of nitrogen, with the majority of the nitrogen in wastewaters being in one of the other forms), and Keldjahl nitrogen (organic forms of nitrogen)
- Turbidity
- Alkalinity
- Phosphorus (P)
- Pathogens

Depending upon the characteristics of the land and soil receiving the pretreated wastewater, some characteristics of the waste stream may be of more concern than others due to the land's potential inability to effectively remove those pollutants. Those have been referred to as land-influencing or "land limiting" constituents (LLCs) in the technical

---

*Wherever BOD is used in this book, it is assumed to mean BOD_5 (5-day biochemical oxygen demand), unless otherwise noted.

literature. More recently, the U.S. EPA's *Process Design Manual* on "Land Treatment of Municipal Wastewater Effluents" refers to those as "limiting design parameters" (LDPs).[3] These influence both the amount of land of a given type needed for safe and sustainable final disposition of the wastewater effluent, and the level of treatment needed prior to final land/soil dispersal of effluent. Utilization of the LDP design approach assures that the wastewater constituent requiring the largest land area is used for the design.

The concept of land-limiting constituents has historically most often been used in association with the land application and beneficial reuse of municipal scale and industrial waste streams, and for projects of somewhat larger scale than discussed in this book. However, the same basic concept and definition applies. The following is from an article published in *Journal of Environmental Quality* (March 2004):

> Land application systems, also referred to as beneficial reuse systems, are engineered systems that have defined application areas and permits for operations, based on site and waste characteristics that determine the land limiting constituent (Overcash and Pal, 1979).[4]

And, the following is excerpted from an article entitled "Beneficial Reuse and Sustainability: The Fate of Organic Compounds in Land-Applied Waste" (M. Overcash, R. Sims, J. Sims, and J. K. Nieman, *Journal of Environmental Quality*, 2005)

> The land limiting constituent (LLC) approach is central to the design and sustainable operation of land treatment systems (Brown et al., 1983; Loehr et al., 1979; Overcash and Pal, 1979).[5]

Because of LLC's common use as an acronym in business settings, LDP will be used here to refer to those land-limiting constituents.

As an example of the LDP concept, pathogen reduction may be the pollutant of greatest concern for a site due to that soil's lesser ability to filter and naturally remove pathogens. In that case, the wastewater would need to be treated for pathogen reduction prior to soil dispersal. For other sites such as those with deep clay soils, the infiltrative capacity (or "percolation rate") of the soil, and water-loading rates would tend to be a more critical factor for final land treatment and final dispersal of the wastewater effluent. Wastewater flow (or water) would be the LDP. Pathogen reduction would be of lesser concern for subsurface effluent dispersal there because of the higher organic content and, in general, better biological treatment capabilities of clay soils. For clay soils, much of the improved treatment capabilities have to do with the slower percolation or infiltration rates, thus providing longer time periods for treatment as the effluent moves through the soil. A greater amount of land/soil would, however, be needed in clay soils to prevent overloading and saturating the soil, as compared with sites having soils with higher infiltration rates. Saturation of the soil would not only result in surfacing effluent with accompanying health and environmental risks, but would fail to maintain aerobic conditions for the bacteria needed with natural treatment processes.

Some soils have lower organic content and drain too rapidly (such as a very sandy or gravelly soil), and fail to retain the percolating effluent long enough in the presence of enough bacteria to remove nitrogen and other wastewater constituents adequately. Ammonium-nitrogen in percolating effluent tends to convert to the nitrate form of nitrogen fairly rapidly in aerobic soil conditions (nitrification). The nitrate form of nitrogen is very mobile in the soil, and moves with percolating water. If soils do not have enough organic content and retention capacity (drain too rapidly), there is not enough residence time for subsequent conversion of nitrate to N_2 gas (denitrification) or uptake of nitrogen by vegetation. Nitrate-nitrogen is regulated in U.S. drinking water supplies as a pollutant of concern because when existing in higher concentrations in groundwater or other water supplies, studies have found it to contribute to an infant health condition called methemoglobinemia, or "blue baby syndrome." The United States limits nitrate in drinking water supplies to 10 mg/L as nitrogen (45 mg/L measured as nitrate). For sites having nitrogen as an LDP, it's therefore important to treat the wastewater for greater levels of nitrogen removal prior to final soil/land dispersal. Using lower soil loading rates also offers more opportunity for further natural treatment to occur in limiting soil conditions.

Nitrate and other forms of nitrogen are also of concern for inland surface waters and in coastal areas because it is a key nutrient for algae growth. Adverse effects of eutrophication, or excessive plant growth and decay, include lack of oxygen and reductions in water quality, fish, and other animal populations. Wastewater with significant concentrations of ammonia tends to deplete dissolved oxygen levels in waters for fish and other aquatic species. That's a concern for surface-applying or irrigating wastewater effluent in areas with relatively steep slopes draining fairly quickly into watersheds.

Figure 2.1 shows three basic levels of predispersal treatment, and forms of nitrogen associated with each as long as certain physical and biochemical conditions are present. Some of those critical conditions are indicated in the figure. The first major step in domestic wastewater treatment is called "primary" treatment, and is intended to remove larger and readily settleable solids from the waste stream along with flotation of oils and greases through natural gravitational processes. Primary treatment for most decentralized wastewater systems occurs in a septic tank, or multiple septic tanks in series. Depending on the soil and site conditions, just primary treatment may be sufficient prior to final disposition and natural treatment of the effluent in the soil. Just a teaspoon (about 5 mL) of productive soil may have from hundreds of millions to billions of naturally present bacteria that play a crucial role in onsite biological wastewater treatment processes. The soil matrix provides treatment through a combination of physical (e.g., filtration) and biochemical processes.

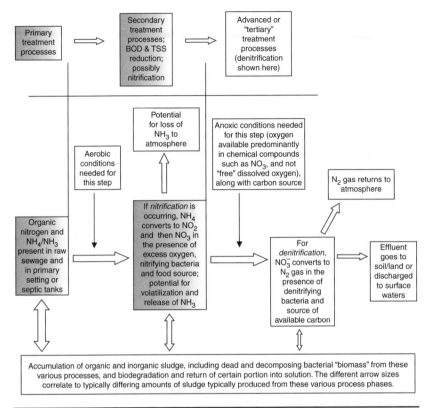

FIGURE 2.1   Basic levels of predispersal treatment and forms of nitrogen at each level under given conditions.

The removal of BOD and TSS to certain levels is called "secondary" treatment. BOD is a measure of the organic content in the waste stream that would be available as a food source for bacteria, and is typically considered a measurement of the overall pollutant level of a wastewater. Bacterial respiration occurs as the dissolved organic molecules in wastewater are processed, and this exerts an oxygen "demand," or oxygen depletion in the wastewater. It's important to have sufficient oxygen available for bacteria to process the dissolved organics in the wastewater, as well as have sufficient settling time and capacity for settling out solids. As a microbiologist colleague once quipped during a public meeting, "Bacteria don't have teeth." Therefore, suspended organic and inorganic solids that bacteria can't readily process or consume as a food source need to be settled out of or filtered from the waste stream. The bacteria need both enough time and available oxygen to process the dissolved organics in the wastewater, which is important to treatment process sizing and designs.

Because BOD is considered a measure of the food supply available for bacteria, it is of concern for potential clogging of subsurface dispersal fields and trenches along with suspended solids, or TSS. When solids and excessive bacterial populations, or "slime layers," build up along the interface between subsurface dispersal system pipes and the surrounding soils, a clogging layer develops at that interface that prevents effluent from percolating as readily into the soil. There are varying technical views and studies that have looked at this phenomenon, with some concluding that it's actually of benefit to treatment processes when this occurs. Clogging layers would increase the time during which treatment of effluent could occur before infiltrating, and the conditions in and around the clogging layer would likely be favorable to certain natural treatment processes. It seems reasonable to conclude that there would be some treatment benefits for slime layers previously thought to be solely a negative development. However, it's important to maintain conditions in the dispersal field that allow for enough effluent percolation to avoid field saturation or backup for the range of usage and flow conditions typical of most decentralized wastewater systems.

Most soils have such an abundance of bacteria capable of processing dissolved organics in wastewater as a food source, so that BOD is most commonly not found to be an LDP. As long as designs take into consideration the need to prevent buildup of slime or clogging layers in subsurface dispersal fields and in piping, most soils are capable of receiving primary treated wastewater effluent with relatively high levels of BOD (up to several hundred mg/L) without environmental or public health concerns. However, higher BOD concentrations can effectively be a treatment "process-limiting" constituent as related to the various forms of nitrogen and pathogen levels.

Sufficiently low levels of BOD are needed for the subsequent removal of nitrogen and pathogens in predispersal treatment processes, if removal of one or both of those pollutants is needed prior to soil dispersal. "Advanced" or "tertiary" treatment follows "secondary" treatment (BOD and TSS reductions) as shown in Fig. 2.1.

Because nutrient loading to watersheds (and especially nitrogen for so many geographic areas) is an environmental concern and is oftentimes regulated for decentralized systems, some discussion of the "nitrogen cycle" is presented here to explain the various forms of nitrogen and the conditions under which each tends to exist. Figure 2.2 below is a conceptual illustration of the nitrogen cycle today, which includes natural sources and increasing amounts of nitrogen introduced from human activities, including industry, auto emissions, and so on. Nitrogen continuously cycles through the world's ecosystems in various organic, dissolved, and gaseous forms.

The chemical formulas for the various forms of nitrogen in domestic wastewater can offer some insights into the predominant forms nitrogen would assume in certain biochemical conditions in wastewater

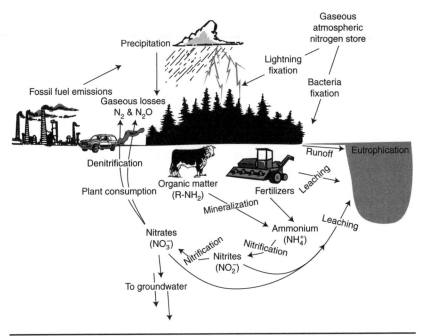

FIGURE 2.2   Nitrogen cycle.

treatment systems and the environment. Forms that are of principal concern to the domestic wastewater industry are

- Organic forms of nitrogen, such as urea
- NH_4^+ = ammonium ion (water soluble, and so is dissolved in wastewater)
- NH_3 = ammonia nitrogen (the degree to which it reacts in solution to give NH_4^+ depends on pH; the gaseous form of nitrogen that tends to volatilize at varying rates from fluids, depending temperatures, and other factors). Ammonia is commonly associated with odors from wastewaters)
- NO_2^- = nitrite ion (water soluble), a very unstable and transient form of nitrogen that tends to be present in very small concentrations, and which can be an indication of problems with a nitrogen conversion cycle if occurring in larger concentrations
- NO_3^- = nitrate ion (water soluble), a very stable anion that is highly mobile in subsurface/soil conditions, in that it tends to move with the water and not bind to the soil
- TKN = total Kjeldahl nitrogen = $NH_4^+ + NH_3$ + organic nitrogen
- Total nitrogen = TN = TKN + NO_2^{-2} + NO_3^-

It's not the focus of this book to cover in detail the biochemistry of wastewater. However, selection of suitable treatment processes relies on a basic understanding of at least some of that science.

A variety of bacterial species play a vital role in the treatment processes. Two important elements of bacterial molecular composition are carbon and nitrogen (as will be briefly discussed in Chap. 5 for composting operations). Pollutants in domestic wastewater can provide both a carbon and a nitrogen source for bacteria, and depending on the concentrations of those along with the availability of oxygen, certain types of bacteria will tend to thrive more than others. Other wastewater characteristics, including pH, temperature, and alkalinity must be such that necessary biochemical processes can occur for particular steps in treatment processes. Alkalinity will be discussed further in Chap. 6. "Resources and Helpful Links" on this book's website (http://www.mhprofessional.com/paisows) has technical references covering a wide range of wastewater science and engineering topics in more detail.

Biological wastewater treatment processes basically follow the same natural cycle if conditions in the treatment processes lend themselves to the conversion of and/or presence of nitrogen in those forms. For raw wastewater and wastewater in septic tanks where conditions are predominantly "anoxic" (absence of air/oxygen), nitrogen is predominantly in the form of TKN (organic nitrogen and ammonium/ammonia). To achieve significant pre-soil-dispersal removal of total nitrogen, aerobic conditions must be provided (nitrification step), followed again by anoxic conditions and an available source of carbon (denitrification step). That results in the release of N_2 gas back to the atmosphere. Natural treatment processes such as wetlands rely on a combination of vegetative uptake and a limited amount of nitrification/denitrification. They are limited in their ability to nitrify (due to fairly anoxic conditions depending on size and depth), and thus limited in their nitrification/denitrification nitrogen removal capabilities. Wetlands are discussed in Chaps. 6 and 8 in more detail.

Where oxygen available to bacteria may be limited (such as in a tank reactor when aerators/blowers entraining air into the wastewater aren't able to keep up with the oxygen demand), the conversion of ammonium nitrogen to nitrate (nitrification) can likewise be limited. For most biological wastewater treatment processes, BOD levels of approximately 20 mg/L or higher in the treated effluent or following certain processes in the overall treatment system would typically mean that relatively little nitrification had occurred.[6] That's a problem where either the effluent ammonia/ammonium-nitrogen concentration needs to be low, or if a denitrification process follows and the goal is total nitrogen removal. The bacteria responsible for nitrifying wastewaters tend to be present in much fewer numbers when BOD levels are higher, with the bacteria fed by the organic pollutants associated with higher BOD levels in greater numbers.

Some treatment processes are much more effective than others for nitrification, as discussed in Chap. 6.

For biological denitrification to occur following the nitrification step, a variety of conditions must be provided as noted in Fig. 2.1 including an anoxic environment and an available source of carbon, along with acceptable temperature and pH ranges. Sources of carbon that have been used successfully for denitrification processes in decentralized wastewater systems include septic tank effluent, settled greywater, and methanol.

Nitrogen has generally been considered a nutrient of greater concern for coastal waters as compared with phosphorus, with coastal waters most commonly found or assumed to be more nitrogen-limited as compared with phosphorus. The extent to which any nutrient is "limiting" (and thus may need to be more limited in terms of watershed pollutant loading) is based on the availability and demand for other essential nutrients (in terms of what algae and other species contributing to eutrophication require for growth). If, for example, there's ample phosphorus available in a watershed for cell growth, and under natural conditions not enough nitrogen for that growth to occur, nitrogen is "limiting." Each receiving water body should be evaluated to determine which nutrients should be controlled for maintaining water quality.

While both phosphorus and nitrogen can be nutrients of concern for freshwater and watersheds in general, many soils have a relatively high capacity for retaining, or "binding" phosphorus. That is, the phosphorus chemically converts to a form in the soil that binds and retains it. Those soils would, however, over time reach a "saturation" point, beyond which the phosphorus retention capacity would be significantly reduced. As compared with more mobile nitrate, phosphorus therefore tends to be a nutrient of somewhat less concern for many geographic areas and sites. Phosphorus in soil systems is discussed further in Chap. 3.

Pathogens are removed from percolating wastewater effluent through a variety of mechanisms, including filtration and natural die-off, or attrition. Natural attrition of microorganisms is time-dependent, and requires that organisms be retained in the soil matrix for varying amounts of time, depending on the species. For soils/sites having pathogen reduction as an LDP's, or where surface application of the treated effluent onto land is to be used, some method of disinfection or pathogen removal must be provided.

In general, the type(s) of pretreatment used prior to final land disposal should be based upon the land-limiting constituents determined to apply to a particular site and land type. Chapter 3 discusses key wastewater parameters of concern in the context of site evaluations, soil types, and system selection based on a soil's natural treatment capabilities.

2.2 Sustainability Factors for Onsite/Decentralized Wastewater Systems

Each of the sustainability factors identified at the beginning of this chapter is briefly discussed below. Consideration of the specific geographic setting and conditions where a project is located are critical for determining the most sustainable approaches for each project. Particular sensitivities or vulnerabilities associated with local conditions and/or resources must be considered, along with the availability of resources needed for the project. For systems requiring electrical power, the source and reliability of those energy sources are important factors related to long-term sustainability. However, because there are rarely ideal circumstances for achieving the "greenest" projects, choices typically must be made regarding the priorities given certain planning factors.

1. *Ability of the system to reliably meet appropriate treatment levels*

Applicable regulatory authorities in many parts of the world have adopted standards which systems are obliged to meet at a minimum. In other countries and regions no such standards have yet been adopted. However, even where minimum standards are adopted and being enforced, regulations typically do not include requirements for long-term sustainability as mentioned previously. In many and perhaps most cases requirements for water quality protection have been adopted based on local "political will." That is, policy makers have determined that there is sufficient acceptance by the public for the adoption of those requirements. Those decisions, however, may be based on local socioeconomic conditions and demographics, and costs associated with certain requirements, rather than necessarily being based entirely on optimal public and environmental health protection. Therefore, to achieve truly sustainable levels of treatment and avoid adverse long-term impacts, property owners may need to decide whether or not the locally applicable minimal standards are sufficient. That can of course be somewhat subjective, but is something that can be discussed with an engineer or designer who is informed on those technical considerations.

The specific watershed and any receiving waters, and geophysical conditions of the site where a project is located need to be considered relative to potential impacts and long-term sustainability and protection of public health. This planning factor relates to the previous discussion on LDPs, or land-limiting wastewater constituents regarding the level of treatment needed. Based on those considerations, the engineer/designer of the system would need to help guide the owner toward those candidate technologies capable of meeting the necessary treatment levels (including both pretreatment and any natural treatment provided through final land/soil application of the effluent).

However, if no effective management entity is available, nor a means of otherwise managing a fairly complex wastewater system, then a less complex and more manageable system should be used. Some basic technical concepts associated with these types of considerations and judgments will be discussed in Chap. 3.

   2. *Consideration of local climatic and seasonal conditions, including how they may impact performance of the system*

   The importance of climate relative to sustainable water and wastewater planning cannot be overemphasized. For example, in some regions of the world, annual and/or seasonal rainfall levels are low enough that maximizing reuse of wastewater effluent for irrigating vegetation is considered an essential element of sustainable planning. However, alongside of that in those regions, potable water supplies may also be of critical concern so that recharge of groundwater supplies with sufficiently purified effluent is considered an important part of overall sustainability. Rainwater collection and storage for water supplies may be necessary along with use of available groundwater and treating surface waters for potable water supplies. Rainwater harvesting can also contribute significantly to good stormwater management plans by helping reduce peak surface run-off during rainfall events, and thus reduce erosion/sedimentation problems and stormwater pollution for receiving waters.

   In other regions of the world there may be sufficient rainfall year round that there is less concern about water supplies, and more concern about adequately treating wastewater to prevent adverse impacts to surface and groundwater supplies. Rainwater harvesting may also be an important element of stormwater management in regions such as the Midwestern United States. However, its use as a method of potable water supply in these regions has less to do with availability of water, and more to do with the availability of and costs associated with treating surface water to a potable level, and distributing it to end users. In general, sustainable water and wastewater management practices depend on local climatic and geophysical conditions.

   In U.S. states such as Iowa, because of the ability of more organic soils to treat domestic wastewater to levels higher than is typically cost-effective for centralized systems using direct/stream discharge of effluent, sustainable decentralized wastewater planning would tend to focus more on optimizing both treatment and return of water to the hydrologic cycle. That is, because there may be an excess of rainfall and water to deal with on a site, it may be most important for designs to ensure as rapid groundwater recharge as possible without polluting those water supplies. Chapter 3 discusses those concepts in greater detail as related to different soil types.

   Climate can also have a significant impact on the performance of certain types of systems, particularly those reliant on biological treatment processes or when severe weather conditions impact systems' physical integrity. Climates with significant seasonal temperature variations may adversely affect performance for some systems more than others during colder months. In settings such as the Caribbean region and U.S. Gulf States that are exposed to tropical storms and hurricanes, physical conditions including periods of intense rainfall and high winds, and the potential for short- or longer-term power outages should be considered in system planning. Treatment processes that are open to ambient conditions and may take in significant added water during storms, or be structurally vulnerable and/or subject to corrosion from coastal conditions may not be the best options for that type of setting. The overall sustainability of a system will depend on its ability to continue to perform sufficiently well during stressed weather conditions.

3. Initial materials and installation costs for the system

The most sustainable and cost-effective systems on a long-term basis will not necessarily be the least costly to install initially. As with so many types of products and services, purchasing the least expensive products may include the use of lower quality materials of construction, labor practices, and so on. Some systems such as intermittent sand filters, for example, may have more labor and materials construction costs initially, but much lower operational costs as compared with many "packaged" treatment units.

Some further inquiries and product research may be needed by engineers/designers and/or project owners to evaluate costs versus quality and benefits of different wastewater system options. Since most persons must deal with limits on available resources and spending, it's important to obtain realistic estimates of installed costs for different wastewater service approaches determined to be viable for the given conditions and project needs. Engineers/designers, with input from experienced local contractors, should be able to advise project owners on estimated costs for options being considered. Initial capital expenditures can then be combined with long-term operation and maintenance costs, along with reasonable estimates of useful service life and the potential need to repair or replace significant parts of the system, to determine the most cost-effective approach overall. This factor also relates to LDPs for sites and the level of treatment needed, with those associated costs.

4. Land area needed for a particular system

Several factors associated with each type of system potentially used relate to this consideration. With regard to treatment systems, some natural treatment systems (e.g., subsurface flow wetlands and buried sand filters) continue to provide habitat for terrestrial fauna, though the habitat likely differs from the natural conditions prior to the system's installation. Other systems may take land area out of use entirely. Likewise, natural treatment systems/processes tend to occupy more land area than manufactured or "packaged" treatment units, even when utilizing essentially the same treatment process. For example, although both methods of treatment system are categorized as *recirculating biofilters*, based on average accepted loading rates for these two systems, recirculating sand/gravel filters require about 3 to 6 times the footprint (or land area) as synthetic textile media filters. There are advantages or benefits associated with each as discussed in Chap. 6. However, on sites where space to install a system is very limited, use of the manufactured treatment unit may be the best option.

Land area is also a consideration for methods of final effluent dispersal. In many areas of the world, and in particular where annual rainfall exceeds averages found in arid or semiarid regions, the land area needed for surface application of effluent typically exceeds that needed for subsurface effluent dispersal. That is particularly so if comparable levels of treatment area provided. Surface irrigation application rates should be such that surface runoff of effluent is prevented, even during extended wet weather periods. For most parts of the world, that realistically means that added storage capacity is needed for treated effluent during those periods.

Experience with both surface and subsurface dispersal systems over time has shown that because of the great capacity of many soils to provide for further natural treatment, subsurface application rates can oftentimes be safely much higher than for surface application systems. And for properly designed, installed, and operated subsurface systems there is not the concern for surface runoff during wet weather periods as there is for surface dispersal systems.

In short, different types of treatment and dispersal technologies have varying land area requirements for each set of site conditions that may be considered as related to overall system sustainability.

   5. *Energy consumption needed for meeting target treatment levels during normal/design operating conditions*

   Power consumption for various processes used in decentralized wastewater systems today can vary widely, and is a principal concern given increasing energy costs and reliability of service issues. Some treatment processes are very reliant on a continuous supply of power to the system, whereas other processes operate intermittently during the day and can easily withstand a certain amount of downtime. For the latter, it's also often possible to incorporate added storage capacity to accommodate longer power outages. For most decentralized wastewater systems, comparisons of energy usage would consider number and size (horsepower or kilowatts) of pump(s) and/or blowers/compressors used for the system, and run times needed for those electromechanical devices to operate the system and maintain acceptable conditions. Controls (panels, floats, etc.) use a relatively minor amount of power as compared with pumps or blowers.

Systems using activated sludge, or in general a "suspended growth" treatment process tend to consume significantly more power as compared with, for example, packed media filter treatment systems. The reason for that is inherent to the major differences between the two processes. Suspended growth treatment units use either "continuous flow" or "batch" tank reactors into which air (oxygen) must be introduced into the tank of liquid to achieve and maintain sufficiently aerobic conditions for certain biological treatment processes to occur. Examples of continuous flow treatment systems are a number of different manufacturers' aerated tank units, and batch tank reactors include sequencing batch reactors or SBRs, also produced by several manufacturers.

In contrast, aerobic packed media filters are dosed near their surface with effluent trickling down by gravity through the media. They are able to maintain aerobic conditions naturally as long as they're designed and built properly, and not flooded. Examples of those include intermittent sand filters and recirculating sand/gravel filters, recirculating trickling filters (e.g., Bioclere™), and the AdvanTex® synthetic (plastic) media recirculating filter. Bioclere and AdvanTex proprietary treatment systems, and other "fixed film" processes are discussed further in Chap. 6. Such media filter treatment units can be dosed intermittently during the day, maintaining a moist environment for the bacteria attached to the surface of the media, and continue to exhibit good treatment following periods of at least a certain duration when they are not being dosed. The time over which they are able to sustain bacterial populations responsible for treatment during nondosing periods depends on the media type and surface area.

6. Expected useful service life

Some project owners may view a long useful service life of their decentralized wastewater system as less of a priority due to the likelihood of their shorter-term ownership of the property. In general, however, the importance of quality of construction methods and materials to sustainability for wastewater systems can't be overemphasized. The period of time over which a particular system would be expected to operate reliably without problems or failure, and any major repairs or replacement of major components has major health, environmental, and resource implications. The expenditure of energy, materials, and money associated with repair, replacement costs, and lesser performance over time tends to far outweigh any short-term savings associated with using inferior quality materials or installation methods.

The conditions in which a particular type of system is used can greatly affect its useful service life, and it's therefore important to review operating histories of systems used under comparable conditions. Such evaluations should be made based on systems that have been designed and installed properly. Factors affecting useful service life include materials of construction used such as pipe quality and thickness, selection of materials as related to susceptibility to corrosion or other adverse impacts from local conditions, and so on. For example, Schedule 40 PVC pipe exposed to similar operating conditions might be expected to last longer than thinner-walled pipe or tubing. Concrete or inappropriate types of stainless steel*[,7] exposed to corrosive wastewater or coastal conditions would likely not last nearly as long as fiberglass, though concrete might be needed in some cases due to structural considerations. Such things need to be considered on behalf of long-term use and sustainability of the system.

  7. *Recycling and/or reusing some or all of the treated waste stream*

    While returning water to the hydrologic cycle is "recycling" or "reusing" water in a regional and global sense, the terms "recycle" and "reuse" are used here in a somewhat narrower and immediate way. Depending on local climate and rainfall patterns, there may be a need to try to reuse at least some of the treated effluent from onsite systems for watering lawns or gardens. While there is often a wish to water plants via surface application, certain aspects of that approach should be considered when comparing it with other reuse approaches.

    For systems reliably and consistently producing a high enough quality of effluent, the treated wastewater might be used for a nonpotable indoor water use such as toilet flushing water. This type of use is considered recycle, in that it replaces water supplies that would otherwise be needed for those uses. Currently most recycling of treated effluent occurs with larger wastewater systems with recycled water used predominantly for offices and commercial facilities for toilet flushing, HVAC cooling water, and other nonpotable uses. An example of this is a large high-rise apartment building (the Solaire) in New York City's lower Manhattan area, where treated effluent from the wastewater plant located in the basement of the building is recycled for toilet flushing, HVAC water and subsurface irrigation of an adjacent park. As technologies improve over time, recycle of treated effluent will likely become increasingly possible with minimal risks to humans or pets in households.

    Depending on any applicable regulatory requirements, public and environmental health considerations and subsurface conditions, surface applying effluent usually necessitates having to treat the wastewater to a higher level, including disinfection, as compared with certain subsurface dispersal options. For systems located at residences or where there would be public exposure to surface-applied effluent, it is important to select a treatment process that ensures consistently low levels of effluent ammonia for odor control and to prevent adverse impacts to local watersheds in surface runoff, as well as

---

*See literature for stainless steel alloy 2507, used for more corrosive and seawater environments.

efficient and reliable disinfection. Subsurface dispersal methods are now commonly used that dose the treated effluent into the upper soil horizons, and still within root zones of vegetation, resulting in a beneficial reuse of that effluent. Low pressure dosing is such a method, and is described in detail in Chap. 10. Further treatment may be provided naturally through biological processes in the soil, resulting in lower overall treatment and operational costs as compared with surface application systems.

Potential water quality impacts that may occur with surface application systems, particularly following rainfall events above sensitive watersheds should also be considered when comparing options for reusing effluent. Relatively shallow subsurface application of effluent places the effluent in the root zone for uptake of nutrients and moisture by plants while preventing surface runoff of the effluent during wet weather periods. Further, certain household chemicals tend not to be broken down in the predispersal treatment processes typically used for onsite systems. Many types of soil have the ability to retain and provide the opportunity for the biological degradation of chemicals that can build up in the environment over time if allow to combine with stormwater runoff.

8. *Expected repairs or replacement of components over time, and their costs*

This factor relates to long-term overall costs of systems, the useful service lives of systems components and resources needed to keep systems functioning properly over time. An example might be the type of pump selected for a system. While pumps might be considered a relatively minor part of a system due to its (typically) lesser cost as compared with other components, it is certainly important to the system's ongoing proper functioning. Experience has shown that high head effluent pumps, which are vertical turbine well-type pumps modified for use with effluent, tend to have longer service lives than most lower head centrifugal effluent pumps on the market today. Based on that experience, some manufacturers offer 5-year prorated warranties for vertical turbine submersible effluent pumps as compared with 1-year warranties typical for centrifugal submersible effluent pumps. Manufacturers of grinder pumps recommend sharpening cutters/blades on those types of pumps at least about once every year or so, and due to the wear on pumps having to macerate solids as well as move liquid at certain operating pressures, grinder pumps would be expected to require repair or replacement more frequently than effluent pumps of comparable quality.

For suspended growth treatment systems relying on aerators/blowers to introduce air/oxygen most of the time or continuously, those components will require replacement from time to time. The U.S. EPA estimates that compressors and aerators for aerated tank units last about 3 to 5 years, costing from about $300 to $500 to replace in the United States, and more where parts and service providers aren't locally available.[8]

Regardless of the types of components used for a system, realistic consideration should be given to the need for repairs or replacement over time, and those costs and component accessibility considered in the planning and design.

9. *Energy needed to operate the system, and maintain "ready for use" treatment conditions during and following periods of lower or no usage*

Systems serving vacation or seasonal use homes and businesses tend to be subjected to operational stresses that need to be considered when selecting technologies and developing the details of designs. Biological treatment processes important to the proper functioning of systems need to be sustained well enough for variable or seasonal occupancy homes and businesses, so that when the system is next used it will perform acceptably. Some processes are much more vulnerable to performance problems than others in that type of situation.

As mentioned above for packed media filter treatment systems, aerobic bacteria attached to the filter media can survive in sufficient numbers as long as even a small amount of effluent is dosed over the media periodically. That is due to the unsaturated and naturally aerobic conditions in packed media filters, and to the much greater surface area in the media serving as bacterial "habitat," as compared with tank reactor systems. For tanks of liquid, aerobic bacteria must have air/oxygen mechanically added to the fluid to maintain aerobic conditions for the bacterial population to survive. The bacteria in biological treatment systems are responsible for treating the wastewater, so it's important to maintain healthy enough bacterial populations. Without a sufficient source of food and oxygen, the suspended bacteria simply die and accumulate in the bottom of the tank as "sludge."

These types of natural phenomena inherent to different methods of treatment are important considerations in the planning process, and for ensuring long-term reliable performance.

10. *Sludge/septage production associated with the treatment process(es)*

Essentially all types of centralized or decentralized wastewater treatment systems produce at least some amount of "residuals" needing removal and proper handling. The amount produced, however, may vary significantly, depending upon the particular processes involved in the complete system. For decentralized systems, there is almost always primary settling or a "septic" tank needing periodic pumping to remove buildup of settled solids at the base of the tank and scum/grease accumulating to some depth at the surface of the tank. The time interval between pumping and hauling is very dependent on the design and sizing of the primary settling tank(s) or compartment(s). For secondary and advanced methods of treatment the sludge/biosolids accumulation rate is also very dependent on the design and the particular treatment process used. Waste sludge/septage needs to be hauled away and properly treated off-site. Many geographic locations do not have nearby facilities capable of dealing with that waste.

These factors are very important to consider when selecting a method of treatment, and sizing system components for particular projects. It is rarely cost-effective in the long term to skimp on tankage and other process capacities when the added incremental cost for sizing components in a more sustainable way is typically minor in comparison with significantly shorter pump and haul intervals. In cases where there may be long transport distances to suitable septage and sludge "drop-off" facilities, it may be appropriate to provide added capacities in the design to further increase time intervals between pumping. Designers should determine approaches that best serve the project and don't compromise important aspects of system performance, while balancing various cost and sustainability factors. Sludge and septage are discussed further in Chap. 5.

11. Treatment by-products from the system resulting in immediate or long-term environmental impacts

A common method of disinfection used for decentralized and small-scale treatment systems is chlorination. While larger (and in particular municipal) scale systems have increasingly provided for dechlorination following disinfection using chlorination, many small-scale systems do not. Chlorinated organics, which can readily form from chlorinated wastewater effluent, are known human carcinogens. Chloride ion is also, like nitrate, a very mobile ion in soils and subsurface conditions and migrates with water and moisture. It is therefore frequently measured and used for water-quality-monitoring activities.

While it may take some period of time for the effects of discharging chloride species and other chemicals into the environment, it seems intuitively clear that there will eventually be adverse impacts. Therefore, if disinfection of wastewater effluent is needed based either on site conditions or the method of final effluent disposition to be used, options such as ultraviolet irradiation (UV disinfection) might be considered, along with treatment processes that lend themselves to effectively using UV. Filtration with sand or other types of fine media can naturally provide very high levels of pathogen reduction if designed and built using the proper type and gradation of media. Maintenance requirements and power usage for each disinfection method should be considered, along with other factors such as performance reliability and costs.

12. Local availability of materials and resources used for the system

The use of local materials and resources for projects to the extent that is feasible is important for a variety of reasons ranging from energy consumption to ecological and environmental considerations.

Using local materials for construction aids local economies and avoids energy consumption and use of resources needed to transport materials long distances. It also tends to enhance the economic feasibility of recycling certain materials that cannot compete economically when considering marketing and transport costs from elsewhere. An example of this would be the use of ground glass for sand filters in areas of the world where there may be a shortage of suitable natural filtration media. This is discussed briefly in Chap. 7. Another example would be the use of chipped tires for dispersal field media where it may either be costly to transport suitable aggregate to a construction site, and/or where there are chipped tires needing to be recycled or disposed of properly. The use of chipped tires for subsurface flow wetlands and dispersal field trenches is discussed in Chaps. 8 and 10, respectively.

The use of locally collected and propagated native vegetation for onsite wastewater dispersal fields is also very important not only for the health of the vegetative cover, but also for local and regional ecological reasons. The introduction of exotic or nonnative vegetative species can adversely affect certain terrestrial environments that can in turn impact local watersheds and wildlife habitat. As an example, certain dry forest climates may not have naturally adapted turf-grass species that might be used for vegetative coverage over effluent dispersal fields. In such cases a mixture of native bunchgrasses and sedges might be used to accomplish a good coverage rather than to import nonnative turf-grass species that may become invasive and/or might not be compatible with the needs of local fauna. Local faunal species around the world have evolved depending on the presence and health of certain floral species for their continued survival and propagation.

13. *Transport costs and resources expended for shipping products to the project site*

This factor is related to numbers 3 and 12 above, and primarily considers financial realities and use of resources associated with different options. For example, the construction of a sand filter may cost more due to the added labor for the installation as compared with the use of a manufactured treatment system that produces comparable effluent quality. However, if the sand filter can be constructed using locally available filter media, and only a minimal number of products shipped from suppliers, it may be more cost-effective than shipping a full treatment system to the site from perhaps a longer distance.

Proprietary treatment units and major system components, unless manufactured locally, must be transported to the project site with those associated costs and energy consumption. Some proprietary units can be shipped in large numbers in a very compact manner, making possible much more efficient energy and resource expenditures. Housings of some modular units can be stacked inside each other and their smaller internal components assembled by a single trained technician at a local facility, or at jobsites for larger projects. Familiarity of engineers/designers with these types of possibilities offers greater opportunity for more efficient use of energy and material resources where products must be shipped over long distances.

14. *Labor and material costs associated with operating the system*

Some system types and designs require more frequent visits, and in general a greater presence of trained service providers or operational adjustments to ensure proper continued operation of those systems, as compared with others. For example, if a system needs certain valves or controls adjusted fairly frequently for processes, and perhaps an operator on hand to make those adjustments on short notice, the sustainability of that system might be questionable if the site is a significant distance from trained service providers or operators. Examples would include remote rural settings or islands.

Other potential operational costs to consider include any monitoring and analytical costs that might be needed or required for certain types of systems. An example of this would be systems surface applying or irrigating effluent, and the need to monitor the quality to verify adequate pathogen reduction is occurring.

15. *Labor and material costs associated with routine maintenance of the system*

As with other factors, some systems require more frequent routine maintenance and servicing than others. Properly designed and built sand filters, for example, should require only one routine servicing per year, as compared with 4 to 6 visits per year by a trained service provider recommended by the EPA for aerated tank units.[8]

While designs should be configured so as to minimize the potential for clogging as much as possible, for packed media filter (or "fixed film") treatment systems, comparisons of maintenance and replacement costs for filter media in the event of clogging over time should be considered. Some types of filter media better lend themselves to easy maintenance/servicing and/or periodic replacement, as described in Chap. 6.

Facilitating easy access to system components for routine maintenance or for periodic replacement of certain components should be considered in the design and installation of the system. This is something that is all too often neglected by designers, since they are often not the person(s) returning to sites later to perform that servicing. As a result, it is not uncommon for installers to make adjustments to the original designs when they anticipate problems later on with access or servicing, if designers or regulators don't check all of the details of the installation. And because the installer is not typically privy to the thinking and calculations leading to the design as planned, this can lead to unintended and undesirable consequences with the system's operation. Examples can be things as simple as where a check valve is located in discharge piping following a pump. It's therefore important for designers to be involved in the construction process and communicate with installers as needed, inviting comments and questions early on and during the installation to prevent problems.

16. *Aesthetic and social considerations, including potential for odors, or other detracting features of the system*

An example of aesthetic considerations would be odors that might occur with residential onsite systems using surface irrigation of effluent, and particularly those using certain treatment methods. Some treatment processes tend to nitrify effluent much more efficiently than others, thus producing much lower ammonia concentrations in the effluent that would cause odors at higher concentrations. Properly designed packed media filters tend to nitrify wastewater much more efficiently than residential scale aerated tank units, based on data for those two categories of treatment systems. Sand filtration produces a very clear effluent with low levels of suspended solids that is suitable for disinfection using ultraviolet irradiation. If an effluent is to be surface applied, odors associated with either chlorine from disinfected effluent or ammonia from inadequately nitrified effluent can be an aesthetic issue, and a concern to surrounding property owners.

Costs associated with most of the above factors can be included directly in long-term cost analyses to compare alternate system options. Monthly, annually, or periodically recurring operation, maintenance, repair, and replacement costs should be included in a long-term cost analyses (or "net present value" analysis) using reasonable assumptions for interest rates, power costs, sludge/septage removal costs and service lives for system components. Expected useful service lives and major component repairs or replacement costs at those intervals should be included in those analyses. For example, some systems have major components that may only last 15 to 20 years, while other system types may have 30- to 40-year service lives for their major and most costly components. For a reasonable comparison of such options, replacement costs at that 15- to 20-year interval would need to be included for the former.[9]

Essentially all of the factors discussed above, except possibly the last (aesthetic considerations), contribute in one way or another to the "carbon footprint" of wastewater service approaches, methods, and materials. The extent to which construction and use of a certain wastewater treatment system…

- Consume nonrenewable energy resources
- Deplete, conserve, and/or reuse water and material resources
- Produce harmful discharges (liquid, gas, or solid)
- Produce harmful by-products from the manufacturing of products needed for the system

...are all factors central to its long-term sustainability and carbon footprint. Production of lesser or greater amounts of greenhouse gases depends on the aggregate of factors associated with implementing specific wastewater service approaches.

The use of subsurface effluent dispersal fields for final disposition of wastewater effluent tends to be highly compatible with other sustainable and low impact development (LID) practices. Impervious covers, including roads, driveways, sidewalks, rooftops, and so on have shown to be directly related to adverse impacts to watersheds. Impervious covers eliminate otherwise natural land areas from infiltrating rainwater and natural degradation and assimilation of nutrients and sources of both human and animal pollutants. Natural plants and vegetative land covers continually go through organic decomposition, and produce nutrients and organic pollutants that if not assimilated into the soil can run off with rainwater and have adverse impacts on watersheds.

Concrete and other impervious land covers can also significantly raise temperatures for rainwater running off of those surfaces. Even slight temperature increases in receiving waters can have serious impacts for many aquatic species. While some types of engineered controls have shown to be very effective for minimizing erosion and sedimentation from disturbed areas during and following construction, most measures used for attenuation of dissolved stormwater pollutants such as nitrogen have shown only limited success.[10,11] It's therefore particularly important for watersheds with sensitive aquatic species and resources, such as coral reefs, to maintain significant buffer zones or green belts/strips between developed areas and receiving waters.

Surface application of treated wastewater may contribute to pollutant loading for sensitive watersheds during wet weather periods unless sufficient storage is provided to accommodate wastewater flows during those periods. However, effluent dispersal methods such as low pressure dosing in subsurface trenches, as discussed in Chap. 10 can accomplish both sustainable final disposition of wastewater effluent, while leaving vegetated surfaces available for infiltration of rainfall and uptake of nutrients or other potential pollutants. It is usually helpful to "step back and take a bird's eye look" at projects, considering how well each element is integrated with others, and whether it contributes favorably or unfavorably to a sustainable project overall.

2.2.1   LEED Rating System

The LEED rating system is used by architects, engineers, interior designers, landscape architects, developers and real estate professionals, and governments to encourage the use of more sustainable construction practices around the world. Rating systems continue to be developed by LEED committees that assign point scoring to project elements considered consistent with specific sustainable building goals. The LEED program doesn't award scoring credits to particular products, but to the ability of projects and their elements to meet certain performance standards established in the rating systems. It is left to the project design and implementation team to determine which products to use in meeting those goals. More information about the LEED program can be found at the following U.S. Green Building Council website: www.usgbc.org.

Some manufacturers, such as Orenco Systems, Inc. have developed LEED *profiles* for some of the collection and treatment systems they promote that use their product lines. Aspects of collection, treatment, and effluent dispersal approaches for which LEED ratings could be sought by designers will be discussed in the context of sustainability. But again, it is up to project owners, managers, and designers to determine which products will achieve certain LEED rating and sustainability goals targeted for specific projects.

Determining the most appropriate and sustainable system is usually somewhat of a balancing act among potentially competing factors. In some cases a system may require less operation and maintenance over time and have less energy usage, but may cost more initially to install. With the help of the engineer/designer the owner can consider long-term and overall costs for different approaches, resulting in more informed project decisions.

References

1. "Pharmaceuticals and Personal Care Products: An Overview," *Pipeline: Small Community Wastewater Issues Explained to the Public*, Winter **18**(1) 2007, Marilyn Noah (NESC staff writer/editor), West Virginia University, National Environmental Services Center (NESC), Morgantown, WV.
2. "Traces of Drugs Found in Drinking Water," ABC News Report, Oct. 15, 2008, Gigi Stone, Research by Dr. Brian Buckley, Rutgers University.
3. *Onsite Wastewater Treatment Systems Manual*, United States EPA 625/R-06/016; Sep. 2006.
4. M. Overcash and D. Pal, *Design of Land Treatment Systems for Industrial Wastes*, Ann Arbor Science, Ann Arbor, MI, 1979.
5. M. Overcash, Ronald C. Sims, Judith L. Sims, and J. Karl C. Nieman, "Beneficial Reuse and Sustainability: The Fate of Organic Compounds in Land-Applied Waste," *J. Environ. Qual.* **34**:29–41, 2005.
6. S. C. Reed, R. W. Crites, and E. J. Middlebrooks, *Natural Systems for Waste Management and Treatment*, 2nd ed., McGraw-Hill, New York, 1995.
7. *Corrosion and Corrosion Control in Saltwater Environments: Proceedings of the International Symposium*, Electrochemical Society Corrosion Division, Electrochemical Society Meeting, Phoenix, Arizona, 2000; Edited by D.A. Shifler, P.M. Natishan, T. Tsuru, and S. Ito; Published by the Electrochemical Society.
8. U.S. EPA *Onsite Wastewater Systems Treatment Manual* (EPA/625/R-00/008), Feb. 2002.
9. J. L. Riggs, *Essentials of Engineering Economics*, McGraw-Hill, New York, 1982.
10. M. E. Barrett, "Comparison of BMP Performance Using the International BMP Database," *J. Irrig. And Drain. Eng.*, ASCE, 134(5):556–561, Sept./Oct. 2008.
11. *Final Report of the Retrofit Pilot Program*, Report No. CTSW-RT-01-050, CALTRANS, 2004, California Department of Transportation.

CHAPTER 3

Project Planning and Site Evaluation

One of the most important elements of implementing cost-effective and sustainable wastewater systems designs is early planning and coordination of the system design with the overall project plan. Very early in the planning process, site planners, architects and/or developers should consult with the engineer/designer of the wastewater system(s) (assuming none of those other planners has expertise in that discipline). The wastewater planning best begins alongside of, and possibly before, conceptual layouts of structures to avoid potential conflicts, since relatively large areas may be needed for it. Available areas of sites should be identified that are the most likely to offer the best natural soil treatment conditions. Once preliminary estimates of wastewater flows have been made, those potential effluent dispersal sites (and estimated areas) can be physically evaluated to determine limiting design parameters (LDPs). After that is established, methods of treatment and dispersal capable of meeting the necessary effluent quality can be identified. Candidate approaches can then be compared to determine the most cost-effective, appropriate, and sustainable options capable of meeting project needs and any applicable regulatory requirements. Replacing wastewater systems on sites where facilities to be served already exist is often more challenging, since areas best suited for treatment and dispersal components may already be used for structures or utilities.

Figure 3.1 is a flow chart outlining the basic steps in the planning and implementation of systems, with each of those steps discussed below. As can be seen from the dashed lines on the left side of the chart, early on in the planning process there may be points at which it's necessary to adjust the scope of the project. Those adjustments may be needed based on such findings as excessively high preliminary cost estimates for candidate wastewater service options, or land area requirements and limitations associated with projected wastewater flows from the system. At the point where detailed design of the system commences, all of those factors should have been considered and any necessary changes made to avoid potentially costly revisions later in the planning process.

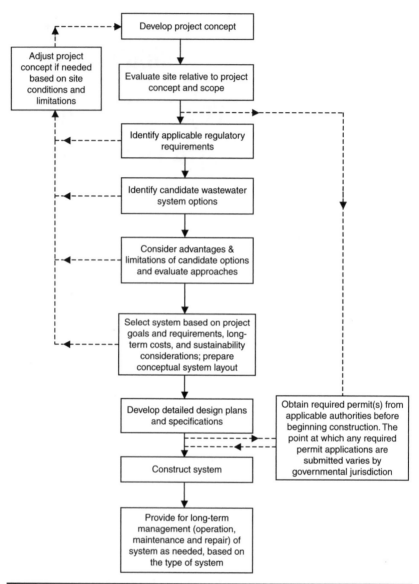

FIGURE 3.1   Basic steps in the planning and implementation of decentralized wastewater systems.

3.1   Project Concept Development and Implementation

It's important, during the earliest parts of a planning process needing an onsite or decentralized wastewater system, to involve an engineer/ designer who'll be responsible for that portion of the project. That person may be a subconsultant to the architect or builder on the project, or may work directly for the property owner or developer. If the property owner or developer contracts directly with the waste-water engineer/designer, it's important for coordination to be maintained between that person and the architect(s) and all other key persons involved in the project planning process.

The focus of and needs for cost-effective and sustainable waste-water systems can be somewhat different from those associated with planning other property improvements, and it's important early on to effectively integrate those various project elements. For example, a particular site may have very rocky areas with steeper slopes and other areas with lesser slopes and deeper soils. If those flatter-sloped areas are found to be suitable for final disposition of wastewater effluent, that would likely result in significant cost-savings and a much more sustainable system overall as compared with using more difficult areas of the site. Locating buildings and other above-ground structures in rocky conditions may actually be preferred for structural stability and foundation construction, whereas it can result in substantially higher installation and maintenance costs for onsite wastewater systems providing acceptable levels of treatment to address LDPs in those conditions.

The wastewater planning and implementation elements of both small and larger projects can generally be broken down into four principal phases. Those include the following:

Preliminary/conceptual planning: In this earliest phase of the planning, site and soils evaluations are conducted; applicable rules and requirements are identified; viable alternatives are identified and evaluated; preferred options are selected based on the project needs and goals (including sustainability considerations), and site conditions; conceptual layouts and preliminary cost estimates are developed as needed to evaluate and compare options; and a selection is made for a particular approach. Depending on any local applicable regulatory requirements, coordination with local permitting authorities may be needed at this point to confirm acceptability of the wastewater system for the particular project.

Detailed design: Detailed plans and specifications are developed for the system in this phase. In many cases, and depending upon local requirements, a permit application is submitted along with the plans and system specifications to regulatory authorities for review. Regulatory review processes frequently require response by the engineer/designer of the system to specific questions about the design, construction and/or long-term management of the system.

Construction: This phase of the project begins following detailed design and issuance of any required preconstruction permits. The timing of the system installation should be coordinated with other construction activities to ensure that (1) wastewater system components aren't damaged if installed before other construction activities are occurring and (2) that there is enough access to areas of the site(s) by earth moving and any other equipment needed for installation of the various wastewater system components.

In many cases system engineers/designers work with project owners/managers to select a qualified contractor for installing the wastewater system. Because it is a specialized area of the construction industry, most U.S. permitting authorities require that wastewater systems installers have certain training and certification/licensure. Whether that is or isn't the case where a project is located, the engineer/designer can be helpful in reviewing the qualifications of candidate installers, and making recommendations accordingly. Contractors should be evaluated and selected based on their experience and performance history installing the type of system planned.

In cases where no local contractors are experienced with the installation of a particular type of system, one of several approaches can be taken depending on the size of project. For small-residential-scale projects, the project owner or manager might arrange with the system engineer/designer to be on hand during construction to respond to questions by the contractor as they arise. For larger projects, both the engineer and technical field support staff for manufacturers of products used for the installation might coordinate with the contractor prior to and during the installation to make sure all questions are answered by someone experienced with the particular system component. For some larger projects, due to either their complexity, difficult or sensitive site conditions or other factors, it may be best to try to arrange for a contractor from elsewhere who's experienced with installing that type of system in comparable conditions to either (1) supervise and coordinate with a local contractor, and effectively train the local contractor for future installations or (2) bring enough experienced construction crew personnel to the project site to do the installation with the help of local laborers. A determination of the best approach depends of course on project budgetary, scheduling, and any other logistical constraints.

Because the system engineer/designer has the best understanding of the design, and the reasons for certain system elements, it's important for owners or project managers to plan on the engineer visiting the site to observe important parts of the installation as it proceeds. There is rarely a "perfect" set of detailed drawings and specifications, and it's important that the contractor maintain a clear understanding of the designer's intent as the work proceeds. Even seemingly minor deviations can adversely impact the functioning of the system (e.g., placement of a valve in a slightly different location due to plumbing challenges in tight spaces, or not maintaining positive grade on a pressure line in a critical stretch of pipe, and potentially creating air pockets/blocks that would show up after the installation was complete). Contractors should communicate and coordinate with the system engineer and any inspectors to be sure that all parts of the installation are examined before being covered/backfilled. Because most decentralized wastewater systems are underground utilities with mostly buried components, "a small amount of prevention is worth many pounds of cure." It may take several years for certain problems to become apparent, but by then otherwise preventable conditions may have resulted in serious damage or failure to very expensive parts of the system. The contractor may no longer even be available (and willing) to take responsibility for and correct the problem at that point.

Long-term system management: Regardless of how simple a system may be, all wastewater systems need some degree of long-term maintenance and care. Those elements should be considered prior to implementing a given type of system. If it is not feasible or cost-effective to provide the level of care needed to ensure long-term effective and reliable performance of a particular system or approach, other options should be considered.

Where skilled operators and/or technicians will be needed for the ongoing functioning and care of a system, if no one is available locally to provide those services, arrangements will need to be made for one or more persons to be trained. That can often be arranged during construction and start-up of the system if an experienced contractor has traveled to the site to construct the system or be a consultant to the construction process. Most contractors experienced with the installation of certain types of wastewater systems are also involved in their long-term maintenance and/or repair calls, and may be willing to train local persons to provide basic ongoing maintenance and repair services.

Regardless of the approach taken for providing ongoing system maintenance and any needed servicing, the importance of implementing a sound management plan cannot be overstated relative to sustainable wastewater systems. Problems that would otherwise arise can have very expensive consequences environmentally and financially, and on public health.

3.2 Wastewater System Flow Determination or Estimation

To identify the most suitable candidate effluent dispersal areas for a site, and to make sure that a large enough total area will be evaluated for that use, it's important to have reliable estimates of wastewater flows early in the planning stages for decentralized systems.

For existing systems needing modifications or replacement, determining or estimating flows may be easier and more reliable utilizing water usage records, if available, for the facilities served by the system. For new systems, local usage patterns should be considered to the extent that such records are available for review. In reviewing usage records and patterns it's important, however, to try to distinguish between water entering the wastewater system and water used for outdoor purposes such as car washing or lawn and/or garden watering. For geographic regions having significant seasonal temperature differences, using water usage records for the coldest months of the year during which there's much less outdoor water usage can yield more accurate estimates than averaging all months together. For example, in southwestern U.S. states, water usage records for the months of December, January, and February may be averaged and used to determine average monthly water usage, which can then be divided by numbers of days during that period to estimate average daily flows.

For individual residential and commercial buildings, it's important to consider wastewater production from specific plumbing fixtures and appliances to be used. Those water conservation efforts and the use of water-saving fixtures may not be reflected in general water usage estimates for a local population. The use of water-saving appliances and fixtures can greatly reduce daily water usage. Discussions with those who'll be living in the home and relying on the system or architects planning homes for developments are an essential part of achieving relatively accurate estimates of wastewater flows. For both commercial and residential facilities, potential changes in usage over time should also be considered. For example, a couple without children might occupy a relatively large residence and generate a relatively low wastewater flow for that size of home. If in the future the same house is occupied by a family of four to five persons, and if the wastewater system was only designed to handle the flows typical of the previous owners, problems and ultimate failure of the system would almost surely occur. If the original system and site planning were done with the lower flows in mind, there might not even be a suitable area left on the site to expand the system.

For these reasons, most U.S. regulatory permitting authorities specify design flow criteria for residences that are based on both square footages of living spaces as well as numbers of bedrooms. Using both square footages and numbers of bedrooms for existing or planned structures takes into consideration the fact that remodeling may occur within the exterior walls of the structure that might add to the number of bedrooms and/or number of persons the house would accommodate. Different formulas are used to assign flows based on numbers of bedrooms and square footages, depending upon the particular state and permitting authority. For example, in Texas, if water-saving fixtures are planned or being used for a residence, the minimum daily per capita flow to be used for designs is currently 60 gallons per person per day (227 liters per person per day). It is further assumed that the number of persons in a dwelling for a certain number of bedrooms is $N + 1$, where N is the number of bedrooms. The presumption there is that two persons (e.g., two parents) would occupy a bedroom and one person would occupy the other bedroom(s) (e.g., children, roommates, or other family members). So, for a three-bedroom dwelling, the design criteria for flow would call for a minimum design flow of $(3 + 1)$ persons $\times 60$ gpcd $= 240$ gal/day (908 L/day). The dwelling square footage associated with that daily flow would be required to be less than 2500 ft^2 (232.5 m^2) of living space.[1]

Obviously there's a significant potential for either over- or underestimating flows, depending upon actual circumstances that might vary greatly from such assumptions. It's therefore important to be realistic about both current and potential future usage patterns. For projects that will be phased in over time, and for which it may not be possible to accurately anticipate occupancy, usage patterns and wastewater flows several years into the future, it may be most appropriate and cost-effective to use wastewater service approaches that enable phasing-in of the wastewater treatment system and tracking of flows over time. A nationwide study of large-/community-scale wastewater systems in the United States showed that this approach could result in considerable cost-savings for projects. Cost data obtained in that study showed that systems' cost per average gallon of actual (measured) flow was in many cases much greater than cost per gallon of design flow. Costs per gallon of design flow ranged from $6 to $140 per gallon, while costs per actual treated/measured flow ranged from $18 to $494 per gallon.[2]

Water usage patterns depend greatly on climate and geographic location along with local cultural patterns and socioeconomic conditions. Daily per capita wastewater production from plumbing fixtures might range from an average of just a few gallons per day per person up to 150 gallons per person per day (568 liters per person per day) for some resort hotels and vacation homes. While the higher end of that flow range might not be considered a sustainable or ecologically sound level of water usage, it's important to be realistic with variations in daily usage by persons from different regions of the world for facilities serving tourists and vacationers.

It's also important to consider any seasonal or daily/weekly variations in use for buildings or facilities that would result in variable wastewater flows. Examples include churches, restaurants, vacation homes, schools, resorts and hotels, and day or overnight use parks that might have much greater use during certain seasons. Several of these types of facilities may have little to no usage during certain months. The wastewater system designer should take such usage patterns into consideration on behalf of a variety of critical design factors, including storage and flow equalization capacities, and loading rates to treatment processes and soil dispersal areas. Based on the projected duration of peak usage periods and possible costs associated with adding storage or equalization capacity, it may be necessary for the system to be designed based on flows during those peak usage periods. These design issues are discussed further in later chapters.

Using more conservative design flows tends to be more important for systems serving single dwellings, smaller multifamily units or relatively small commercial projects because of the greater proportional impact each change in usage patterns and occupancy may have on the system. For larger decentralized systems, with flows produced from larger numbers of homes and/or businesses, experience has shown that variations in use and occupancy tend to better attenuate variations in flows, with differences in flows generated from individual residences tending to better balance each other out. Exceptions to this, however, may be found in geographic areas where there are significant seasonal changes in tourism, such as in the Caribbean region. Occupancy for certain vacation resorts and developments tends to be much lower overall between the months of July and October. Each specific project should, therefore, be evaluated for realistic daily and seasonal use patterns to determine sound flow estimates, along with meeting any applicable regulatory design flow requirements.

A variety of onsite and remote flow-monitoring approaches are available today in control panels, along with improvements in flow monitoring/measurement devices. Control panels can be ordered with event counters that record the number of times, for example, that a pump dosing a dispersal field activates. This information can be used in combination with flow measurements from draw-down tests performed on a system during start-up (or at some time thereafter) to determine the total flow dosed to the field during a given time period. Such event counters can also be used to track flows through treatment units if they are dosed with a pump, or a siphon equipped with those tracking capabilities. Direct flow measurement devices have varying accuracies depending on their design, size, and specific use. Flow tracking methods and materials appropriate to the specific system need to be determined by the design engineer.

The table Typical U.S. Residential Wastewater Design Flows (among several downloadable tables available online at this book's website) shows typical flows for systems serving dwellings of varying sizes in the United States and elsewhere in the world where wastewater production would not tend to be limited by the availability of water supplies. The values shown in the table are, however, based on residential water usage prior to the more common use of water-saving fixtures today in the United States. For example, U.S. toilet manufacturers may no longer make toilets for home use with flows greater than 1.6 gallons per flush (as of January 1, 1994). Toilets operating with higher volumes per flush could continue to be manufactured, but only for certain commercial applications and only until January of 1997.[3]

The table Commercial and Multi-Family Residential Wastewater Usage Rates (downloadable from this book's website) provides per capita flows for a variety of nonresidential uses typical of those used for designs within the United States. As can be seen from those tables, the use of water-saving fixtures can reduce water usage and wastewater production by 20 to 25 percent or more. That translates into a variety of significant savings that contribute to more sustainable projects, including reduced treatment capacities, conveyance needs, and soil dispersal areas, all with their associated costs. For existing buildings and facilities served by systems, the cost-effectiveness and feasibility of replacing fixtures and appliances with water-saving ones might be well worth exploring.

The two tables offer some guidance for establishing design flows for conditions typical of the United States. The per capita flows those estimates are based on however are *averages*, and may need to be adjusted to account for periods of significantly higher flows, consistently higher flows, or in cases where water usage may be consistently lower. *Average* is distinguished from *design* flow in that sense, with design flows needing to be established based on actual daily, weekly or seasonal, and peak usage patterns.

3.3 Wastewater Characterization

3.3.1 Characterizing Wastewater Strength and Constituents for Each Project

There are many variations in wastewater quality that are all considered "domestic" sources, with potentially significant differences in levels of key pollutant concentrations and LDPs for given sites. Even within certain categories of use, such as restaurants and schools, the wastewater strength and qualities can vary greatly depending on the specific activities in that facility.

For example, elementary schools in the United States tend to have fewer showers as compared with junior-high and high-school facilities, due to increases in sports programs in the latter. Depending upon the numbers of specific types of plumbing fixtures for which there would be different uses and frequency of use, there may be substantially more greywater produced as compared with black water, and thus less total nitrogen concentration most likely. Laundry services might introduce more phosphorus into the waste stream with lower than average nitrogen and pathogen levels as compared with other domestic wastewaters. Solids, concentrations can also vary greatly for different sources of wastewater, such as kitchens where garbage grinders are frequently used. Garbage grinders can contribute to a variety of problems in onsite system treatment processes, as discussed further in Chap. 5. Larger solids are more easily settled out of wastewater in primary settling tanks. Therefore, if it is an option to compost or otherwise dispose of organic wastes, that's preferable to the use of garbage grinders. Some types of restaurants use much more grease for cooking as compared with others, with this almost invariably contributing to higher-strength wastewater despite grease traps and settling devices being used. Chapters 5 and 6 discuss some of these design considerations in greater detail.

The table Reported Raw Wastewater and Septic Tank Effluent Levels for Key Constituents (downloadable from this book's website) shows statistical quality medians, averages, and ranges for key wastewater constituents in residential and some categories of commercial wastewaters (both raw and septic tank effluent) in the United States. The data reported in the table was developed in a 2006 WERF study.[4] As can be seen in that table, average concentrations of certain parameters tend to vary based on types of facilities served. Food preparation and service systems, for example, show significantly higher averages for BOD and suspended solids. Those facilities would also tend to produce wastewater higher in fats, oils, and greases, with that increase dependent on the types of foods prepared. Total nitrogen levels were found to be somewhat lower on average for medical facilities, as compared with other commercial sources. While a significant amount of information has been compiled on raw and primary treated wastewater quality from residential sectors, more detailed evaluations and judgments are often needed for characterizing waste streams for commercial projects.

The table Raw Wastewater and Septic Tank Effluent Levels for Key Constituents: Combined Study Results (downloadable from this book's website) presents combined residential raw wastewater and septic tank effluent averages and ranges from other studies. Data from all of these studies were gathered from U.S. systems, for which average per capita wastewater production was found to be about 60 to 70 gal/day, depending on the study. With increased water conservation practices or limitations on available potable water supplies, the concentrations for some parameters would tend to increase. Local wastewater quality data for comparable projects and socioeconomic conditions, if available, is helpful to better estimate levels for the various constituents.

For determining typical mass loadings from those concentrations based on flows from facilities, the following calculations would be used:

$$\text{Mass loading in lb/day} = \text{concentration (mg/L)} \times \text{flow (gpd)} \times 8.34 \times 10^{-6}$$

$$\text{Mass loading in gram/day} = \text{concentration (mg/L)} \times \text{flow (L/day)} \times 10^{-3}$$

The table Breakdown of Residential Wastewater Constituents by Plumbing Fixture or Appliance (downloadable from this book's website) shows averages and ranges for the same residential wastewater constituents broken down by source within households. Typical percentages of flows associated with those fixtures or activities are included. The potential water conservation benefits of using low-flow toilets can be seen in the significant drop in percent contribution to total per capita flow from that source. Other low-flow fixtures and conservation measures contribute to reduced flows, but tend to maintain a similar overall percentage contribution.

"Resources and Helpful Links" on this book's website references resources and links to information that may be helpful in characterizing the wastewater to be produced from a variety of nonresidential sources. Information compiled by the U.S. EPA and various wastewater industry scientific and professional organizations that may be helpful in this respect are referenced in that section of the book's website.

For properties relying on decentralized wastewater systems, whether onsite or clustered/collective systems, it is very important that users of the system exercise care with wastes put into the system. It is critical that materials such as paints (whether latex or oil-based) not enter onsite/decentralized wastewater systems. It is not uncommon for residents in urbanized areas connected to large-scale municipal treatment systems to wash paint brushes in sinks, and empty a variety of household waste chemicals into the sewer system. The large numbers of users of municipal scale system tends to lessen the individual effects contributed by each of those events, though such practices should not be used even for large centralized wastewater systems. During home or business remodeling and other activities occurring on properties served by decentralized wastewater systems, it's all the more important to avoid introducing those wastes into the system. Doing so can result in many different types of problems, including toxicity to microorganisms needed for treatment and clogging of filters and piping.

3.3.2 "Greywater" and "Black Water"

Because of common perceptions and occasional misunderstandings about what constitutes "greywater" versus "black water," some discussion of those two definitions and concepts is presented here. "Black water" consists of toilet wastewater, along with any other domestic wastes included in (or combined into) that waste stream. It is sometimes thought that "greywater" consists of residential wastewater treated to a level that can be reused or discharged onto lawns or gardens. Under U.S. definitions of greywater, that is not the case. Greywater, whether treated or not, would never have included toilet wastes, or "black" waters. U.S. states vary in their definitions of "greywater," though all U.S. states agree on that one point. The U.S. EPA defines greywater, or "graywater" as "Wastewater drained from sinks, tubs, showers, dishwashers, clothes washers, and other non-toilet sources."

Because greywater from these various plumbing sources may contain significant levels of bacterial or pathogenic contamination, and especially depending on whether kitchen wastes or such things as washed baby diapers might have contributed to the waste streams, care should be taken to provide enough treatment prior to applying greywater to gardens or other areas where there may be human or animal exposure. If greywater drained from showers, sinks, and clothes washers that exclude any human excreta is to be discharged through some type of distribution piping into a lawn or garden area, it will still need adequate settling to remove such things as hair and lint to avoid clogged pipes or tubing. Primary settling requirements discussed in Chap. 5 are applicable to both greywater and black water.

As discussed in Chap. 2, sustainable wastewater planning and implementation includes recycle and reuse of wastewater to the extent practicable given the specific project circumstances and location. The use of greywater to replace potable water use for lawns or gardens is certainly of benefit from a resource conservation perspective, as long as adverse impacts don't result from the practice. Care should be taken to consider constituents in the waste stream to make sure the receiving environment is capable of assimilating the waste. At least one study has examined effects of greywaters consisting of different household sources of wastewater, including kitchen and laundry waters, on a variety of ornamental plant species. Laundry waters were observed to have no adverse effects on the plants. However, when kitchen wastes were included in greywaters, the waste was found to severely damage every species included in the study.[5]

3.4   Site Evaluation

Evaluation of the site conditions relative to a project's conceptual plan is one of the most critical steps in the planning process, and one which sometimes receives insufficient attention from planners and engineers. While most U.S. permitting authorities regulating onsite/decentralized wastewater systems require a physical site evaluation that includes soils analyses, in many parts of the world and even in some U.S. territories systems designers continue to rely on general soil surveys for detailed design of onsite wastewater systems. While such surveys can be very useful for preliminary planning for larger projects covering a variety of conditions, geophysical conditions can vary significantly across a single site without those changes being reflected in a soil survey.

The area(s) selected for evaluation on a site should take into consideration whether subsurface or surface application of effluent will be used, to make sure that there are adequate setbacks from features as needed for ensuring necessary protections. Considerations for effluent dispersal method selection are presented and discussed in Chap. 9, including advantages and limitations for different methods as related to costs and long-term sustainability. Achieving more sustainable wastewater systems depends on optimizing the use of natural treatment capabilities of the soils and other site conditions.

Depending upon where a project is located, the list of features and conditions to assess may vary due to either specific regulatory requirements or geographic conditions. In addition to an evaluation of soils and subsurface conditions, a list of basic site factors affecting system selection and design would include the following:

Site topography: A topographic survey of the planning areas is needed, and is helpful during even the earliest phases of planning for developing conceptual layouts. The survey should be sufficiently detailed for the wastewater system engineer/designer to have a good understanding of site elevations and changes in grade relative to structures served as well as any drainage-ways, or ground or surface water levels. Making the best use of the natural grade in the wastewater system planning can result in substantial lifetime cost-savings for systems, and reduced site disruption.

Providing adequate setbacks from environmentally sensitive features and site improvements: The project's conceptual layout should be developed so as to maintain adequate setbacks from features and project elements needing protection. This is best done when the site layout is first being developed, based on any applicable regulatory requirements and/or engineering judgments on sufficient setback/separation distances between site features and treatment effluent dispersal components.

Geophysical features and site improvements for which setbacks are either commonly required or should be considered include

- Potable and nonpotable water wells and cisterns, and potable water lines
- Streams, "guts," ponds, lakes, drainage-ways, drainage structures, floodways, wetlands and other freshwater natural or man-made water features
- Swimming pools
- Pastures used for large animals
- Buried utilities
- Man-made excavation "cuts," escarpments, steep slopes, and significant grade breaks
- Foundations, buildings/structures, and other surface improvements

- Property lines and easements
- Marine shorelines
- Trees, vegetation, and other natural site features and improvements

Adequate distance should be maintained between any edible vegetation on the site and wastewater system components, with those distances dependent upon the specific system and type of plant. Due to the potential for direct contact with plants from surface-applied effluent, care should be taken to consider prevailing and seasonal wind and site drainage patterns. For subsurface effluent dispersal systems, soil and subsurface conditions should be considered, and enough distance provided between dispersal fields and any food crops to ensure that pathogens or any other harmful wastewater constituents are not available for plant uptake. Although these components should be watertight, at least a certain separation distance should be maintained between food crops and wastewater conveyance lines and treatment units, due to the potential for leaks, overflows or root intrusion into those components over time.

In tropical regions where hurricanes and significant storm swells may occur, or in seismically active regions where water surges toward shorelines (tsunamis) may raise water levels well above high tide levels, it's important to anticipate the eventual likelihood of these events and plan accordingly with the wastewater system to minimize adverse shoreline impacts and replacement costs over time. Sufficient structural protection should be provided for vulnerable components exposed at the surface.

3.4.1 Soil Textural and Structural Analyses

A critical part of evaluating sites for decentralized wastewater systems is a physical examination of the soils. Soils analyses are also important for determining suitable locations for onsite treatment units and components, on behalf of construction costs and needed equipment. Where treatment units are to be installed, avoiding excessively rocky or shallow groundwater areas can yield significant installation cost and energy savings.

The land's natural ability to attenuate wastewater pollutants, and potential limitations in that respect (LDPs), must be determined for all sustainable designs using soil dispersal of the effluent. The level of treatment needed must be determined, along with acceptable methods of effluent dispersal for that level of predispersal treatment. Textural and structural soil evaluations are needed to make informed judgments on the effectiveness of several factors governing a soil's natural treatment capabilities. Those include

- Hydraulic residence time in soil horizons in which treatment will be dependent

- Soil's infiltrative capacity and rate for movement of water
- Soil particle size and surface area
- Organic content of the soil
- Presence or absence of physical conditions contributing to preferential flow patterns for effluent as it moves through the soil

A suitable soil application method and rate are also determined based on the soil's textural and structural evaluation, and any final vegetative cover to be used. Most U.S. onsite wastewater systems permitting authorities have prescribed criteria for soil loading rates associated with levels of treatment provided for specific categories/ types of soil as well as other subsurface and overall site conditions (e.g., vertical distance to groundwater or rock). However, no such criteria currently exist for many geographic regions of the world.

As discussed in Chap. 2, the biochemical conditions and travel time associated with wastewater effluent percolating through soil and subsurface conditions are key to the natural treatment capabilities of a site. Historically, "percolation testing" was used to evaluate infiltration rates for soils and subsurface site conditions. Conducting a percolation test properly, however, requires having sufficient water at or taken to the site to conduct the test, and readings before and after the test over about a 24-hour period. Percolation testing by itself may also fail to distinguish between certain important soil conditions, and their relevance to potential LDPs.

For example, percolation testing is typically done with the test hole dug to the likely excavation depth for subsurface dispersal trenches or bed(s). That depth might be to only about 1 to 2 ft (0.3 to 0.6 m). If the test showed a very slow infiltration rate, there might be no way of knowing whether it was due to a fractured rock lens or rock strata of substantial depth, or some depth of underlying clay soil. In the case of the former, the need for added predispersal treatment might be indicated to address wastewater pollutant LDP(s) along with perhaps a lower application rate, whereas the latter would indicate the need for a reduced effluent loading rate with less concern about adverse groundwater impacts from percolating effluent.

While percolation testing can provide helpful information about site infiltration rates and suitable effluent loading/application rates for soils along with the site's natural ability to treat wastewater effluent, the examination and classification of site soils has become increasingly used for this. In many U.S. states, percolation testing is no longer used at all for onsite wastewater systems planning and permitting. Soil "profile" holes are instead typically used to sample and evaluate textural and structural characteristics of soils at each of the soil horizons down to a particular depth.

Some soils are, however, difficult to classify and characterize for their wastewater treatment capabilities. For sites having soils with potentially problematic structures with preferential flow patterns, or where soils may be of a type that can't be texturally classified using the field method discussed later in this chapter (e.g., caliche-type soils), percolation or infiltration testing may help with determining appropriate treatment and dispersal designs. In cases where percolation testing may help determine appropriate treatment and dispersal designs, guidance for performing percolation testing may be found on this book's website under "Supplemental Information," Chap. 3.

If percolation testing is performed, test holes should be located so as to be sufficiently representative of the potential dispersal field area. For small projects (i.e., single family residences), as few as two test holes may be sufficient, as long as conditions do not appear to be too variable or with steep slopes. In those cases and for larger projects and field areas, holes should be located similarly to profile holes as discussed below. Percolation test results, if performed, are correlated to recommended effluent loading rates and treatment levels presented in Chap. 9.

Systems permitting authorities often specify required profile hole depths, but they should at a minimum be deep enough to evaluate conditions as they relate to appropriate soil loading rates and levels of predispersal treatment. Those depths may vary from about 3 to 6 ft (1 to 2 m) or possibly deeper, depending upon the soil type and conditions encountered. The depth of the profile hole should in essence be considered a soil treatment "boundary," below which no adverse water quality or health impacts would be expected from percolating effluent. As well as evaluating the type of soil at each horizon change along the depth of the hole, it's important to know if there is evidence of seasonal, "perched" or constant groundwater (and if so at what depth), or if rock (fractured or solid/massive) is present and at what depth. Local water well logs, if available, can be used to approximate depths to groundwater table(s). If geotechnical borings are used for structural design on the site at depths greater than profile hole excavations, these may also be useful for further examining subsurface conditions and depths to any shallow or "perched" water tables.

For both profile holes and percolation or infiltration testing on sites, it's important to locate the test holes so that truly representative results are obtained. This can sometimes be challenging due to the variability of subsurface conditions over even very short distances for many sites. In flatter areas where a subsurface dispersal system is to be located for a relatively small (e.g., residential) project, a total of two holes may be sufficient. Those might be located as shown in Fig. 3.2*a*. Figure 3.2*b* shows profile holes located for a steeper site. Figure 3.2*c* shows possible profile holes locations for a larger system.

(a) Small dispersal field—mild slopes

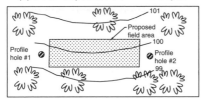

(b) Small dispersal field—steeper slopes

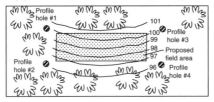

(c) Larger dispersal field—mild slopes

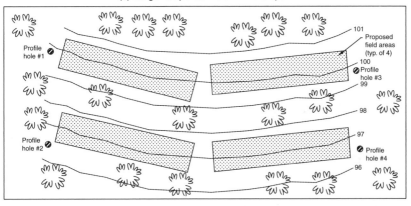

Figure 3.2   Locating soil profile holes for subsurface dispersal fields.

Profile hole excavations can be done using a backhoe or mini-excavator if one is available and can be taken to those areas of the site needing evaluation. Otherwise, hand-digging or auguring may be necessary. Auguring is not the best approach since it doesn't result in the ability to sample and evaluate each soil horizon independently. Soil boring and sampling is an option, but might again involve getting heavy equipment on site depending on how penetrable the soil is. Care should be taken not to excavate or otherwise disturb the soil in areas to be used for effluent distribution, as that can contribute to short-circuiting and preferential flow patterns into the excavations for percolating effluent. Leaving approximately 4 to 5 ft (1.2 to 1.5 m) from those areas might be enough distance unless subsurface horizons have very pronounced lateral or horizontal bedding planes, in which case 8 to 10 ft (2.5 to 3 m) might be more appropriate.

Note in Fig. 3.2 that locations were selected for the profile hole excavations that offered representative soil sampling around the dispersal field areas, yet avoided excavation under tree canopies to better protect root systems. Avoiding damage to or removal of trees is important to preventing erosion from sites (and downstream sedimentation), even where slopes are very mild.

Figure 3.3 shows a "soil texture triangle" that has traditionally been used to classify soils into textural categories based on the percentages of sand, silt, and clay in the soil. Soils have varying proportions of sand, clay, and silt, which is used to classify them as type Ib, II, III, or IV. Those soil classes are used by the U.S. Department of Agriculture's National Resource Conservation Service (NRCS, formerly the Soil Conservation Service or SCS) to evaluate their suitability for certain uses and behavior relative to physical phenomena such as moisture retention and rainfall runoff. Type Ia soil has a sufficient portion of gravel or coarse fragments so as not to be included with the other textural soil categories.

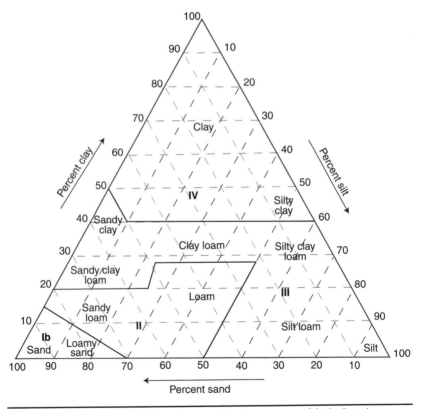

Figure 3.3 Soil textural triangle. (*Source: U.S. Department of Agriculture.*)

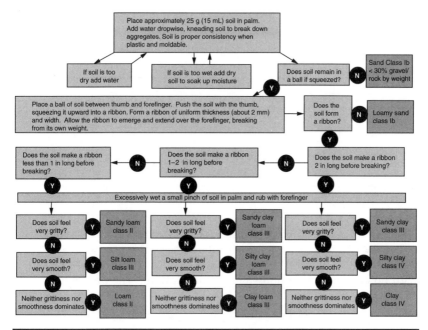

FIGURE 3.4   Field method of soil texture determination. (Modified from Journal of Agronomic Education. 8:54–55, *A flow diagram for teaching texture-by-feel analysis*, S. J. Thein, KSU 1979.)

While soil samples can be obtained from each visible horizon in profile holes and analyzed at soils labs (if one is available). Figure 3.4 provides a relatively simple procedure for classifying many soils into the four types shown in Fig. 3.3. The test can be done in the field with a small amount of water, or samples can be labeled, bagged, and taken elsewhere for analysis.

To use the procedure outlined in the flowchart in Fig. 3.4, a soil sample should be obtained from each visible soil horizon (distinguishable layers) down to the lowest depth being evaluated, and any rock pieces or gravel removed from the sample prior to performing the test. Representative samples should be taken from about the middle of each horizon. To distinguish between soil types Ia and Ib, it may be necessary to determine the weight of gravel/rock in the sample, and verify whether it constitutes less than or greater than 30 percent of the sample by weight. Figure 3.5 provides a particle size scale to help distinguish gravel versus sand. Approximately one ounce, or a heaping tablespoon (25 g or 15 mL) of soil is placed in the palm of the hand, and water is added drop-wise and kneaded between the fingers and thumb to break down any clots. The soil should be kneaded until it's plastic and moldable (like moist putty or "playdough"). The stepwise procedures in Fig. 3.4 are then followed to determine the soil class.

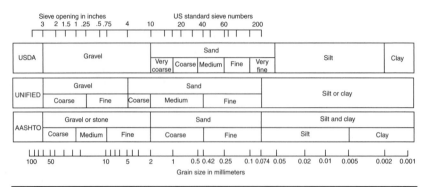

FIGURE 3.5 Soil classification systems and particle size distributions. (*Source: U.S. Department of Agriculture.*)

Figures 3.6 and 3.7 show profile holes dug in two different types of terrain. Figure 3.6 shows the site where the profile hole was excavated and was found to have at least 6 ft of easily excavated soil, with no evidence of groundwater or soil mottling. The profile hole

FIGURE 3.6 This profile hole was dug to a depth of approximately 6 ft (1.8 m). The soil was found to be type III, consisting of sandy clay and clay loam, with pockets of powdery white "caliche" deposits.

FIGURE 3.7   This profile hole could only be dug to a depth of approximately 10 in (25 cm) before very hard limestone rock was encountered. Soil above the rock was found to be very rocky/gravelly type III soil.

shown in Fig. 3.7 showed relatively shallow depth to fractured rock, with no sign of groundwater at that shallow depth. Project examples covered in Chap. 11 describe systems planning results that were based in part on the presence of these two different types of site conditions, among others.

As soil types change from type Ia to IV, several things are observed that affect a soil's ability to treat percolating wastewater effluent. In general, there tends to be an inverse relationship between the water-loading capacity of a soil and other LDPs, as related to soil type and decreasing tendency for preferential flow patterns.[6] For example, with clay soils water loading capacity (and thus wastewater flows) are an LDP due to lower infiltration/percolation rates, and the need for greater land areas to receive the same flows in a sustainable manner. However, because of the increased retention time for biochemical processes to occur in clay soils, they tend to facilitate greater removal of key pollutants such as nutrients, as long as the soil isn't overloaded and becomes saturated (exceeding the water-loading capacity).

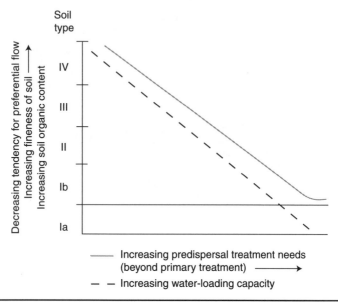

FIGURE 3.8   Conceptual relationships between soil types, water-loading capacity and added treatment needs.

Nutrients would not tend to be an LDP for clay soils as compared with other much more rapidly draining soils.

The tendency for that inverse relationship to occur is illustrated conceptually in Fig. 3.8. No scales have been included on the chart due to the difficulty in quantifying these phenomena, given the variability of conditions on sites, nonhomogeneity of most soils and other factors such as level of treatment needed for protection of different watersheds. While the relationships shown in the plots appear to be linear, soil biochemical and hydraulic behavior are certainly not linear. The figure is intended to show general trends and tendencies in the relationships between water infiltration and pollutant removal capacities for different soil types. The conceptual plot assumes that a sufficient depth of each type of soil is present to accomplish its respective treatment needs/goals. Also, the solid line conceptually represents all pollutants of concern in a wastewater, and those would vary by waste stream. Nontextural/nonstructural factors also affect a soil's ability to remove some constituents as compared with others. For example, soils within certain pH ranges tend to have greater phosphorus retention capabilities, as discussed further below.

The overall concept illustrated in Fig. 3.8 may be generally useful as applied to soils evaluations and the level of treatment that should be considered for sustainable wastewater systems planning. A basic premise here is that a planning goal applied to all sustainable systems would be to prevent ground or surface water impacts from pollutants of concern (e.g., nutrients and pathogens). It further assumes that sustainable wastewater planning attempts to optimize the use of a soil's natural treatment capabilities.

Figure 3.8 illustrates that as the hydraulic/water loading capacity increases significantly for sandy soils as compared with clay soils, so does the need to provide treatment (beyond primary treatment) for certain wastewater constituents of concern. Examples of those constituents, or LDPs, would include nutrients and in particular nitrogen and pathogens. As the fineness of the soil increases to silts and clays, water-loading capacity increases as an LDP design factor, while there is less concern about nutrient and pathogen removal as long as sustainable hydraulic loading rates are maintained.

Where the solid "treatment needs" line meets the horizontal line differentiating soil types Ia and Ib, essentially no further treatment of certain constituents of concern such as nitrogen and pathogens would be occurring as effluent is applied to the soil. Soils of that type (Ia) are texturally coarse with a high degree of preferential flow patterns. Pre-dispersal treatment of those constituents would need to be provided before soil/land dispersal. Similarly, if soils are very fine, such as clays, but the structure of the soil is such that there are still significant preferential flow patterns (e.g., columnar or prismatic soil structure), treatment of pollutants of concern would also need to be provided in that case ahead of the soil dispersal system. Soil structural classifications and descriptions are provided later in this chapter.

In general, as the tendency for preferential flow patterns *decreases*, natural soil treatment capabilities for key wastewater constituents such as nutrients tend to *increase*, thus necessitating less predispersal treatment. This is discussed further in Chap. 9. Systems planned for all of these soil types would, however, need sufficient primary treatment/settling capacity to remove settleable solids and fats, oils and greases.

Several wastewater treatment constituents of potential concern for domestic wastewater systems relying on land/soil-based treatment processes and final land disposition of effluent are discussed below. The extent to which any is an LDP will depend on the specific site conditions.

3.4.2 Total Nitrogen Removal in Soils

As discussed in Chap. 2, a soil's natural ability to remove nitrogen from infiltrating wastewater effluent depends on the presence of conditions that contribute ultimately to the release of nitrogen back to the atmosphere or uptake by vegetation. For nutrient management in watersheds, it must be remembered, however, that plants ultimately decay and return a certain amount of nutrients to the soil, so harvesting or removal of decaying or cut vegetation may be needed. Nitrogen in septic tank effluent is mainly in the form of organic nitrogen and ammonia.

For most unsaturated and well-drained soils, nitrogen tends to convert to the nitrate form of nitrogen within relatively short distances
and retention times. The availability of oxygen in the soil greatly
affects the efficiency of that conversion. Sand filters have been shown
to nitrify wastewater effluent very efficiently, as is the case with
sandy soils.

The results of a case study included in the U.S. EPA's *Onsite
Wastewater Treatment Systems Manual* (2002)[7] show ranges and average levels of soil treatment for percolating septic tank effluent at distances of 2 ft (0.6 m) and 4 ft (1.2 m) through a *"fine sandy"* soil. The
results of that study showed that on average only about 50 percent
total nitrogen reduction had occurred after 2 ft (0.6 m). At that depth,
almost all of the measured nitrogen had been converted to the nitrate
form. At a soil treatment distance of 4 ft (1.2 m), about 20 percent
more total nitrogen reduction was reported.

For "fine sandy" soils, and average total nitrogen levels of approximately 45 mg/L for effluent applied to a subsurface dispersal system,
the results of that study show that total nitrogen levels would on average still be above 10 mg/L at a depth of 4 ft (1.2 m). As mentioned in
Chap. 2 for groundwater and drinking water supplies, the U.S. EPA
limits nitrate-nitrogen levels to less than 10 mg/L. As nitrate-nitrogen
migrates to lower soil depths where there is less organic content, bacteria and sources of carbon for denitrification processes, nitrogen
tends to remain in the nitrate form as it moves with the soil moisture
and to ground or surface water supplies. The results of the above
study, therefore, show that the soil type in this study would not tend
to provide adequate treatment of septic tank effluent at those percolation distances. Added predispersal treatment would be needed for
lesser soil treatment (separation) distances to critical features such as
fractured rock or groundwater for protection of ground or surface
water supplies and/or aquatic resources.

For soils having higher organic content than sands (e.g., type III
soils such as clay loams), and fewer preferential flow patterns, higher
soil retention times offer the opportunity for greater levels of treatment
for key wastewater constituents such as nitrogen.[8,9] Natural layering
of upper soil horizons that offers both aerobic and relatively anoxic
conditions along with adequate retention times in those conditions
tends to result in better total nitrogen reduction.[10] Certain methods of
subsurface effluent dispersal can also greatly enhance natural treatment
capabilities, as discussed in Chap. 9 with respect to alternate dosing and
resting cycles and achieving relatively uniform soil distribution.

3.4.3   Pathogen Removal in Soils

Due to the expense and available lab testing facilities and equipment/
materials needed to test for the myriad of different viral and bacterial pathogens that may be present in wastewater, several types of

"indicator organisms" are used as statistical tests for the likely presence of pathogens. While testing for indicator organisms is a very useful tool for evaluating the quality of wastewater as related to pathogen levels, there is no assurance that very low levels of fecal coliform or *E. coli* would mean the absence of certain pathogenic species.

Pathogen removal processes for percolating wastewater effluent in soils include filtration, natural attrition due to retention times (particularly in aerobic soils), temperature and solar irradiation, and adsorption onto particle surfaces in the soil. Pathogen reduction capabilities for different types of soils, as statistically measured using indicator organisms, tend to vary less than for other pollutants of concern such as nitrogen as long as there's not a significant tendency for preferential flow patterns. Various studies have shown that as little as 1 ft (0.3 m) of many soil types (though without preferential flow patterns) provides high levels of pathogen reduction. For the same EPA-referenced study as cited for nitrogen removal above, fecal coliform levels were mostly nondetectable at a soil treatment depth/distance of about 2 ft (0.6 m).

Due to the unpredictable fate and transport of pathogens, and particularly viral species in sandy soils where preferential flow patterns are likely to exist in soils, use of more conservative and longer setback distances (both horizontal and vertical) may be needed for long-term protection of public health and environmental resources. Viral species have been observed to migrate further distances as compared with bacteria, as well as to desorb from sandy soils and continue migrating with soil moisture.[11,12] Chapter 9 discusses common requirements for vertical separation distances from rock and ground water tables for different types of subsurface dispersal methods in U.S. states as related to pathogen reduction.

3.4.4 Phosphorus in Soil Systems

While concerns about phosphorus concentrations in wastewater effluent are typically not on behalf of public health, some soils and site conditions offer pathways to nearby surface waters where eutrophication may occur. Phosphorus is a major plant nutrient along with nitrogen, which contributes to algae growth in surface waters. As discussed in Chap. 2, receiving watersheds potentially affected by projects should be evaluated to determine sustainable nutrient levels that won't degrade water quality over time.

Unlike the nitrogen cycle, the phosphorus cycle is a part of aquatic and terrestrial natural processes, and doesn't tend to involve exchanges of forms of phosphorus between the atmosphere and land or water. Elemental phosphorus is very unstable and reactive, and tends to readily combine with oxygen when exposed to the air. In water and land ecosystems, phosphorus exists as a form of phosphate, in which a phosphorus atom is surrounded by four oxygen atoms (PO_4^{3-}). This molecule is called orthophosphate, which is the simplest form of phosphate, and has a negative charge that invites combination with oppositely charged atoms or molecules. The tendency to combine with different atoms or molecules depends on the pH of the soil or aquatic conditions.

In soils, phosphorus may exist in many forms, but in general, may be stored in one of three principal reserves: (1) living mater (plants and animals), (2) dead/decaying organic matter ("biomass"), and (3) inorganic matter or mineral forms. Phosphorus cycles between these various reservoirs, or "stores." Plants only uptake phosphorus in its inorganic form, with plant roots absorbing $H_2PO_4^-$ or HPO_4^{2-}. These are called "available" forms of phosphorus. Figure 3.9 is a conceptual illustration of the distribution of inorganic phosphorus in soils at different pH's.[13,14] Orthophosphate exists mainly in the HPO_4^{2-} form at pH's of 6.5 or greater, and in the form of $H_2PO_4^-$ at lower pH's. The "Available P" portion of the figure shows this trend, with the peak of that section of the distribution occurring at a pH of 6.5. The release of inorganic phosphate from organic phosphate is called *mineralization* (similar to mineralization in the nitrogen cycle).

In addition to plant uptake of available phosphorus, another important portion of the phosphorus cycle in soils is the reaction and combination of phosphate with other inorganic molecules in the soil. Those reactions are referred to as "phosphorus fixation." In soils with pH less than about 6.5, phosphorus is predominantly fixed by aluminum, iron and manganese. At pH's above about 7, phosphorus is fixed mainly by calcium and magnesium present in soils. The opposite reaction, or release of fixed phosphorus, occurs naturally at a much lower rate than fixation.

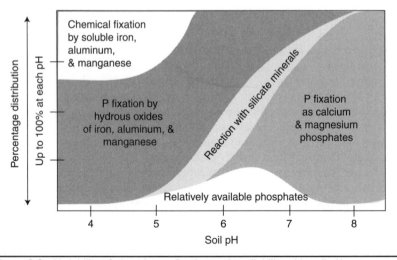

FIGURE 3.9   Variability of phosphorus fixation and availability with soil pH.

The same EPA case study referenced for both nitrogen and pathogens reported that on average about 95 percent of total phosphorus was removed at a travel distance of 2 ft (0.6 m) through the sandy soil. The average total phosphorus concentration in the septic tank effluent applied to that fine sandy soil was 8.6 mg/L, with the concentration measured at 2 ft (0.6 m) averaging 0.4 mg/L for 35 subsurface monitoring samples collected. At a soil treatment distance of 4 ft (1.2 m), total phosphorus was further reduced to an average concentration of 0.18 mg/L.[14]

Clay and organic content appear in general to favorably affect phosphorus retention in soils.[15] Mechanisms controlling the sorption of phosphorus in soils in clays and other soils can be somewhat complex, but appear to be driven by soil pH along with molecular structure and charge of the "host" mineral sites. "Exchangeable cations" (positively charged ions) have been found to influence phosphorus adsorption by affecting the availability of receptor sites for negatively charged phosphate ions along the surfaces of clays.[16] Due at least in part to the higher surface areas and greater numbers of inorganic "receptor" sites for sorption of phosphate ions in finer textured soils such as clays and silts, phosphorus retention in these soils has been observed to be significantly greater than for sands and more coarse soils.

Laboratory scale tests can be set up to estimate the phosphorus removal capabilities of specific soils to check for the need for any pre-dispersal treatment of phosphorus that may be needed. However, the actual phosphorus retention would be expected to be several times greater than bench-scale testing due to the exchange of ions and availability of new sites made available in soils.[17,18]

The removal processes in soils for phosphorus in percolating wastewater effluent consist basically of plant uptake and retention through fixation and sorption processes. Leaching of phosphorus from soils would mainly be a concern for "phosphorus-saturated" soils (depletion of receptor molecules for sorption and fixation processes) due to high or prolonged phosphorus loading, and/or where preferential or rapidly draining flow patterns would facilitate movement of phosphate with soil moisture and reduced soil retention times.[19]

3.4.5   Other Organic and Chemical Pollutants of Concern

The conceptual relationships between soil types and pollutant attenuation capabilities shown in Fig. 3.8 tend to hold true for many household pollutants and pharmaceuticals that may be present in domestic wastewaters. The higher the organic content and clay content of the soil, and the lower the tendency for preferential flow patterns, typically the better the soil's ability to remove those pollutants. The nature of the chemical compound will, however, affect that trend. Some inorganic chemicals, depending on their molecular structure and "charge" may be removed less efficiently. Molecules with a positive charge would tend to bind more readily in most clay and organic soils, and larger molecules would tend to be removed more efficiently than smaller molecules from infiltrating effluent through filtration processes.

Significant percentages of rock or gravel in the soil can reduce that capability (and increase preferential flow patterns), and are therefore limited to 30 percent by weight in the USDA "texture by feel" soil classification approach presented in Fig. 3.4. Type 1a soils have a sufficiently high percentage (>30 percent) of gravel and/or coarse fragments that they tend not to offer enough treatment retention time for percolating effluent. Therefore, the "added treatment" line in Fig. 3.8 tapers quickly to the right as soils approach type 1a. Caution might, therefore, be needed for soil types 1a and 1b, with increased setback distances between effluent dispersal areas and critical environmental features, water supplies, and aquatic resources.

3.4 6   Preferential Flow Patterns and Soil Structure Considerations

Figure 3.8 illustrates the importance of preferential flow patterns in soils/subsurface conditions and soil textural classification along the vertical axis, as related to level of treatment needed. An evaluation of soil structure should, therefore, be made alongside of textural analyses. Several basic types of soil structures are described in Table 3.1. With regard to wastewater treatment capabilities, "blocky" or "granular" soils are preferable to "columnar" or "platy" soils due to their combined infiltration and treatment capacities. Columnar and prismatic soils would tend to have potentially severe downward preferential flow patterns and thus reduced treatment capabilities, while platy soils would tend to be very limited with respect to water-loading capacities. Two of the basic types described in Table 3.1 (massive and single-grained) are considered to be without structure.

3.5   Matching Systems to Site Conditions

Based on careful consideration of the project needs, goals, and site evaluation, the system designer should identify candidate methods of treatment and final effluent disposition that would likely be best suited to those conditions. Chapter 6 presents comparative information on a variety of treatment methods for those sites needing some level of treatment greater than primary (septic tank) predispersal treatment. Chapter 9 presents information on dispersal methods that might be considered in combination with those treatment processes, along with advantages and limitations associated with those dispersal methods.

Soil Structural Type	Illustration	General Description
Granular		Roughly spherical, resembling "grape nuts" or cookie crumbs; is usually less than 0.5 cm in diameter. Commonly found in surface ("A") horizons where plant root growth and decomposition have occurred.
Blocky		Roughly cube-shaped and irregular blocks with more or less flat surfaces; If rounded, it's called "subangular blocky;" Usually 1.5–5.0 cm in diameter, and typical of "B" horizons with higher clay content.
Prismatic		Vertically elongated columns of soil, often with 5 sides and commonly about 10–100 mm across; commonly occur in fragipans.
Columnar		These shapes are similar to prisms and are bounded by flat or slightly rounded vertical faces; they may have a salt "cap" at the top, as found in soils of arid climates.
Platy		Flat platy structures that lie horizontally in the soil. Platy structure can be found in A, B, and C horizons. It commonly occurs in an A horizon as the result of compaction.
Single-grained (lacks structure)		This soil is considered to have "no structure." Every grain acts independently, with no binding agent to hold the grains together into peds. Permeability is rapid, with low fertility and water holding capacity.
Massive (lacks structure)		This soil is also considered to be without structure; compact, coherent soil not separated into peds of any kind; in clayey soils usually have very small pores, slow permeability, and poor aeration.

Source: USDA, National Resources Conservation Service.

TABLE 3.1 Soil Structural Descriptions

All of these systems, however, require adequate primary settling capacity prior to either subsurface soil dispersal of that effluent or further treatment of some type. Chapter 5 presents detailed information on the design of primary treatment components suitable for use in small to larger decentralized wastewater systems.

Certain dispersal methods, as discussed in Chap. 9, have been found to more effectively provide soil treatment for key wastewater pollutants (site-specific LDPs) such as pathogens and nutrients. Appropriate methods and levels of predispersal treatment combined with those methods of dispersal can jointly provide very high levels of overall treatment that often can't be as cost-effectively provided by systems relying on direct discharge of treated effluent. That is due to the natural ability of bacterial populations existing in most soils to supplement predispersal treatment methods. Recommended soil loading rates coupled with certain levels and methods of treatment for specific soil types are presented in Chap. 9. Recommendations for minimum depths of soil beneath the bottom (infiltrative layer) for methods of dispersal coupled with treatment levels for the basic soil types are also provided in that chapter.

References

1. *On-Site Sewage Facilities*, 30 TAC Chapter 285 state rules, Texas Commission on Environmental Quality, Austin, Texas, 2008.
2. S. M. Parten, *Analysis of Existing Community-Sized Decentralized Wastewater Treatment Systems*, Water Environment Research Foundation, Alexandria, VA, July 2008.
3. U.S. EPA, *How to Conserve Water and Use it Effectively*, On-line Nonpoint Source Pollution program information (http://www.epa.gov/owow/nps/chap3.html). Last accessed June, 2009.
4. K. Lowe , N. Rothe, J. Tomaras, K. DeJong, M. Tucholke, Dr. J. Drewes, Dr. J. McCray, et al., "Final Report: Influent Constituent Characteristics of the Modern Waste Stream from Single Sources: Literature Review," Water Environment Research Foundation, Alexandria, VA, 2007.
5. S. L. Warren, A. Amoozegar, W. P. Robarge, C. P. Niewoehner, and W. M. Reece, "Effect of Graywater on Growth and Appearance of Ornamental Landscape Plants," *Proceedings of the 10th National Symposium on Individual and Small Community Sewage Systems*, ASAE Proceedings, Sacramento, CA, 2004.
6. K -J. S. Kung, "Funnel-Type Flow and Its Impact on Contaminant Transport in Unsaturated Soil," *Proceedings of the 7th International Symposium on Individual and Small Community Sewage Systems*, ASAE, Atlanta, GA, 1994.
7. *Onsite Wastewater Treatment Systems Manual*, EPA/625/R-00/008, USEPA, February 2002, Table 3-18.
8. "Final Report: Alternative Wastewater Management Project", Report on Phase I Monitoring Results; City of Austin, TX Water Utility; Prepared by Community Environmental Services, Inc., 2005.
9. D. F. Weymann, A. Amoozegar, and M. T. Hoover, "Performance of an On-Site Wastewater Disposal Systems in a Slowly Permeable Soil," Proceedings of the 8th National Symposium on Individual and Small Community Sewage Systems, ASAE, Orlando, FL, 1998.
10. J. C. Converse, M. E. Kean, E. J. Tyler, and J. O. Peterson, "Nitrogen, Fecal Coliforms and Chlorides beneath Wisconsin At-Grade Soil Absorption Systems," *7th Northwest On-Site Wastewater Treatment Short Course and Equipment Exhibition*, University of Washington, Seattle, WA, 1992.
11. *Onsite Wastewater Treatment Systems Manual*, EPA/625/R-00/008, USEPA, February 2002.
12. L.A. Nicosia, J.B. Rose, L. Stark, and M.T. Stewart, "A Field Study of Virus Removal in Septic Tank Drainfields," *Journal of Environmental Quality*, 30: 1933–1939, 2001.

13. Sources for modified diagram: UT-Austin Engineering School, Dept. of Civil, Architectural and Environmental Engineering, graduate course notes, Dr. Raymond C. Loehr, Land Treatment of Municipal Wastes; Nova Scotia Agricultural College, Department of Plant and Animal Sciences, "Management of Soil Phosphorus Fertility" on-line technical information, 2008; and USDA-CSREES technical literature.

14. *Onsite Wastewater Treatment Systems Manual*, EPA/625/R-00/008, USEPA, February 2002, Table 3-18.

15. B. Lalljee, "Phosphorous Fixation as Influenced by Soil Characteristics of Some Mauritian Soils," Food and Agricultural Resource Council, Reduit, Mauritius, 1997.

16. Mohamed El-Nennah, "Phosphate Adsorption by Na-, Mg- and Ca-Saturated Soil Clays," *Journal of Plant Nutrition and Soil Science*, 138(1):33–37, 1975.

17. R. Crites and G. Tchobanoglous, *Small and Decentralized Wastewater Management Systems*, WCB/McGraw-Hill, New York, 1998.

18. "Soil Testing for Phosphorus: Environmental Uses and Implications", Southern Cooperative Series Bulletin No. 389, a publication of SERA-IEG 17, USDA-CSREES Regional Committee, 1998, http://www.sera17.ext.vt.edu/Documents/Soil_Testing_Uses_Implications.pdf, last accessed on August 20, 2008.

19. C. Hyland, Q. Ketterings, D. Dewing, K. Stockin, K. Czymmek, G. Albrecht, and L. Geohring, "Phosphorus Basics—The Phosphorus Cycle," *Agronomy Fact Sheet Series*, Fact Sheet Number 12, Cornell University Cooperative Extension, Nutrient Management Spear Program, 2005, http://nmsp.css.cornell.edu/publications/factsheets/factsheet12.pdf, last accessed on August 20, 2008.

CHAPTER 4

Wastewater Collection and Conveyance Methods

For onsite wastewater systems serving individual properties, wastewater "collection" usually consists of just the gravity sewer line(s) leaving the house and any other structures served, draining to the septic/primary settling tank(s). For larger projects with multiple buildings spread out across the site, or wastewater systems serving multiple properties where longer runs of collection lines are needed, more complex methods and materials may be needed. Where site topography prevents flow by gravity either to primary settling tanks or to treatment facilities, some method of pumping will be needed.

4.1 Basic Methods of Decentralized Systems Collection and Conveyance

There are four basic types of collection and conveyance system approaches commonly used for decentralized wastewater systems, each with its advantages or limitations depending on specific site conditions. These include

1. Conventional gravity sewer systems

2. Septic tank effluent gravity (STEG) collection systems

3. Septic tank effluent pumped (STEP) systems

4. Grinder pump pressure sewer systems

For conventional gravity sewers and grinder pressure sewers, primary treatment (as discussed in Chap. 5) needs to be provided separately prior to any further treatment process(es) and/or final land/soil effluent dispersal. Primary treatment is provided as an integral part of effluent (STEG and STEP) collection systems.

Hybrid STEP/STEG configurations are also frequently used, with that approach used in two of the project examples discussed in Chap. 11.

There is a fifth type of collection system, vacuum sewers, which is used for some small community and larger collection systems. That method of collection, however, necessitates the construction of at least one central vacuum station, with those costs contributing enough to the overall collection system costs that the number of users of the system needs to be at the small-community-scale or larger. Vacuum sewer systems have been used more in Europe than in North America, though there are a number of U.S. communities served by them today. Because of the minimum size of system that tends to be needed because of costs, vacuum sewers will not be covered here. Good sources of information about vacuum sewers as well as further information on each of the other methods of collection listed above are the U.S. EPA's manual: *Alternative Wastewater Collection Systems* (EPA/625/1-91/024), and the more recently published manual of practice FD-12, *Alternative Sewer Systems*, 2nd ed., published by the Water Environment Federation (2008).

For conventional gravity sewers, main/street collection pipes are typically 8 in (200 mm) or larger, and laid at depths ranging from 3 to 30 ft (1 to 10 m). Pipes are sized and designed to have straight alignments and uniform grades to maintain adequate pipe flushing velocities for solids. Manholes are installed between runs of pipe for accessing and clearing blockages, and are located at changes in slope or pipe direction, and usually no more than about 400 ft (122 m) apart. Because conventional gravity sewer lines must maintain steady grades and alignment, burial depths can become quite deep in areas with hills and variable topography in the direction of flow. Because of their larger size to handle solids, excavation widths are greater, and lines tend to be less watertight than smaller effluent or grinder collection lines. Each of these factors affects costs and environmental and public health considerations.

STEG sewers convey effluent by gravity to a common/community treatment location, or to an effluent pump station that pumps the effluent to the treatment location or into another portion of a collection system. A septic or "interceptor" tank is located near the building or facilities served by the collection system, with settleable solids, oils, and grease removed by this tank before the effluent drains to the common/main collection lines in the street or easement. Collection mains are most commonly Schedule 40 PVC and 3 to 6 in (80 to 150 mm), depending on the number of homes and/or businesses served. Burial depths tend to be shallower, and trenches much narrower than conventional gravity sewers. Cleanouts are used rather than manholes. Because the collection pipes convey clarified effluent, alignments, and slopes are much more flexible than conventional gravity sewers. Valve boxes housing PVC ball-type shutoff ("isolation") valves are located at the edge of each property up-flow from the main effluent collection line.

STEP sewers are very similar to STEG sewers except that a pump is used at homes and businesses served to convey effluent to the common collector lines. The pump, which is typically a 1/2-hp (0.37-kW) high head effluent pump, is located either in a screened vault in the last compartment of a septic tank, or in a separate tank or basin. STEP collection is used where topography doesn't allow gravity flow of effluent from lots, and is often used in combination with STEG service connections. STEP collection mains are typically 2 to 4 in (50 to 100 mm), depending on the number of homes and/or businesses served, and can follow natural grade with minimal burial depth. Since STEP collection lines convey clarified effluent, flow velocities are much less critical than for grinder pressure sewers. Figure 4.1 illustrates some of

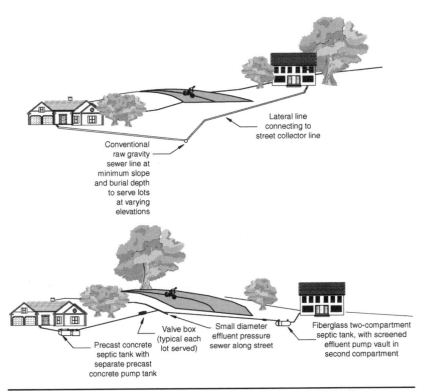

FIGURE 4.1 Conceptual illustration of conventional gravity and STEP effluent pressure sewer connections.

the key advantages of effluent sewers, particularly as related to several sustainability factors. The conventional raw wastewater gravity sewer line shown at the top requires relatively deep burial depths for undulating terrains, to maintain proper slope/grade for the main collection lines running along streets or easements. Unlike effluent gravity sewers (STEG) that convey septic tank effluent, conventional raw wastewater sewers need to maintain minimum slopes to move both solid and liquid waste through the lines. STEP sewers (bottom layout) can follow grade, using minimal pipe burial depths, and narrow trenches for small-diameter piping. The ability to follow natural grade much more closely results in far less site disturbance and accompanying adverse environmental and water quality impacts. Two different types of STEP on-lot tank configurations and materials of construction are illustrated in the figure (bottom), and either or both could be used for the same system. Properly installed and well-built fiberglass tanks tend to last longer than concrete tanks due to corrosive wastewater conditions. Costs will depend on the availability of each to the project, and site conditions. Fiberglass tanks can be more easily installed in tight conditions as compared with concrete tanks delivered by heavy trucks.

Grinder pressure sewers convey slurries of macerated solids and liquids in pressurized lines. Homes and buildings served by this type of collection system have a pump vault equipped with usually at least a 2-hp (1.5-kW) grinder pump. Lateral/service and main collection lines, typically Schedule 40 PVC, must be sized to ensure enough velocity to scour lines and prevent excessive buildup or clogging (2 to 7 ft/s or 0.6 to 2.1 m/s). Line sizes are usually 1-1/2 to 3 in (40 to 80 mm). These pressurized collection lines are laid in relatively narrow shallow trenches that can also follow natural grade. Street collector lines can maintain minimal burial depths where there's variable topography, since all connections and main lines are pressurized. Grinder pressure sewer pump stations on lots served typically use a fiberglass basin with watertight grommets used for inlet and outlet piping. Basins are commonly 30 to 36 in (0.77 to 0.92 m) in diameter.

Where there are relatively minor undulations in the terrain, but with overall favorable drainage from each lot in the desired direction, pure STEG or conventional gravity collection systems may be used. The narrower/shallower trenches and flexibility of grade associated with STEG/effluent sewers offer advantages over conventional gravity sewers, along with other advantages listed in Table 4.1. Where some lots are lower and some higher than the hydraulic grade line of the main collector line, a combination STEP/STEG system might be used.

As seen in Fig. 4.1, small-diameter effluent sewer connections (both pressure and gravity) typically have shutoff or "isolation" valves to stop flow from the system for system maintenance and/or repairs.

Collection Technology Element	Conventional Gravity Sewers with Lift Stations	Septic Tank Effluent Gravity Sewers (STEG)	Septic Tank Effluent Pumped Sewers (STEP)	Grinder Pressure Sewers
Socio-cultural considerations	• Public, engineers, and contractors tend to be most familiar with this collection method	• Less familiar to public, engineers, and officials in some parts of world as compared with conventional gravity sewers	• Less familiar to public, engineers, and officials in some parts of world as compared with conventional gravity sewers	• Less familiar to public, engineers, and officials in some parts of world as compared with conventional gravity sewers
On-lot components (including primary settling/ interceptor tank and service lateral)	*Advantages:* • No power is required for normal functioning • Minimal amount of, and use of space for on-lot components (no onsite primary treatment) • No power required for on-lot components (as compared with grinder and STEP systems)	*Advantages:* • No power is required for normal functioning • Discharges wastewater to common parts of system free of solids and grease (primary treatment provided onsite) • Lateral line/service connection is typically 1-1/2–2 in (40–50 mm)	*Advantages:* • Discharges wastewater to common parts of system free of solids and grease (primary treatment provided onsite) • Lateral line/service connection is typically 1–2 in (25–50 mm)	*Advantages:* • Grinder pump stations occupy less horizontal space than STEP/STEG systems (though more than conventional gravity systems) • Lateral/service lines typically 1-1/4–1-1/2 in (32–40 mm)

TABLE 4.1 Comparison of Wastewater Collection Technologies

Collection Technology Element	Conventional Gravity Sewers with Lift Stations	Septic Tank Effluent Gravity Sewers (STEG)	Septic Tank Effluent Pumped Sewers (STEP)	Grinder Pressure Sewers
	Potential disadvantages: • No primary treatment provided to wastewater before entering common collection system, so harmful or caustic substances will combine directly with full treatment/dispersal system • Primary treatment required at the common treatment site • Minimum diameter of 4 in • Must maintain steady grade and minimum slopes/fall	*Potential disadvantages:* • Uses certain amount of on-lot space (typically 6 × 15, or 1.8 × 4.6 m) for primary settling tank • Requires periodic tank inspection and cleaning	*Potential disadvantages:* • Requires power (115 V) • Uses certain amount of on-lot space (typically 6 × 15, or 1.8 × 4.6 m) for primary settling tank • Requires periodic tank inspection and cleaning	*Potential disadvantages:* • Requires 230-V power (2-hp or 1.5-kW pumps) • Provides no primary treatment onsite • Inorganic solids are macerated and blended with organic matter • Have very little to no emergency storage or hydraulic retention time in the event of power outages or pump failures, thus requiring service personnel to respond immediately to alarms • Harmful or caustic substances are passed directly into the collection system

72

Main/common collection lines			
Advantages: • Larger diameter lines able to pass raw wastewater and normal sized solids entering collection system **Potential disadvantages:** • Typically 8-in (200-mm) diameter and larger • Grade/slope of pipe is critical, potentially resulting in very deep cuts/excavations in hilly areas to maintain pipe slopes • Larger excavation equipment is usually needed • Manholes used instead of cleanouts • Designs assume certain amount of inflow and infiltration in line sizing • Potential for leaks, or exfiltration, from lines with accompanying water-quality impacts	**Advantages:** • Typically 2–6 in (50–150 mm) diameter lines for systems serving up to several dozen homes (or equivalent commercial flows) • Lines may be laid in narrow, shallow trenches, with less site disruption • No manholes are required (just cleanouts for longer pipe runs) • No minimum slope must be maintained, and less topography-sensitive (effluent lines can have variable grades and better follow natural drainage patterns) • Small-diameter lines are installed to be watertight, with	**Advantages:** • Typically 2- to 4-in (50 to 100 mm) diameter lines for systems serving up to several dozen homes (or equivalent flow from nonresidential sources) • Lines can follow grade using minimal burial depths • No manholes are required (possibly just cleanouts for very long pipe runs) • Lines may be laid in narrow, shallow trenches, with less site disruption • Small-diameter collection lines are installed watertight, without opportunity for infiltration and inflow	**Advantages:** • Small-diameter lines that can be laid in shallow, narrow trenches • Pipe can follow grade/contour of land • No added line capacity is needed for infiltration and inflow • Small-diameter lines are installed to be watertight, with much less potential for infiltration and inflow, or exfiltration **Potential disadvantages:** • Lines must handle all liquids and macerated solids, and tend to need more maintenance/clearing than effluent sewers (though possibly less

TABLE 4.1 Comparison of Wastewater Collection Technologies (*Continued*)

Collection Technology Element	Conventional Gravity Sewers with Lift Stations	Septic Tank Effluent Gravity Sewers (STEG)	Septic Tank Effluent Pumped Sewers (STEP)	Grinder Pressure Sewers
	• Not suitable for high-groundwater or flood-prone conditions • Greater potential for odor problems (at manholes)	• much less potential for infiltration and inflow, or exfiltration • Interceptor tanks followed by isolation valves between tanks and main collection lines allows for disruption of service (for road work, etc.) with a certain amount of storage capacity provided in the interceptor tank (usually at least a day's flow) • No added line capacity is needed for infiltration and inflow	• Interceptor tanks followed by isolation valves between tanks and main collection lines allows for disruption of service (for road work, etc.) with a certain amount of storage capacity provided in the interceptor tank (usually at least a day's flow) • No added line capacity is needed for infiltration and inflow	than conventional raw gravity sewers) • Pipe sizing is critical for maintaining adequate scouring velocities (to prevent clogging from slurry being pumped) • Very little on-lot storage capacity for main collector line maintenance or repairs

Pumps	Potential disadvantages:	Potential disadvantages:	Potential disadvantages:	
• Pump stations, where needed, require larger and higher horsepower pumps with solids handling capabilities (3 in or 100 mm, or more) • Pumps are very heavy and require lifting equipment for removal and/or servicing • Pumps tend to be expensive, and require more maintenance	• Pure gravity effluent collection lines must have adequate overall slopes for lines to drain • Where needed, effluent lift station pumps are lower horsepower and smaller than either grinder or raw wastewater pumps • Relatively easy to service/maintain and rebuild if needed • Relatively low replacement costs • Long service lives are typical (15–20 years is common)	• Due to potential for odors from septic tanks, requires air-tight lids and appropriate venting • High-head effluent pumps used for individual properties are much lower horsepower and smaller (typically 1/2 hp or 0.37 kW) than either grinder or raw wastewater pumps • Relatively easy to service/maintain and rebuild if needed • Relatively low replacement costs • Long services lives are typical (15–20 years is common)	• Requires more powerful pumps than STEP systems (minimum 2 hp or 1.5 kW) • Pump lifetime estimated at 3–10 years • Pumps are heavy and require lifting equipment (commonly mounted on rail systems in pump vault) • Corrosion problems often observed for grinder pump stations	

TABLE 4.1 Comparison of Wastewater Collection Technologies (*Continued*)

Collection Technology Element	Conventional Gravity Sewers with Lift Stations	Septic Tank Effluent Gravity Sewers (STEG)	Septic Tank Effluent Pumped Sewers (STEP)	Grinder Pressure Sewers
	• Redundant pumping capacity is often needed due to greater tendencies for lift station failures/problems • Greater power usage for pumps capable of solids handling	• Effluent pumps offer very high head capabilities (wide ranges of flows and heads for different models of effluent pumps) • Constructed of noncorrosive stainless steel and thermoplastic parts	• High head effluent pumps operate reliably over a wide range of pressure/head conditions • Constructed of noncorrosive stainless steel and thermoplastic parts	• Higher cost pumps to replace • Manufacturers recommend sharpening of grinder blades at least annually, to prevent problems with solids maceration
Effects on treatment facilities	• Must provide primary treatment in addition to any other treatment processes prior to soil/land disposition • Infiltration and inflow impact treatment performance • Excess capacity needed for treatment plant designs to accommodate inflow and infiltration	• Primary treatment already provided by tanks at facilities served, with BOD and TSS reductions those reported for septic tank pretreatment (see tables, Chap. 3, this book's website) • Some flow attenuation and	• Primary treatment already provided by tanks at facilities served, with BOD and TSS reductions those reported for septic tank pretreatment (see tables, Chap. 3, this book's website) • Some flow attenuation and equalization occurs with on-lot interceptor tank	• Must provide primary treatment in addition to any other treatment processes (unless extended aeration without primary treatment is used) prior to soil/land disposition • Solids settling and separation of oils/greases (scum) tends to be more difficult

• No attenuation of potentially harmful substances that may enter collection system	• equalization occurs with on-lot interceptor tank • Significant dilution and attenuation of any harmful chemicals or caustic substances that might have adverse impacts on common treatment and dispersal system • No added treatment capacity is needed for infiltration and inflow	• Significant dilution and attenuation of any harmful chemicals or caustic substances that might have adverse impacts on common treatment and dispersal system • No added treatment capacity is needed for infiltration and inflow	for grinder waste as compared with other collection methods for primary treatment processes • No attenuation of potentially harmful substances that may enter collection system • No added treatment capacity is needed for infiltration and inflow
Construction cost considerations	• Larger and more expensive common/main collection lines • For projects and developments to be phased-in over time, primary treatment capacity must often be built even where portions of the system	• Common collection system/mains tend to be much less expensive to install than conventional gravity lines with manholes • Long service life for materials (typically 40–50 or more years)	• Common collection system/mains tend to be much less expensive to install than conventional gravity lines with manholes • Long service life for materials (typically 40–50 or more years)	• Common collection system/mains tend to be much less expensive to install than conventional gravity collection lines • On-lot portions of collection system may be installed when that home or commercial

TABLE 4.1 Comparison of Wastewater Collection Technologies (*Continued*)

Collection Technology Element	Conventional Gravity Sewers with Lift Stations	Septic Tank Effluent Gravity Sewers (STEG)	Septic Tank Effluent Pumped Sewers (STEP)	Grinder Pressure Sewers
	served by the collection lines won't be online for long periods of time. Developers/owners are in turn not able to defer those costs that might otherwise be paid later by individual property owners (as with STEP/STEG systems)	for properly installed systems) • On-lot portions of collection system may be installed when that home or commercial building is constructed, thereby either deferring or transferring those costs (depending on whether developer is also home-builder)	for properly installed systems) • On-lot portions of collection system may be installed when that home or commercial building is constructed, thereby either deferring or transferring those costs (depending on whether developer is also home-builder)	building is constructed, thereby either deferring or transferring those costs (depending on whether developer is also home-builder)
Operation and maintenance	• Lines and manholes need periodic flushing/cleaning to clear solids and build-up • Greater sludge production at centralized primary treatment facility as compared with septic tanks with natural	• Good reliability with low level of maintenance and repairs • Septic/interceptor tank inspections may be scheduled well in advance, with relative flexibility of sludge/septage	• Good reliability with low level of maintenance and repairs • Septic/interceptor tank inspections may be scheduled well in advance, with relative flexibility of sludge/septage pumping	• More line maintenance and lift station maintenance than effluent collection systems • Much greater on-lot power usage than for STEP pumps (2 hp or 1.5 kW used for grinder pumps as

anaerobic digestion processes occurring • Inability to stage sludge/septage pumping through use of distributed primary settling capacity at individual facilities served • Higher power usage for lift stations	pumping intervals (as contrasted with centralized/common primary treatment facilities that may reach critical capacities, and have less reserve storage capacity for sludge and scum) • No pumps to service or replace at residences	intervals (as contrasted with centralized/common primary treatment facilities that may reach critical capacities, and have less reserve storage capacity for sludge and scum) • Reliable pumps that use much less power than grinder or raw wastewater pumps	compared with 1/2 hp or 0.37 kW for STEP systems pumps) • Greater sludge production at centralized primary treatment facility as compared with septic tanks with natural anaerobic digestion processes occurring • Inability to stage sludge/septage pumping through use of distributed primary settling capacity at individual facilities served

TABLE **4.1** Comparison of Wastewater Collection Technologies (*Continued*)

Grinder and STEP pressure sewer service laterals have a PVC swing check valve on the tank/pump side of the isolation valve, before tying into the street collector line after the shutoff valve. Shutoff valves are best located in a public easement or right-of-way for accessibility to those managing the collection system. Shutoff valves are typically ball-type valves for effluent gravity and pressure sewers, and either ball or plug valves for grinder pressure sewers. For pressure sewers, both the check and shutoff valves may be located in a single covered valve box, or two separate boxes. Check valves should also be located in the discharge piping after the pump in STEP and grinder pump tanks or basins to prevent drain-back of liquid in the line when pumping stops. Regardless of the collection method used, pipe burial depths should be sufficient to provide enough protection from surface loads, and freezing conditions where applicable. Pipe sleeving may be needed where burial depths or locations are such that surface loads are an issue.

As well as serving new homes or businesses, conventional sewers, effluent collection systems (STEP and/or STEG) and grinder pressure sewers can all be used to serve properties with existing failing onsite wastewater systems, and connect them to clustered decentralized or centralized wastewater systems. The suitability, cost-effectiveness and long-term sustainability of each would, however, depend on the conditions at the properties served, local topography, and the wastewater system to which the properties would be connected. For both conventional gravity sewers and grinder pressure sewers, primary treatment is typically still needed at the community/common treatment site.

4.2   Factors Affecting Sustainability for Collection Systems

Table 4.1 presents comparative information for the above basic wastewater collection types, along with some advantages and limitations associated with each. Characteristics of each type of system are compared in the context of overall sustainability.

Several key considerations around which method of collection is used can significantly affect the sustainability of the overall wastewater treatment approach, and are discussed below. This discussion pertains mostly to projects larger than individual dwellings or businesses. Single buildings would tend to use gravity sewer lines from the structures leading to a septic tank, followed by any further treatment needed and final soil dispersal. For larger projects serving multiple sites or buildings spread out over a larger area on the same site, the specific geophysical conditions will tend to govern which wastewater collection method(s) is(are) the most cost-effective and appropriate.

Topography of the planning area(s): Where site or planning area topography prevents building sewer lines from draining with enough slope (at least 1/8 to 1/4 in per linear foot, or 10 to 20 mm/m of sewer line) to a location where treatment and/or final soil dispersal would occur, some method of pumping the sewage to the treatment unit location must be used. For the size ranges of systems covered in this book, raw wastewater lift stations are not discussed because they tend to be used for larger municipal scale collection systems with full-time operators available to respond to trouble calls. Combinations of collection methods described above might be used, although grinder pump stations would not tend to be used in conjunction with effluent sewers (gravity or pressure). The most cost-effective and suitable approach will depend on the specific conditions of the physical area and facilities served. Figure 4.2 shows a large primary treatment tank located next to the facilities served, with a third hydraulically separate compartment in the tank used for a pump station delivering effluent to the small-diameter effluent pressure line shown in Fig. 4.3. The small-diameter effluent pressure line can easily follow grade and the curvature of the road, using a shallow burial depth.

Large three-
compartment fiberglass
septic and pump tank

FIGURE 4.2   A large three-compartment fiberglass septic tank provides primary treatment, with the third compartment housing a screened effluent pump vault. Primary treated effluent is then pumped through the small-diameter effluent pressure line to a more remote area where treatment and final soil dispersal occur. Closer areas were historically protected and/or reserved for other public uses.

2-in (50-mm) effluent pressure line

Conduit

FIGURE 4.3   Note the shallow narrow trench to be used for both the electric conduit and the small-diameter effluent pressure line, and the curvature in the alignment possible for smaller lines. This trench was cut with an 8-in-wide (200-mm) rock trencher.

Watertightness of the collection lines: The use of watertight gravity and pressure sewer lines is critical to long-term sustainability for wastewater systems on behalf of both environmental health and operational concerns. As mentioned in Table 4.1 for conventional gravity sewer lines, both infiltration and exfiltration of wastewater is a significantly greater issue for that method of collection as compared with the other three types of collection systems. That is due to the larger pipe sizes and methods/materials typically used for pipe connections, along with the need for manholes for changes in grade or direction and longer pipe runs.

PVC sewer lines greater than 4 in (100 mm) in diameter tend to use "bell and spigot" joints with elastomeric gaskets. Smaller-diameter PVC pipes (4 in or 100 mm, and less) are typically solvent welded, which creates an actual chemical bond between the two lengths of pipe (and their fittings). The ability to achieve more watertight lines for effluent gravity and pressure sewers provides much greater water quality and environmental protection, and prevents excess water from entering the treatment system and overloading processes. Their watertightness enables small-diameter effluent gravity and pressure lines to be used in areas with high groundwater or periodic flooding, and in other environmentally sensitive areas. Smaller-diameter lines can also follow gentle curves in trench lines, enabling narrower trenches and less excavation as compared with stiffer larger-diameter piping for which trenches either have to be cut along very straight lines or cut wider to align the piping.

The use of solvent-welded *pressure* PVC couplings and joints (rather than drain fittings) doesn't add that much incremental cost to projects, and provides much more protection against leaks or water infiltration. Also, since Schedule 40 PVC pipe is manufactured in 20-ft (6.1-m) sections and with bell-and-spigot ends that can be inserted into each other and solvent-welded, piping layouts that enable use of 20-ft (6.1-m) lengths of pipe with fewer joints and couplings offers fewer opportunities for leaks over time. In some geographic areas, pipe may only be available in 10-ft (3-m) lengths without bell-and-spigot ends, in which case added care should be taken to use only pressure couplings and fittings for solvent-welding joints together, even for gravity effluent lines. PVC pressure couplings and fittings are significantly longer and offer much more surface area for the solvent-welded connection at the joints than drain couplings/fittings. Some basic installation tips and photos for installing PVC piping are included in Chap. 5.

High-density polyethylene piping may also be used to achieve watertight lines, although the joints are normally heat-fused using special equipment at the installation site along with persons specifically skilled with the installation of that type of piping. Fusion-welding of the joints, however, creates very watertight connections.

Energy consumption: Energy consumption is a very important consideration in selecting the most sustainable method of wastewater conveyance and collection for projects. Grinder pumps expend considerably more energy than effluent pumps because they must macerate solids as well as pump the resulting slurry. The minimum horsepower pump used for grinder lift stations is 2 hp (1.5 kW), with 230-V service. Effluent pumps used for residential scale STEP systems are typically 1/2-hp (0.37-kW) pumps capable of pumping to high-pressure heads. Maintaining scouring velocities (and energy losses associated with those velocities) is not nearly as much a factor for effluent pressure lines for which somewhat lower velocities can be used.

Maintenance: Maintenance is another key consideration, with grinder pressure sewers and conventional gravity sewers tending to require significantly more maintenance than effluent collection systems. Manufacturers of grinder pumps recommend that the cutting blades on the pumps be sharpened at least once a year. Because of the solids suspended in grinder pressure sewer wastewater, lines tend to need clearing ("pigging") from time to time. Conventional raw gravity sewers also tend to need blockages

FIGURE **4.4**   This effluent lift station basin is constructed of precast concrete, and sealed with epoxy for added corrosion resistance. The duplex pump station is approximately 4 ft (1.2 m) in diameter, and 8 ft (2.4 m) deep. Fiberglass basins are also commonly used for effluent lift stations.

cleared at times. Effluent lift station pumps are much smaller and lighter weight as compared with grinder pumps handling the same flows. Effluent pumps, which handle only the liquid portion of the waste stream, tend to require much less ongoing mainte-nance and have fewer problems as compared with solids handling pumps (whether grinder or raw wastewater). Figure 4.4 shows a duplex effluent lift station serving a commercial facility, with total daily design flows of approximately 5000 gal/day (approximately 19,000 L/day).

Burial depth of lines and site disturbance: Conventional gravity sewer lines tend to cause significantly more disturbance as com-pared with any of the other three collection types, due to the larger line sizes and need to maintain steady grades. As illustrated in Fig. 4.1, the need to maintain certain pipe slopes can result in fairly deep earth cuts, depending on the local terrain. STEG sewers must maintain enough slope to drain, but can have inflections in the lines since there are lesser concerns about solids build up in those areas. Both types of pressure sewers (effluent and grinder) are best able to follow natural grade with minimal burial depths. STEG, STEP, and grinder pressure sewers can be installed in narrow trenches (6 to 8 in or 150 to 200 mm for smaller-diameter lines), resulting in less site disturbance as compared with conventional gravity sewers.

Presence of shallow groundwater: Because STEG, STEP, and grinder pressure sewers all use small-diameter pipe that is more reliably watertight than larger-diameter conventional gravity piping, effluent or grinder pressure lines are considered more suitable for areas with high groundwater or that may be flood prone. The need for manholes with conventional gravity sewers also makes that option less appropriate for those conditions.

Shallow depth to rock: Smaller-diameter lines tend to be significantly less costly to install than conventional gravity sewers in areas with shallow depth to rock. Larger diameter rock-saws or hammers used to remove rock are very expensive to operate and use large amounts of fuel as compared with other less energy-intensive excavating equipment. While STEP and STEG systems require installation of primary settling tanks near buildings, in rocky, hilly areas, they can often be "tucked" into hillsides near buildings with minimal added rock removal. Lower profile tanks can also be used that are only about 5 ft (1.5 m) in overall height. Pump basins used for grinder pump stations are typically at least 6 ft (1.8 m) deep, but deeper if more wastewater reserve storage capacity is needed. Manholes used for conventional gravity lines can also be relatively deep, depending on the burial depth for the sewer line.

Effects on treatment processes: STEP and STEG collection systems provide primary treatment before the wastewater enters the combined collection system, thereby reducing costs, use of land area and site disturbance for treatment at that location. While grinder pumps have historically been used in some cases ahead of activated sludge treatment processes without first providing primary settling to remove settleable solids, this is considered a less efficient and sustainable wastewater treatment approach as compared with others. Secondary and advanced treatment processes will be discussed in Chap. 6, with comparisons made for the various processes.

Based on the above considerations, effluent sewers are considered a particularly sustainable sewering approach for decentralized systems, where found to be cost-effective and when designed and installed properly. There are cases where the use of grinder lift stations may be needed (e.g., locations along hills where there isn't enough space for a septic tank), but those cases are most often associated with existing onsite systems needing replacement or being taken off-line and connected to a collection system. For new residential or commercial construction, sites can usually be laid out to accommodate the necessary tank capacity and location for connection to effluent sewers. However, early coordination of the various planning disciplines is needed to ensure those wastewater service elements are integrated into the site plan.

CHAPTER 5

Primary Treatment

The simplest types of decentralized wastewater systems consist of primary treatment in a septic tank followed by subsurface soil dispersal of that effluent. As discussed in Chap. 2, primary treatment is needed before most secondary or advanced treatment processes, and always before final soil disposition of wastewater effluent.

Historically, grinder pumps or comminutors have sometimes been used in place of primary treatment for grinding larger solids ahead of certain types of treatment systems such as packaged activated sludge system. However, macerating/grinding solids can significantly lessen primary settling efficiencies of raw wastewater, potentially increasing tank capacity needs for longer retention times for suspended solids to settle. The power usage and routine service needs of grinder pumps and comminutors are also very high on a long-term basis. For those reasons, along with other treatment factors to be discussed further in Chap. 6, those approaches would typically not usually be considered the most cost-effective or sustainable in the long term unless needed for special circumstances.

A more passive approach relying on natural solids settling in appropriately sized tanks is therefore considered more sustainable for essentially all of the systems discussed in this book.

5.1 Design Capacity Needed for Primary Treatment Tanks

On average, properly designed primary treatment tanks would be expected to provide the following levels of treatment:[1]

- About 60 to 80 percent suspended solids removal
- About 50 to 60 percent BOD_5 removal
- Fats, oils, and grease removal to about 80 percent
- Minimal nutrient removal

Figure 5.1 is a simplified illustration of a septic tank used for residential-scale primary treatment, showing some basic elements critical to sound designs. Important features affecting performance of primary settling/septic tanks include

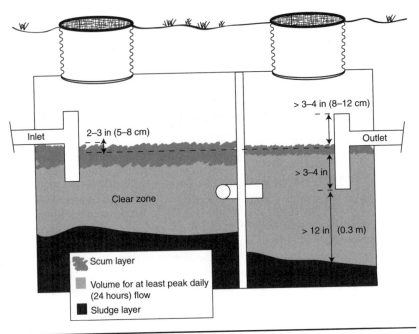

Scum layer

Volume for at least peak daily (24 hours) flow

Sludge layer

Figure 5.1 Basic septic tank elements.

1. *Sufficient volume in the tank where settling can occur for incoming suspended solids, and flotation of fats/oils/greases (FOG)*

The rule of thumb for this is that the "clear zone" (middle zone in the tank) provides at least 24 hours of hydraulic retention time based on design flows for all operating conditions (including just prior to the point when pumping of the tank is needed). So, if a small tank such as the one shown in Fig. 5.1 were serving a residence with a daily design flow of 300 gal/day (1136 L/day or 1.14 m³/day), the total clear-/middle-zone volume should be at least that amount.

2. *Multiple compartments or multiple tanks in series*

Figure 5.1 illustrates the effects of compartments by showing greater sludge and scum buildup in the first compartment, which helps prevent short-circuiting of solids and FOG toward the outlet of the tank. The tank shown in the figure has a PVC "tee" placed in the baffle wall between the two compartments at about the middle (vertically and horizontally across the width) of the clarified effluent zone. This helps prevent solids from settling in the direction of the inlet to the second compartment. Larger fiberglass tanks often use baffles with two to three 3- to 4-in-diameter (80- to 100-mm) holes placed at intervals across the baffle at a depth of approximately the middle of the clear zone.

3. Watertightness of the tank

It's very important that the inlet and outlet pipes be the only way that water can enter or exit the tank with the access riser lids in place. Excess water, from either surface ponding around the tank or groundwater, will overload and reduce settling performance. Absence of watertightness can of course also result in leakage of wastewater *from* the tank. Tank construction and installation practices contributing to watertightness are discussed later in this chapter. Notice in Fig. 5.1 that the access risers extend from the top surface/lid of the tank to the ground surface, with a watertight lid bolted down to the access riser. This enables access to the tank for periodic checks, pumping, and any other servicing needed.

4. Structural integrity

This aspect of septic tanks will be discussed later in this chapter, but will be mentioned briefly here due to its importance to watertightness. Several factors affect the structural integrity of tanks, including:

- Very corrosive conditions in the septic tank, with deteriorating effects on concrete and exposed metal over time.
- Exterior pressure, or loading on the sidewalls and lid of the tank from earth/soil/backfill.
- Interior pressure on the sidewalls of the tank if it's water tested prior to backfilling (which is the preferred way of ensuring no leaks).
- Potential fatigue of inlet/outlet fittings during periods when the tank is filled and emptied (for periodic pumping/maintenance).
- Whether just foot traffic, or small to heavier vehicular traffic occurs above and around the tank.
- Bedding and proper placement of the tank.

Hydrogen sulfide gas associated with wastewater has an opportunity to form sulfuric acid in moist conditions, which is highly corrosive. Some materials and methods of construction hold up much better under those conditions, including well-built fiberglass tanks, reinforced concrete tanks constructed with high-strength sulfide-resistant (type II or V) cement and thick enough walls, and well-built medium or high-density polyethylene and polypropylene tanks. Each of these materials of construction has advantages and limitations, with care needed to specify requirements for each to ensure long-term watertightness. Materials of construction for tanks are discussed later in this chapter.

5. Inlet tee in place

The inlet tee directs incoming wastewater downward and below the water surface, to reduce disturbance that would decrease settling, and to prevent solids from being directed toward the outlet of the tank or compartment.

6. Outlet tee in place

The outlet tee prevents settling solids and scum from directly exiting the tank. This tee should extend down to a depth that would prevent scum from exiting the tank under all operating conditions, and well into the clear zone. A vertical length of pipe of about 12 to 14 in (0.31 to 0.36 m) from the "invert" or "flow line" (6 o'clock inside the horizontal pipe leaving the tank) to the tee inlet is usually a sufficient length. The outlet tee should be at least 3 in (80 mm) in diameter, unless the tank outlet is fitted with an effluent filter. If allowed under any applicable regulations, the outlet for an effluent filter could be reduced to 2 in (50 mm) since larger solids would be prevented from exiting by the filter. Effluent filters are discussed in Sec. 5.3 of this chapter.

7. Sufficient length to width ratio for the tank

Because of the basic physical laws governing particle settling and separation of FOG from water, it's important to provide enough tank length relative to its width for wastewater entering from one end and exiting from the other end (lengthwise). The goal is to offer ample settling/flotation time and distance. Most technical guidance literature and regulations call for a length to width ratio of at least 3:1 for rectangular-shaped tanks.[1] The same would reasonably apply to cylindrical, or roughly cylindrical tanks. While higher length to width ratios would tend to offer greater settling distance, for a given tank volume it would also decrease the depth available for settleable solids and scum.

8. First compartment has at least 50 percent, and preferably more, of the total tank capacity; for multiple tanks, it is recommended that the first tank be at least somewhat larger than the second tank

As illustrated in Fig. 5.1, the first compartment of a single tank, or first tank of multiple tanks, tends to build up greater amounts of settleable solids and scum. It's therefore important to provide enough volume there for that buildup, to avoid excessively short tank pumping/cleaning intervals.

9. Adequate depth of tank for settling and sludge/scum buildup

For a certain minimal clear-zone volume under all operating conditions, a minimal length to width ratio results in an average depth of clear zone for a particular design flow. Referring again to Fig. 5.1, the *minimum* clear zone depth in the tank shown would be about 16 in (0.41 m). That allows for a minimum of 12 in (0.3 m) on average across the length of the tank between the top of the sludge layer to the opening of the outlet, and on average 4 in (10 cm) between the bottom of the scum layer and the opening to the outlet tee. *It is critical that sludge not be allowed to accumulate to a point where it exits the septic tank.*

Problems and expenses with that relative to further treatment processes or effluent dispersal systems can greatly outweigh any short-term savings associated with less frequent tank pumping intervals.

For a flow of 300 gal/day (1.136 m³/day), a length-to-width ratio of 3:1, and for a minimum hydraulic retention time of 24 hours (1 day) for the given flow, the following calculation gives the minimum width and length of the tank to provide that average clear-zone depth at the operating point where pumping would be needed (using 16-in or 0.407-m operating depth):

$$(300 \text{ gal/day}) \times (1\text{-day minimum retention}) = 300 \text{ gal}$$
$$300 \text{ gal} = 40.1 \text{ ft}^3 = 3W \text{ ft} \times 1W \text{ ft} \times 1.33 \text{ ft}$$
$$= 3W^2 \text{ ft}^2 \times 1.33 \text{ ft} = 4W^2 \text{ ft}^3$$
$$(40.1 \text{ ft}^3)/(4 \text{ ft}) = 10 \text{ ft}^2 = W^2$$
$$W \text{ (minimum)} = 3.17 \text{ ft} = 38 \text{ in}$$
$$L \text{ (minimum)} = 9.5 \text{ ft} = 114 \text{ in}$$

In SI units:

$$(1.136 \text{ m}^3/\text{day}) \times (1\text{-day minimum retention}) = 1.136 \text{ m}^3$$
$$1.136 \text{ m}^3 = 3W \text{ m} \times 1W \text{ m} \times 0.407 \text{ m} = 1.22W^2 \text{ m}^3$$
$$(1.136 \text{ m}^3)/(1.22 \text{ m}) = 0.931 \text{ m}^2 = W^2$$
$$W \text{ (minimum)} = 0.97 \text{ m}$$
$$L \text{ (minimum)} = 2.9 \text{ m}$$

The above tank length, width, and depth dimensions calculated for a flow of 300 gal/day (1140 L/day) are the *minimum* dimensions needed for the point in time when the tank will need pumping. The total operating depth of the tank would therefore need to be significantly greater to allow for the accumulation of solids and scum (as shown in Fig. 5.1). If that depth is too little, tank pumping frequency will be excessive. Calculations and estimates, along with studies have been done to estimate tank pumping frequencies. However, the frequency will depend on the solids and grease loading to the tank, which in turn depends on a variety of other factors. Those include

- Whether a garbage grinder is used in the kitchen(s)
- Whether the tank receives waste from a grinder pump station
- Food preparation and types of foods commonly prepared
- Occupancy of the building(s) served

Studies have shown that the use of kitchen garbage grinders significantly increases the organic and suspended solids and fats/oils/grease loading to onsite wastewater systems, and creates more rapid buildup of scum and sludge in septic tanks needing removal.[2] As with the use of grinder pumps in collection systems or lift stations, much smaller particles entering a primary settling tank would not be expected to settle as quickly as larger solids, and grinding up FOGs with solids would tend to increase time needed for their flotation and segregation into the scum layer of the tank.

Width-to-height ratios of about 1:1 are typical for septic tanks, with liquid depths ranging from a minimum of about 30 to 36 in (0.76 to 0.92 m) for smaller residential tanks[1,3] to a maximum of about 9 to 10 ft (2.7 to 3 m) for large fiberglass septic tanks. The "head space" (air space) in the tank above the liquid level is typically about 8 to 12 in (0.2 to 0.3 m) above the liquid level, depending on the type of tank. That distance tends to be at the higher end of the range for fiberglass tanks due to the curvature of the tank walls near the top and the need to have inlet and outlet fittings flat/flush against the sidewall. For tanks built to be water-tight all the way to the riser openings, the head space provides reserve storage space in the event that discharge from the system needs to be interrupted for some reason (e.g., pump replacement, service or repair of a further treatment process, or dispersal field maintenance or repair).

For the small residential tank illustrated in Fig. 5.1, if the tank had a total *liquid* depth of 30 in (0.76 m), with an average of 16 in (0.41 m) of clear zone, that would leave about 14 in (0.36 m) for the sludge and scum accumulation on average. With a head space of 8 in (20 cm), the total interior tank depth would be 38 in (0.97 m). Referring to the min-imum width calculated above, that would be consistent with the 1:1 ratio typical of these tanks. However, in that there are typically several inches of scum accumulated at the surface of the liquid, that total liq-uid depth does not leave much space at the bottom of the tank for sludge accumulation. A larger tank, with at least approximately 36 in (0.92 m) of liquid depth along with slightly larger width and length would allow for more reasonable intervals between tank pumping.

Based on the above sizing considerations, Table 5.1 provides sug-gested minimum sizing for domestic wastewater flows, particularly for geopolitical locations where no regulatory design guidelines are in place for septic tank sizing, or where more conservative designs may be needed or desirable.

For systems having flows higher than about 5000 gal/day (19 m³/day), and particularly for those serving multiple residences or commercial establishments that will tend to have variable usage pat-terns, a *working* capacity (volume inside the tank up to the liquid level) of about 2.5 times the daily design flow is likely sufficient. As discussed previously, for larger systems serving greater numbers of homes or businesses, fluctuations in flows from individual facilities served tend to moderate each other. However, design flows should always take into account peak flows to prevent undersizing.

Because sludge/septage pumping can be costly (commonly $400 to $500 in the United States to pump a 1000-gal, or 3.8-m³, tank), it's important to consider pumping intervals in selecting tank sizes for systems. There are also many geographic areas throughout the world

Daily Design Flow, gal/day (L/day)	Minimum Septic Tank Capacity, gal (L) for Liquid Level (or "Working" Capacity)
Equal to or less than 200 (750)	750 (2850)
Equal to or less than 300 (1135)	1,000 (3800)
Equal to or less than 400 (1500)	1,250 (4750)
Equal to or less than 500 (1900)	1,500 (5700)
Equal to or less than 600 (2275)	1,750 (6650)

For flows greater than 600 gal/day (2275 L/day), minimum working septic tank capacity of about 3 times the peak daily design flow.
*For commercial facilities with higher levels of oils/greases such as restaurants, higher suspended solid loadings, or tanks receiving grinder pump waste, additional settling capacity may be needed along with grease trap/separation provisions as discussed at the end of this section.

TABLE 5.1 Minimum Suggested Primary (Septic) Tank Capacities for Small-Scale Systems*

where there are few pump trunks available, and/or they may need to travel long distances and incur high fuel costs to reach tank sites and transport the pumped waste to a facility that can treat and properly handle that type of waste. From both economic and environmental perspectives, the incremental added costs for sizing septic tanks conservatively (large enough) so as to maximize pumping intervals tend to be worth the investment.

It is recommended that water softener backwash waters not be plumbed into onsite wastewater systems, as a general rule. This will be discussed more in Chap. 6, but at least one study has shown that there were adverse impacts to treatment levels achieved for several important wastewater pollutants, including nitrogen and pathogen levels. Water softener brine flows may also increase septic tank pumping frequencies.

In settings such as coastal areas where sand or grit may easily enter sewer lines leading to primary settling tanks, some method of grit removal may be needed ahead of septic tanks. See Ref. 4 at the end of Chap. 6 for more information on grit removal.

5.2 Grease Removal

For restaurants or commercial properties that include restaurants (such as shopping centers or hotels), truck stops and other facilities from which relatively large amounts of oils and greases may enter the wastewater system, some method of grease removal should be used prior to primary treatment in a septic tank. In addition to those measures, it may also be important to up-size the primary settling/septic tank to ensure that fats, oils, and/or greases (FOG) don't enter either further treatment processes or effluent dispersal systems.

There are basically two approaches used for grease removal for decentralized wastewater systems when used ahead of primary settling tanks. Those are each described below:

1. Outdoor grease interceptors

Outdoor (and usually buried) grease interceptors, or traps, may be either small "passive" units, or larger ones sized for higher commercial flows. Larger units are often referred to as grease "interceptors," although grease trap and grease interceptor tank are used synonymously here. Grease traps are constructed similarly to septic tanks, and typically have one to two compartments. For two-compartment units, the first compartment usually has about 60 percent of the total capacity. In contrast to septic tanks and due to the focus on FOG and their tendency to rise to the surface of the waste stream, the opening between the first and second compartments is often lower than for septic tanks to prevent FOG from entering the second compartment or exiting the tank. Grease interceptors may also be fitted with flow diverters that serve to slow the velocity of wastes entering the tank, and better distribute the wastewater across the entrance of the tank. That is particularly helpful for warm or hot wastewaters where emulsified grease needs to cool for better separation before entering the second compartment and exiting the unit. Figure 5.2 below shows the interior of a 500-gal (1900-L) precast concrete grease interceptor with that configuration. Figure 5.3 shows the same grease interceptor tank during testing and start-up. Figure 5.4 shows the basic features of a typical two-compartment commercial scale grease interceptor tank.

FIGURE 5.2 Wastewater enters the grease interceptor tank from the left, with the concrete flow diverter slowing and better distributing the waste stream. The diverter is positioned directly in front of the inlet.

FIGURE 5.3 The same grease interceptor as in Fig. 5.2 is shown here during start-up and testing.

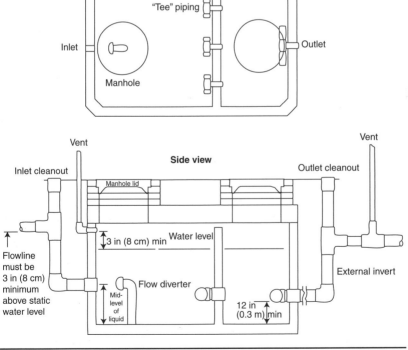

FIGURE 5.4 Commercial grease interceptor tank. (*Source: Austin Water Utility, Special Services Division, City of Austin, Texas.*)

While there are a number of criteria available in the technical literature for sizing grease traps, the Austin (Texas) Water Utility's Special Services Division (Industrial Waste Control/Pretreatment Program) has studied and developed criteria that are cited by municipalities throughout the United States. The basic design steps below have been adapted from the methodology used in Austin's grease trap sizing criteria.[4] The city also has a December 2003 presentation on grease traps that helps emphasize the need for regular inspections, cleaning and necessary maintenance of grease traps. Viewing that presentation is highly recommended for property owners and managers responsible for restaurants, truck stops and other facilities equipped with grease traps. The presentation may be downloaded and viewed using the following web link: http://www.ci.austin.tx.us/water/downloads/wwwssd_iw_greasepres.pdf.

Grease trap sizing steps:

1. Calculate the total number of "fixture units" that will be plumbed to the trap (this should include all fixtures that will carry waste with grease). Fixture units can be calculated using the following:

Fixture Type*	"Pee" Trap and Trap Arm Size, in (mm)	Number of Fixture Units
3-compartment sink	1-1/2 and 2 in (40 and 50 mm)	3 and 4, respectively
2-compartment sink	1-1/2 in (40 mm)	2
Dishwasher	2 in (50 mm)	4
Wok stove	2 in (50 mm)	4
Floor drains, 2, 3, and 4 in (50, 80, and 100 mm)	2, 3, and 4 in (50, 80, 100 mm)	2, 3, and 4, respectively
Floor sinks, 3 and 4 in (80 and 100 mm)	3 and 4 in (80 and 100 mm)	3 and 4, respectively

*Note that garbage disposals/grinders are assumed to be excluded from the calculations, with their use discouraged.

2. Determine the minimum flow rating for the grease trap: Multiply the total fixture unit count times 3 gal/min (fixture count times 3×3.785 L/min).

The grease trap flow rating = fixture unit count × 3 gal/min

The grease trap flow rating = fixture unit count × 3 ga/min × 3.785 L/gal (= flow rating in liters per minute)

3. Calculate the minimum liquid capacity of the trap by multiplying the grease trap. Flow rating times 12 minutes:

Minimum grease trap capacity* = flow rating ×
12 minutes

Two-compartment grease traps are recommended due to their increased flow paths and separation capabilities. Precast concrete grease traps and those manufactured from other materials are available in set capacity increments. So based on the minimum capacity calculated, unless the calculations match one of the sizes available, the next larger size unit should be selected.

The outlet of the grease trap is typically plumbed into (using "wye" type connection) the sewer line leading from the rest of the building(s) to the primary treatment tank's first compartment or to the inlet of the first of multiple septic tanks. Materials of construction are very important for grease traps due to the corrosive nature of the waste in a septic environment. Precast concrete units should be constructed of very high-strength type II or type V (sulfide resistant) with at least 4-in (100-mm) walls so that there is sufficient concrete cover over the reinforcing steel.

2. *Indoor grease separation/interceptor units*

The second basic method for grease removal employs under-sink or under-counter grease separation units. These units are typically much less passive than larger outdoor grease traps, in that they use electro-mechanical skimming components along with baffles internal to the devices. "Point source" units are typically plumbed to a single fixture, such as a multicompartment sink or prerinse station in commercial kitchens. These units are typically sized to handle flows of about 10 to 50 gal/min (1 to 3 L/s). Larger indoor units capable of handling flows from multiple sources/fixtures are often located in basements or mechanical rooms, with plumbing routed from the various sources to that location in the building. Flows for those "centralized" indoor units may range from 50 to 125 gal/min (3 to 8 L/s). Because indoor grease separator units tend to be proprietary with different operational features and materials of construction, it is recommended that manufacturers or their nearest distributors be contacted for aid with design and installation of the units. Figure 5.5 shows the basic configuration of a typical single-source grease separator unit for a commercial kitchen.

---

*Gallons or liters (depending on which units were used for the calculations).

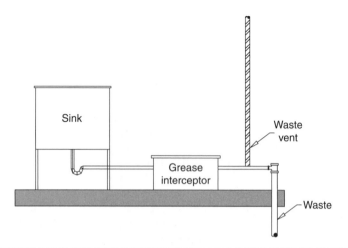

FIGURE 5.5 Basic configuration of a single-source kitchen grease interceptor used for commercial kitchens.

Whether single source or plumbed to multiple fixtures, indoor grease separation units must routinely have grease removed from them. Figure 5.6 shows an under-the-sink unit serving a dining hall at a youth camp. As seen in the photo, there is not a large amount of storage volume available for grease storage. It's recommended that

FIGURE 5.6 Grease interceptor used for a commercial kitchen sink. Grease accumulating in container at front of unit must be checked regularly and emptied as needed into a grease storage receptacle.

commercial kitchens using these units check them on a daily basis and empty them as needed, at least until there is enough operational history based on the amount of grease typically collected by the unit. If not checked daily though, the schedule should be frequent enough to ensure that no overloading occurs.

These units should also be constructed of very corrosion resistant materials. That might consist of stainless steel (proper grade) exterior housing with polyethylene interior materials of construction. Indoor units sized for commercial kitchen flows that are constructed of high quality and noncorrosive materials tend to be significantly more costly than outdoor grease traps. A preference for one approach versus the other may need to depend on the availability, costs, and end use of grease collected by local haulers, along with other factors such as site conditions and potential space constraints. Because under-the-sink or under-the-counter units require plumbing under sinks in a way that is not the usual configuration, care should be taken to make sure that "pee" traps are installed properly.

Oils and grease collected by indoor grease separators have historically most often been either emptied into some type of outdoor storage bin or barrel provided by a local rendering company if those arrangements are made and available, or stored for disposal in a sanitary landfill if allowed. Outdoor traps also need to be checked regularly, with the grease transported for suitable final handling. In recent years, both vegetable and animal fats and oils have successfully been converted into bio-diesel fuels. As long as the methods and materials used result in a net gain of clean energy, if local processing into bio-diesel fuel is available, it would certainly be considered a sustainable approach for dealing with grease and oil wastes.

Composting of grease removed from grease traps has also been carried out as a beneficial use of that waste. In many locations around the world waste sludge or septage combined with grease trap waste is cocomposted with chipped "green" wastes from tree-trimming, landscaping, and operations producing wastes from vegetation. This would also be considered a good end use of grease waste, since good soil amendments for restoring disturbed areas can be produced from well-managed composting operations. In general, some type of beneficial use, or conversion to an energy resource, is considered a much more sustainable approach than delivery to a landfill for disposal.

5.3   Effluent Filters

It is not unusual for property owners with gravity-fed subsurface drain fields to report that their septic system is not draining and sewage is backing up into the house plumbing. Upon digging up sewer lines leading to the field, and portions of the drain field, it's common to find drain field lines impacted by solids over the years, with build-up of solids occurring all the way up the system until there is very little capacity to receive wastewater flow. Such occurrences typically result from a combination of two factors: (1) not checking sludge and scum levels in septic tanks frequently enough, and then removing those accumulations to prevent solids and scum from entering the drain field, and (2) nothing in-line to prevent excess solids from entering the effluent line leading from the septic tank to the drain field. In many cases, owners are not even sure where their septic tank is buried in their yard, and in other cases tanks, risers, and lids are not installed so as to enable checking conditions in the tank.

These things can be avoided using methods and materials available today, including periodically performing certain basic system checks and maintenance activities. Detailed information on operation and maintenance for primary treatment tanks, and treatment and dispersal methods discussed in Chaps. 6 and 9 can be found on the website for this book.

An important element of onsite wastewater systems constructed today is the use of an effluent filter at the outlet of the primary settling/septic tank. Effluent filters are typically one of two types: (1) a gravity flow filter fitted onto the outlet tee of the septic tank, or (2) a screened/filtered vault into which primary treated effluent must flow through to a pump. The latter is commonly used to dose one of several types of system components, including added treatment processes following primary/septic tank treatment, pressure-dosed subsurface soil dispersal fields, or effluent collection systems serving multiple properties (STEP systems).

Effluent filters screen solids of about 1/8 in (3 mm) and larger, with even smaller-sized particles tending to be screened as buildup occurs across the surface of the filter over time. Solids that might pass through these filters are small enough to be flushed from dispersal field lines with relative ease if buildup occurs in field lines over time. In general, an effluent filter is a relatively minor expenditure for even the smallest projects, and can save considerable time, trouble, and expenses over time.

Figure 5.7 shows a photo of a 4-in (100-mm) gravity flow effluent filter installed at the outlet of a concrete septic tank before installing the access riser. The filter cartridge is inserted into a housing plumbed to the septic tank outlet tee. Wastewater in the tank enters the filter housing through some type of opening(s) ("inlet ports" shown on the section view of a typical filter), and must flow through the filtering device before exiting the tank. It's removed for cleaning by pulling it out of the housing with the handle shown in the photo. Some manufacturers produce effluent filters with check balls or other devices that prevent/block flow out of the tank when the filter cartridge is removed. That has a twofold benefit of making sure that someone doesn't simply remove the cartridge without reinstalling it when cleaning is needed, and helps prevent solids from exiting the tank when the cartridge is pulled for normal cleaning. If a septic tank fitted with a gravity-flow

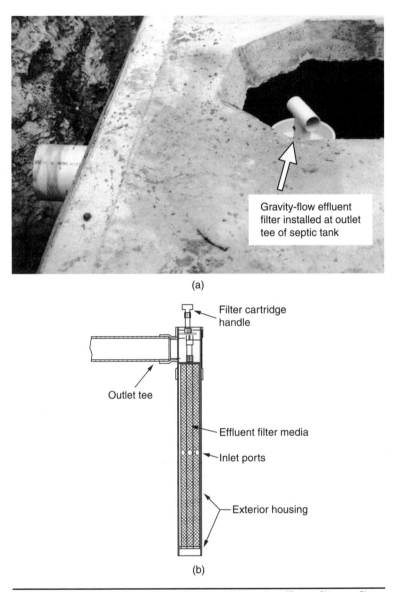

Gravity-flow effluent
filter installed at outlet
tee of septic tank

(a)

Filter cartridge
handle

Outlet tee

Effluent filter media

Inlet ports

Exterior housing

(b)

Figure 5.7 (a) Gravity-flow effluent filter and (b) gravity effluent filter profile.

effluent filter is not equipped with a float and alarm to notify residents or the system manager of excessive buildup and a rise in water level, the filter should be checked at least annually and cleaned if needed. It's best to clean effluent filters when pumping and cleaning the tank, to prevent any solids from washing out of the tank. Sizing effluent filters in accordance with tank cleaning schedules and flows helps in this respect. This is discussed further below.

Figure 5.8 shows photos of an effluent filter installed in a pump vault. These filters should also be checked at least annually to see if cleaning is needed along with periodic checks of other system components. High water alarm floats installed for notification of pump operational problems also serve to notify if water levels rise due to

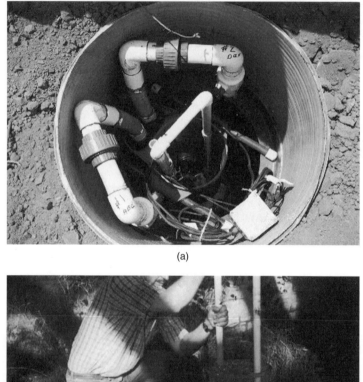

(a)

(b)

Figure 5.8   (a) The photo shows discharge piping from duplex pumps (two pumps installed side by side) in a pump vault with an effluent filter. (b) This photo shows a similar pump vault effluent filter being removed for inspection to see if cleaning is needed.

effluent filter clogging in the pump vault. Because effluent filters/ screens typically pass solids up to 1/8 in (3 mm) in diameter, pumps used in screened vaults should be able to pass that size particle also. Most high head submersible effluent pumps used for decentralized wastewater systems are similar to vertical turbine well pumps, with flow intake screens on their housings. Many of these pumps have screens that will only pass solids up to 1/16 in (1.5 mm), which may cause blinding over of the screen and pump overheating and failure. Detailed technical specifications for pumps should be checked to make sure the pump will pass the maximum size solids that may make their way through effluent filters/screens (typically 1/8 in, or 3 mm).

The pumps in Fig. 5.8(*a*) are dosing different dispersal field zones, so note that the discharge piping is labeled to indicate which zones are dosed by each pump (A, B, and C by #1 and D, E, and F by #2). In Fig. 5.8(*a*) also, note the horizontal members used to suspend the pump vault from the collar of the tank access hole. The manufacturer of this pump vault supplies a PVC pipe for this purpose, but experience has shown that plastic support rods can sag over time. A metal grounding rod, or piece of steel rebar, can be inserted into the plastic length of pipe to provide added support, but must be protected from corrosion in the tank. The white end caps are therefore used to seal the support pipe after insertion of the support rod.

Effluent filters used should allow for easy removal and reinstallation, and be structurally sound and capable of withstanding normal stresses that occur during installation, operation, and servicing. They should be constructed of noncorrosive material, and be sized/designed to handle the system's flows without excessive maintenance. For single-family residential systems, maintenance would be considered excessive if more than one cleaning or servicing is needed per year.

Manufacturers of effluent filters typically each have their own sizing criteria. Those criteria should be referred to for filter sizing and selection. To avoid the need for overly frequent cleaning of effluent filters and/or potential system backups, it's important to make sure that the filter is sized to have enough filter surface area for the given flow and waste characteristics. In general, for residential applications and typical wastewater composition, no less than about 0.008 to 0.01 ft^2 of effective *screen surface area* should be provided per gallon of daily wastewater flow (0.0002 to 0.00025 m^2/L of daily flow), using single or multiple filters. And a minimum effluent *filter flow area* of about 0.0018 to 0.002 ft^2 should be provided per gallon of daily flow (0.000044 to 0.00005 m^2/L of daily flow). Some manufacturers don't use effective surface area per gallon flow, and may instead use historical experience and empirical data with their effluent filters to recommend sizes and models for specific projects. The particular manufacturer or their technical literature about their filters should be consulted prior to selecting an effluent filter based on the design flows and waste characteristics for a project.

5.4   Septic Tank Location and Materials of Construction

5.4.1   Tank Location

During the preliminary/conceptual planning phase of a project, conceptual layouts should be developed for the entire system(s) planned for construction. It's usually helpful to also lay out any future system phases for larger projects that are intended to be built in stages, over time, to identify potential conflicts or complications.

For effluent collection systems, the location and elevation of the septic tanks and street collectors relative to the finished floor elevations of houses and buildings on individual lots will directly influence whether a STEP, STEG, or combination of STEP/STEG is feasible and the most cost-effective. Even if floor plans are not available for homes or businesses on individual lots, elevations and locations for septic/interceptor tanks can be established for each lot as a part of planning for the whole development. These would not typically be installed before building construction begins on the lot, but reasonable finished floor and sewer stub-out elevations should be established for each connection that will allow for gravity drainage of sewage to a suitable tank location. In this way, excessive excavation and site disruption can be avoided, and the use of gravity flow (for STEG systems or portions of systems) can be maximized. Allowances should also be made for possible plumbed cellars or basements in geographic areas where they're included in structures.

Where there's significant rock outcrop or shallow depth to rock, or shallow depth to groundwater, it becomes especially important to locate septic tanks so that burial depths can be minimized. Access to the tank by a pump truck and any water needed for cleaning/flushing the tank or effluent filter should be considered. If the tank is located significantly downhill or a long distance from areas of the site accessible to a pump truck, special pump trucks would be needed with powerful enough equipment. It's also important to avoid excessive velocities with wastewater entering septic/primary settling tanks to avoid disturbance of setting processes. Extra baffling in the tank or some other means of slowing flow velocity without causing blockages should be used for steeper downhill runs of pipe into septic tanks. Structural stability is also an important consideration for hilly sites, where there may be seismic activity, or periodic problems with slope stability or subsidence.

Septic tank inlet elevations must be set based on sewer stub-out elevations from the building(s) served, the distance from the building(s), and the minimum slope needed for the sewer line from the building(s) to the tank inlet. Plumbing codes and most regulatory authorities permitting wastewater systems specify minimum pipe slope for sewer lines leading from buildings to septic tanks. The minimum slope is typically set at about 1/4 in per linear foot (20 mm elevation drop per linear meter).

5.4.2 Materials of Construction for Septic Tanks and Pump Tanks

The material of construction used for septic tanks should be selected based on the availability of local materials and manufactured tanks, the specific site conditions and location where the tank will be installed, and durability of the tank under long-term operating conditions. As discussed previously though, it is very important that the tank be watertight up into the access risers, and that watertightness be maintained under varying operating conditions. That is much more difficult for concrete tanks, with separately cast lids needing to be sealed around that joint. Unacceptable materials would be metal, wood, cinder block, and others that would be very subject to corrosion or leaks.

Three basic categories of materials commonly used today for septic tanks are (1) precast or cast-in-place concrete tanks, (2) fiberglass (FRP, or fiber reinforced plastic) tanks, and (3) polyethylene tanks. Each material has its advantages and limitations, as summarized in Table 5.2 and discussed further below. The cost-effectiveness or suitability of selecting one type of tank over another tends to depend on project location and the availability and costs of one material versus another.

Most U.S. regulatory permitting authorities maintain lists of approved manufacturers of septic tanks constructed of these three materials that will need to be referred to when specifying a tank in those jurisdictions. American Concrete Institute (ACI) and American Society for Testing and Materials (ASTM) both have standards that apply to the septic tank methods and materials of construction. ASTM is an international standards organization, and those applicable standards should be reviewed to ensure that the methods and materials of fabrication used for a specific brand of tank are considered acceptable at least by ASTM standards.

Precast Concrete Tanks—Materials and Methods of Construction

Depending on geographic location, concrete may be more readily available for use in constructing septic tanks, as compared with the other two materials listed above. Cast-in-place concrete tanks should be designed by a structural engineer. Likewise, the design "template" used for precast tanks should be developed and sealed by a structural engineer, with that documentation available from the tank factory. A disadvantage of precast concrete tanks is that, due to their weight, fully built tanks with lids and monolithically poured bottoms and sidewalls are typically limited to capacities of about 2000 gal (7.6 m³).

Cast-in-Place Reinforced Concrete	Precast Reinforced Concrete	Fiber-Reinforced Plastic (FRP or Fiberglass)	Polyethylene
Require structural engineer's design specific to project.	Tanks preengineered by tank manufacturer; ability to better control quality of construction at manufacturing facility, and pour bottoms and walls of tanks monolithically.	Tanks preengineered by tank manufacturer; ability to better control quality of construction at manufacturing facility.	Tanks preengineered by tank manufacturer; ability to better control quality of construction at manufacturing facility.
Requires forming, placing reinforcing steel, and pouring tank on-site in excavation, along with proper bedding.	Requires heavy truck access close to tank excavation for placement of tank with heavy duty boom and lifting equipment.	Relatively lightweight tanks that can be transported strapped-down on a flat bed trailer and moved with equipment typically used for construction such as miniexcavators.	Relatively lightweight tanks that can be transported strapped-down on a flat bed trailer, and moved with equipment typically used for construction such as miniexcavators.
Concrete is subject to corrosion over time.	Concrete is subject to corrosion over time.	Very resistant to corrosion.	Very resistant to corrosion.
Greater variability of quality of construction, and dependent on availability of local labor and engineering skills.	Monolithically poured bottoms and sidewalls limited to tanks with capacities less than about 2000 gal (7.6 m^3).	Available in wide range of sizes, up to very large tanks serving subdivisions and larger commercial projects.	Septic tank sizes typically available up to only about 1500 gal (5.7 m^3) operating capacity, due to structural issues.

If properly designed and built, structurally stable for interior and exterior loading and variable water levels.	If properly designed and built, structurally stable for interior and exterior loading and variable water levels.	Tanks structurally designed primarily for exterior loading/pressures, and care must be taken to ensure that sufficient exterior support is provided to balance interior pressures if/when filling with water before backfilling. Must consult with manufacturer about acceptable filling levels prior to backfilling.	Have historically had problems with flexing/deformation when backfilling with tank empty, or pumping the tank after installation. Improvements in materials and tank designs by at least some manufacturers have reduced these problems.
Top surfaces/lids of tank typically cast separately, requiring sealing between tank walls and lid; larger tanks have to be poured in several sections with greater opportunity for leaks at joints.	Top surfaces/lids of tank cast separately, requiring sealing between tank walls and lid.	Tanks are manufactured in half-shells and assembled either by the manufacturer or by trained technicians in the field (for smaller tanks). For smaller tanks shipped in half-shells and assembled elsewhere, half-shells can fit inside each other allowing for shipment of much larger number of tanks for given container/cargo space.	Tanks are molded in one piece, using either rotational or blow molding processes. Though lightweight, requires much more space for shipping as compared with smaller diameter FRP tanks that may be shipped in half-shells.

TABLE 5.2 Comparison of Septic Tank Materials (*Continued*)

Cast-in-Place Reinforced Concrete	Precast Reinforced Concrete	Fiber-Reinforced Plastic (FRP or Fiberglass)	Polyethylene
Due to weight of tank, flotation of tank and adding ballast are a less concern than for FRP or polyethylene tanks.	Due to weight of tank, flotation of tank and adding ballast are less a concern than for FRP or polyethylene tanks.	Antiflotation ballast, or concrete "deadman" weights of appropriate size must be used for flood prone or high ground water areas.	Antiflotation ballast, or concrete "deadman" weights of appropriate size must be used for flood prone or high ground water areas.
Installation can be difficult along steep slopes or where there's not sufficient access by heavy equipment.	Locations of tank installations must be accessible by heavy trucks.	Can be fairly easily moved to difficult locations on sites, as long as adequate space for size of tank.	Can be fairly easily moved to difficult locations on sites, as long as adequate space for size of tank.
Cannot be moved around site after placement.	Cannot be moved around site after delivery by typical onsite construction equipment.	Can be moved if needed (if empty) around site by typical onsite construction equipment.	Can be moved if needed (if empty) around site by typical onsite construction equipment.

TABLE 5.2 Comparison of Septic Tank Materials (*Continued*)

Precast concrete tanks larger than that are typically delivered to job sites in sections (horizontal) or "lifts," which must then be sealed as well as possible to prevent leaks. Only a certain amount of hydrostatic pressure can be exerted on such seams, or they will leak. Depending upon the method of grouting and sealing, that may only be a few inches up to 1 or 2 ft above the seam. For large concrete tanks and relatively short section heights, that can pose leakage problems for the seams deeper in the tank over time.

The following specifications are an example of those used for the reinforcing steel and concrete mix on projects where precast concrete tanks are to be used.

- Reinforcing steel shall be ASTM A-615 Grade 60, FY = 60,000 psi (413,400 kPa). Details and placement shall be in accordance with ACI 315, ACI 318, and ACI 350, as related to structural tanks containing wastewater or wastewater effluent.

- Concrete shall be ready-mix with cement conforming to ASTM CI50, type II (sulfate resistant). It shall have a cement content of not less than six sacks per cubic yard and maximum aggregate size of 3/4 in (20 mm). Water/cement ratio shall be kept low (±0.35), and concrete shall achieve a minimum compressive strength of 4000 psi (27,560 kPa) in 28 days.

- Tanks walls and bottom shall be a minimum of 3 in thick (80 mm) and lid (top slab) a minimum of 3.5 in thick (90 mm).

- There shall be at least 1-in (25-mm) concrete cover over steel reinforcing bar throughout tank and lid.

- Tanks shall not be moved from the manufacturing site to the job site until the tank has cured for 7 days or has reached two-thirds of the design strength.

- The tank shall be watertight without the addition of seal coatings.

- Tongue-and-groove-type joint with sealer shall be used for affixing concrete lid/cover to tank walls.

- The tank shall be tested for watertightness at the factory, and again after installation.

Some tank casters are able and willing to meet these specifications, while others have indicated they are not able to do so. If a precast tank is to be used, the designer of the wastewater system using the tank will need to determine suitable specifications and acceptability of locally available precast concrete tanks. Because of the costs associated with transporting concrete tanks to a site by large trucks, it's important that water-testing be required prior to moving the tank to a job site. Water may also be more readily available for that at a concrete tank casting factory as compared with a site where construction

is underway. Minor tank repairs, if needed, are more readily performed and reinspected at the factory. The tank caster can also reuse
testing water for other tanks. As long as care is taken with the transport and installation of a watertight tank, testing at the factory will
prevent disruptions to often tight construction schedules due to finding that a tank is leaking after it's installed.

For monolithically poured precast concrete tanks, where available, the bottom and sidewalls of the tank are cast together and the
lid is cast separately. A tongue-and-groove-type joint along with
application of some type of effective sealant between the top of the
walls and the lid is needed to make the tank watertight up to at least
1 or 2 in into the riser. Even with a watertight connection along that
joint, concrete tank lids may still be "floated" (upwardly displaced)
from that seal if the water level in the tank rises as much as just a very
few inches above the top surface of the lid, depending on burial depth
and weight of the lid and soil above it. Some type of mechanical fas-
tening method can also be used to prevent this, and should be used if
groundwater levels would be expected to reach the concrete tank lid.
Again, the details of this should be considered by the wastewater system designer and discussed as needed with the tank manufacturer.

To best achieve watertight conditions at the inlet and outlet of the
tank, Schedule 40 PVC couplings of the correct size can be cast into
the tank walls by the tank caster/factory. This provides for a solvent-
welded joint for the inlet/outlet tees and their connections to exterior
piping. Some precast tank manufacturers supply tanks with beveled
concrete openings, allowing for adjustments to the orientation of
pipes entering and exiting the tank during installation. However,
because of the thinness of the beveled edges of the inlet/outlet holes,
this necessitates grouting around the pipe after it is placed to maintain watertightness over time. Inlet holes are typically 4 to 6 in (100 to
150 mm), depending on the size of the sewer line entering the tank,
and outlet holes typically 2 to 4 in (50 to 100 mm), depending on
flows, whether an effluent filter is used, and any applicable local
requirements for effluent line sizing.

For achieving watertight conditions between concrete tanks and
access risers leading to the ground surface for periodic inspections
and pumping, riser adapters can be cast into the lid (top slab) of the
tank by the precast concrete tank manufacturer. If the tank manufacturer doesn't maintain a supply of the right types of riser adapters, this
requires some coordination among the various persons and entities
involved with the system design and installation, along with the riser
and adapter manufacturer and the tank caster. Manufacturers of
adapters routinely ship such adapters to tank casters. The appropriate diameter of each access hole and riser in the top of the tank
depends on how much space is needed for periodic inspections,
pumping and servicing of that portion of the tank. Figure 5.9 shows

FIGURE 5.9 Plastic riser adapters are shown cast into the tank by the tank manufacturer.

such an adapter cast into the lid of a precast concrete tank. Typically about 15 to 18 in (0.38 to 0.46 m) is sufficient for tank inlets, whereas 24 to 30 in (0.61 to 0.76 m) may be needed for effluent filters and pump vaults. However, the diameter specified to the tank caster must match the available riser adapter sizes available from that manufacturer. PVC risers are affixed to the adapters with either a two-part epoxy or single component adhesive product, providing a watertight bond. The lifting lugs (bent rebar) placed at the four corners of the tank lid shown in Fig. 5.9 are used to place the lid only at the factory, and not to lift the tank. Heavy duty lifting chains are used to lower the tank in place from the transport truck (see Fig. 5.10). A watertight sealant is used between the lid and tank walls. PVC couplings are cast into inlet/outlet walls of the tank by the tank manufacturer at specified tank elevations.

Fiberglass Manufactured Tanks

Fiberglass, or fiber-reinforced plastic (FRP), tanks are very resistant to corrosion and if properly manufactured and installed can remain watertight for many years, and typically longer than concrete tanks. As much as 50-year warranties are reportedly offered by some manufacturers of high-quality septic tanks, as long as they're installed and used in accordance with the manufacturer's instructions. High quality FRP tanks are available in a very wide range of sizes, up to many thousands of gallons (up to +100 m³).

(a)

(b)

Figures 5.10   (a) and (b) As seen in these photos, heavy duty lifting chains and booms are needed for installing precast concrete tanks, along with access to the tank excavation by the transport trucks. Where access is limited such as behind houses/buildings or along slopes, it may be more feasible to use fiberglass or polyethylene tanks.

FRP tanks are cast in halves and assembled at the factory, or for smaller tanks may be assembled by trained technicians following shipment to distributors or installers, depending on the particular manufacturer. For smaller tanks with gasketed/sealed and flanged connections along the midseam (see Fig. 5.11), some manufacturers

FIGURE 5.11 This 1500-gal (5.7-m³) three-compartment fiberglass tank will provide primary settling in the first two compartments, and timed dosing of a further treatment process from a pump vault located in the third compartment.

prefer to assemble the tanks at the factory. Special training is typically required by manufacturers (and needed) for assembling tank half-shells to ensure a watertight seam under all operating conditions. The advantage of shipping tanks in halves is that they can be stacked and occupy much less space for a greater number of tanks during transport by truck, rail, or shipping container. At least one company (Orenco Systems, Inc.) therefore trains installers at their factory on tank assembly.

The FRP tank shown in Fig. 5.11 has three compartments. The third needs to be hydraulically separate from the two primary settling compartments so that the level floats can control the liquid level with the pump in that compartment. Achieving independent hydraulic levels in fiberglass tanks requires special and very careful fabrication methods that may not be possible with field assembly of tank half-shells. In cases where half-shells need to be shipped and assembled away from the FRP-manufacturing facility, it may be more feasible to use two separate tanks, or a pump basin if a pump compartment is needed. FRP tank half shells are joined together to be watertight with a gasketed flange and thick adhesive. The tank in Fig. 5.11 is installed level in both directions, +/− 1/2 in (12 mm), with pea gravel bedding and backfill up to the top surface of the tank. The excavation is cut with a minimum of 12 in (0.3 m) clearance on all sides to prevent any damage to the tank.

High-quality FRP tanks have the characteristic ribbing shown in Fig. 5.11, which provides for structural support for external loading. Fiberglass tanks are not, however, typically designed/manufactured for the same amount of *internal* loading, which must be considered during installation and water testing. It is always best to perform a watertightness test on tanks prior to backfilling so that any problems found can be corrected without having to remove backfilled material, and to have better visibility for any very slow leaks that might not be detected from water level checks during testing. Precast reinforced concrete tanks have the advantage of being more structurally strong for both internal and external loading, and can therefore be filled with water prior to backfilling, with the concrete side walls providing enough structural support for the interior water pressure. More care needs to be taken with FRP tanks, with water-testing of larger tanks and those without flanged midsections being done in stages. The flange along the middle of the tank shown in Fig. 5.11 helps provide structural support for water pressure prior to backfill, with this manufacturer having provided assurances that water-testing up to the top surface of the tank could be done without backfilling.

Unlike concrete tanks, even very large FRP tanks can be transported on flat bed trailers, and lifted with excavators or other equipment already on site. The tank shown in Fig. 5.12 is an 8000 gal (30.3 m³) tank used for a larger-scale onsite system. FRP tanks should be cast with lifting lugs along the top of the tank for lifting and placement. Typical equipment used for decentralized systems installations such as miniexcavators can be used to lift these tanks and set them into place with persons on the ground carefully guiding them into position. Note the inlet piping already cast into the tank by the manufacturer in Fig. 5.12.

Polyethylene Manufactured Tanks

Polyethylene tanks are another type of tank not subject to corrosion by constituents typical of domestic wastewaters. Figure 5.13 shows a residential scale two-compartment polyethylene tank. Polyethylene septic tanks are typically only available in relatively smaller sizes (up to about 1500 gal or 5.7 m³) as compared with FRP tanks, due to structural issues with constructing larger-sized tanks with polyethylene. However, some very long warranties are also provided by some manufacturers for properly installed tanks used in accordance with the conditions of the warranties.

Figure 5.12 This 8000-gal (30.3-m³) FRP septic tank is light enough to be transported strapped down on a trailer.

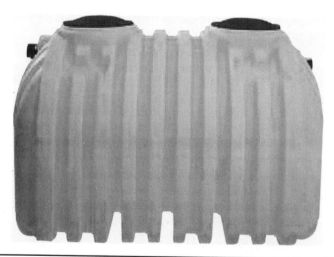

Figure 5.13 This 1500-gal (5.7-m³) two-compartment polyethylene tank has the same type of ribbing as FRP tanks for providing structural support for external pressures from burial. Preplumbed inlet and outlet fittings are shown on this tank. (*Photo courtesy of Norwesco.*)

These tanks are fabricated using either a rotational molding process, or a blow molding process, each of which produces tanks with no seams. The tank in Fig. 5.13 is fabricated using a rotational molding process. Ribbing is also used for these tanks to provide structural strength for external loading/pressure. Because most polyethylene tanks fabricated today for use as septic tanks are seamless, there are fewer concerns about internal loading from static water levels/ pressures when water-testing the tank during installation and prior to backfilling. Since polyethylene is a more flexible material than fiberglass, there have however historically been some problems with polyethylene tanks exhibiting at least slight flexing during filling and emptying the tank (e.g., during tank cleanings) with resulting leaks at the inlets/outlets of tanks. Improvements in the materials and designs used for these tanks have continued to be made with accompanying reductions in those types of problems.

FRP and polyethylene tank manufacturers and distributors typically provide calibrated tables that relate depth in the tank and volume of liquid, based on the shape of the tank. For FRP tanks, inlet- and outlet-hole elevations are commonly specified by the system designer, based on the system configuration and needs. At least 2 to 3 in (5 to 7.6 cm) of water level drop should be provided across the primary settling compartments, with consideration to any applicable regulatory requirements. Inlet/outlet holes can be cut by the manufacturer prior to shipment, along with any inlet/outlet ports that need to be cast into the tank. Because of the curvature of FRP tanks near the top, these holes typically have to be at least 9 to 10 in (22 to 25 cm) from the top surface of the tank to have a relatively flat surface for the tank penetration and inlet/outlet piping connection.

For systems requiring multiple tanks, combinations of tank types (to the extent they're available and cost-effective) can be used to achieve system configurations that best fit the system needs and offer the longest useful service life. Figure 5.18 shows an example of this where a three-compartment FRP tank was used for primary/ septic settling (first two compartments), with the third compartment housing a pump that doses an intermittent sand filter (discussed in Chaps. 6 and 7). A precast concrete tank then receives effluent from the sand filter which, following that process, has negligible amounts of the biochemical constituents that are corrosive to the tank. Fiberglass was used for the septic settling tank portion of that system because the septic wastewater environment is very corrosive.

5.5   Septic Tank and Pump Tank Installation

As with all onsite/decentralized wastewater system components and products, applicable permitting authorities and manufacturer's recommendations and instructions should be referred to for required and/or recommended installation procedures and conditions of use. Information provided in this section is intended to supplement that information. In cases where recommendations in this chapter would result in less stringent or protective methods or materials, the applicable and more stringent practices should of course be used. Manufacturers of FRP and polyethylene tanks (and often even precast concrete tanks) typically have detailed installation procedures, and oftentimes required or voluntary training programs for the installation of their tanks. Manufacturers' instructions should be used to ensure that tank warranties remain valid for their effective periods of use.

While fiberglass and polyethylene tanks are lightweight enough that they can be located in places inaccessible for delivery of precast concrete tanks, greater care is often needed for the installation of those tanks due to their higher susceptibility to damage. For FRP and polyethylene, the details of installation may depend on the availability of certain types of bedding and backfill materials to a greater extent than for concrete tanks.

Tank and system component elevations are established by the system designer and specified on the plans, but should be field verified before placing tanks and treatment units to make sure pipes have adequate gravity fall into and away from system components as needed. For new construction, piping to and from tanks and major components is typically done after installation of the tanks. It's best if the wastewater system installer meets with the system designer/engineer at the site to review the layout of the entire system and identify any potential conflicts before any construction begins. The property owner and/or architect may need to be involved in the meeting also if there are changes or realignments needed that may affect other aspects of site planning, construction, or existing features.

The designer of the system should take into consideration the types of equipment needed for the project and know whether they are used locally by contractors and/or available through rental companies, and if so at generally what costs. Those factors should be considered when specifying the various types of components, methods, and materials used for the project.

1. *Concrete tank installation considerations*

Cast-in-place concrete tank forming, reinforcement, and specific methods of installation should be specified by a structural engineer. Each set of site conditions, size of tank, and other factors will affect the methods and materials used.

Precast concrete tank suppliers should be consulted early in projects (and prior to ordering tanks and beginning construction) to determine what type of truck would be delivering the tank, how close the truck will need to be to the tank excavation for placing it, and whether it can be side or end-loaded into the hole. *In all cases the soil stability around a tank excavation should be considered, along with all other construction safety considerations.* For many concrete tank delivery trucks, it's necessary for the truck to be within a very few feet of the sidewall of the excavation to place the tank. It's therefore important not to excessively overexcavate precast tank holes, while still providing enough sidewall clearance for piping into and out of the tank, and for backfilling.

For precast concrete tanks, a minimum of 4 to 6 in (10 to 15 cm) of added depth of excavation should be provided for the placement and leveling of cushion/bedding sand or pea gravel. A minimum of 6 in (15 cm) of overexcavation should be provided along all precast concrete tank sidewalls *in addition to* whatever space may be needed for inlet/outlet piping and routing of pipe around tanks to their various connections. If rainy weather is expected prior to the time when a tank can be placed and backfilled, it is best to wait for dry weather days if possible to avoid soil and/or rock debris from falling back into the tank excavation prior to backfilling. Bottoms of excavations should be free of rock pieces/fragments, and in very rocky conditions where that may not be possible, a few more inches of cushion sand may be used. Laser construction levels can be used to quickly spot check elevations as excavation proceeds. Cut-ins to the excavation, where inlet/outlet piping or conduit are to enter/exit tanks, may best be dug prior to completing cleanup of the bottom of the excavation and placing bedding material to avoid potential damage to the tanks from heavy equipment after placing the tanks.

To avoid excessive excavation where tanks need to be placed along steeper slopes, terraced excavations may be used in conjunction with retaining walls. Backhoe-mounted rock hammers of the type needed to break out extensive harder rock (Figs. 5.14 and 5.15) require relatively large amounts of diesel fuel for such excavations.

Figure 5.14   Backhoe-mounted rock hammer and miniexcavator used to dig tank hole in rock.

FIGURE 5.15   Rock hammer breaking up rock, which is then removed with the excavator.

Rock broken out from terraced areas can be used for retaining wall construction along the downhill sides of tanks to stabilize backfill and cover material. Figures 5.14 and 5.15 show an excavation for two tanks on a very rocky site being dug using alternately a backhoe-mounted rock hammer, and a miniexcavator to dig out broken out rock. A laser level is used to do quick checks of the depth as the excavation proceeds. In Fig. 5.14 the miniexcavator is positioned over an area where the subsurface dispersal field will be located. Notice that the excavator is tracked, reducing the ground pressure and avoiding overcompaction of the subsurface effluent dispersal field area to be constructed later, and where the equipment needs to operate for excavating the tank hole. It's important to use very low ground pressure equipment over and around dispersal fields to avoid overcompaction before or during dispersal system installation, or damage to field components.

For all tank and trench excavations, trench safety measures should be taken as needed and required, and as determined to be appropriate based on site conditions. In the United States, applicable Occupational Health and Safety Association (OSHA) standards and required methods should be used. To whatever extent the stability of excavated side slopes is a consideration, sidewalls may need to be stepped back or shoring used. U.S. OSHA requirements and standards may be found at the following U.S. government website: http://www.osha.gov/SLTC/trenchingexcavation/index.html.

FIGURE **5.16**   This large tank excavation was stepped back as a means of stabilizing the side slopes, and to avoid the use of shoring to meet U.S. construction safety (OSHA) requirements.

Figures 5.16 and 5.17 show photos of a tank exaction in both rocky conditions (Fig. 5.17) and where slope stability considerations required stepping-back tank excavations (Fig. 5.16). The rocky excavation (Fig. 5.17) was dug at two levels to receive two tanks—one concrete and one FRP as shown in Fig. 5.18. Due to its lesser overall height, the portion of the excavation used for the concrete tank was shallower.

The cushion sand or pea gravel used for bedding should be free of larger aggregate or rock fragments. After placing leveling sand, it is sometimes helpful to moisten it to help with compaction and remove pockets of clumped sand. The bedding material should be leveled in each direction. Laser construction levels can be used to spot check levelness around the excavation as the sand is placed. Either long—4 ft (1.2 m)—hand levels or sides of straight long boards or other materials in conjunction with 24- to 36-in (0.61- to 0.92-m) levels can be used to check levelness across the sand in each direction.

Just after placing a precast tank into the desired position in the excavation, *and while the tank is still secured to the lifting equipment,* the levelness of the tank should be checked in both directions with a long handheld level, straight board and shorter handheld level, or accurate construction laser level. The interior of the tank should be double-checked for any trash or debris, and for soundness of concrete

Figure 5.17 Two tanks of different types and sizes were installed in this excavation, with the deeper and larger FRP tank installed at the near end. Due to the rock all along the sidewalls, no shoring or stepping back of sidewalls was needed.

Figure 5.18 The FRP tank on the left of this photo was placed after the concrete tank to avoid potential damage to it when placing the precast concrete tank. Note that the cutout leading to conduit and pressure piping entering or exiting the tank excavation was dug prior to cleaning the bottom of the excavation and placing pea gravel bedding material.

casting inside the tank. Any divots or "honeycombing," or possible weak spots should be grouted with marine-grade repair material. After determining that the tank is level, inlet and outlet piping should be temporarily capped/sealed, and water-testing performed. Using either water available onsite or trucked to the site, the tank should be filled up to the top surface of the tank, access holes covered to prevent rainfall or significant evaporation from affecting the level, and the water level checked up to 24 hours later. Some permitting authorities require testing over a period of only a few hours. However, slow leaks that should not be occurring with watertight tanks may not be visible within 2 to 4 hours. Unrepaired "honeycombing" of concrete that can occur inside of precast tanks may produce slower leaks. Figure 5.19 shows a concrete tank being checked for levelness prior to water-testing. Once the tank is found to be level and free of any problematic defects that may result in leaking or other problems, it is filled with water to the tank lid (top surface) and water tested by checking the water level before and after a period of time up to 24 hours, and checked along the outside of the tank for any exterior signs of moisture or leaks. Such testing should be performed during dry weather days so that moisture on the outer side of the tank due to leaks can be seen.

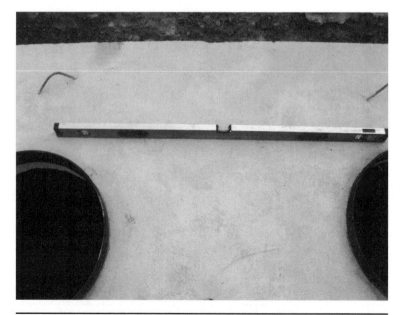

Figure 5.19 A long handheld level, or long straight edge of a board on which a shorter handheld level is placed, is used to check levelness of the tank before the delivery truck with lifting equipment (not shown here) leaves the site.

2. Fiberglass and polyethylene tank installation considerations

Tank excavations for smaller residential scale FRP and polyethylene tanks should provide for at least 8 to 12 in (20 to 30 cm) of bedding for the bottom of the tank, and a minimum of 12 in (30 cm) of over-excavation for sidewalls in addition to space for inlet/outlet piping and placing pipe around tanks. Figure 5.20 shows this for a residential scale FRP tank, and the tank being set into place by a miniexcavator. For larger FRP tanks, at least 18-in (45-cm) space should be provided between excavation sidewalls, along with at least 12-in (30-cm) pea gravel bedding.

Pea gravel free of large or sharp fragments should be used for bedding and backfilling of FRP and polyethylene tanks, or other material deemed by the manufacturer to be a suitable backfilling material (under their warranty conditions). Pea gravel is much more suitable than sand for backfilling these tanks because of their characteristic curvature toward their undersides, and the free-flowing manner that pea gravel tends to fill in voids.

Antiflotation weighting must be used for both polyethylene and FRP tanks where there are concerns about flooding or groundwater levels. Antiflotation "collars" are factory-cast onto some manufacturers' tanks, basins, and other containment units subject to flotation, with concrete poured around these in place. Reinforcing steel should be used for this type of weight (to prevent the concrete from breaking up over time and dislodging from the collar), with the design weight calculated to be as needed for the specific tank and conditions. Concrete "deadmen" weights as shown in Fig. 5.21a may be used where buried FRP or polyethylene containment units are not fabricated with antiflotation collars. The deadmen weights have "anchor points" or lugs on the top for their placement and strapping down tanks, and may either be cast at the site or purchased if/as available from local concrete product suppliers. "Deadmen" should be designed and cast in accordance with American Concrete Institute standards. The straps *and* the connecting lugs in the concrete should be made of materials resistant to biodegradation or deterioration over time. Steel rebar is not an acceptable material for connecting straps to concrete due to the tendency for rebar to rapidly degrade in soil. Wire ropes sized appropriately for the tank and conditions and clipped together at the ends with several stainless steel connecting clips to form an "eye" loop are commonly used for attaching to deadmen anchor points. Wire rope can also be looped around deadmen in appropriate locations along their lengths. Mastic or other type of protective coating may be painted onto wire rope, clips, and anchors to provide further protection against corrosion, to better ensure secure anchoring for long periods of time. In coastal settings, additional measures and/or non-metallic straps and anchoring hardware are needed for those more

(a)

(b)

FIGURE 5.20 (*a*) The photo shows the tank hole overexcavated to allow for about 12 in (30 cm) of space for setting the tank well away from sidewalls and placing backfill. (*b*) The photo shows the FRP tank lowered into place by a mini-excavator. Nylon ropes attached to the tank's lifting lugs on the top and suspended from the digging bucket of the excavator were used for lifting and placing this FRP tank.

(a)

(b)

FIGURE 5.21 (a) and (b) These photos show large FRP tanks strapped to concrete deadmen antiflotation anchors that typically run the length of the tank.

highly corrosive conditions. Whatever methods and materials are specified must last as long as the useful service life of the system for preventing flotation, so enough care should be taken with those types of details.

Concrete deadmen weights should be positioned in the bedded excavation prior to placement of the bedding material and tanks. They should be positioned as shown in Fig. 5.21b, just outside of the tank's shadow (with sun straight above the tank). For tanks placed side by side, at least 18- to 24-in (45- to 60-cm) clear space should be provided between the tanks as shown. Manufacturers of large FRP tanks such as the Xerxes Corporation have very detailed information and installation instructions for a variety of site conditions. Xerxes prefabricates deadmen, anchor straps, and other materials commonly used for large tank installation, and provides information helpful for designing deadmen needing to be built on-site. Their on-line information is referenced in "Resources and Helpful Links" on this book's website.

Some manufacturers of FRP tanks used for septic/primary settling and pump tanks such as Xerxes have established ongoing working relationships with manufacturers of other decentralized wastewater system components, and have developed their products and accessories to fit compatibly with those of other manufacturers.

Riser access ports, adapters, risers, and lids tend to be manufactured in standard and limited numbers of diameters, with those sizes dependent on the tank's material of construction and the specific manufacturer. So in specifying tanks and other products it's important for the designer to make sure that sizing and connections are compatible. Leaving those details to installers of systems may often not result in the most cost-effective or sound project results, since engineers/designers are in the best position to understand the objectives of the design and function of the various components. It is a common complaint among contractors and manufacturers of onsite wastewater systems components that the system designer failed to provide adequate detail for certain basic elements of the system. Those details are critical to basic operation and sound long-term performance, and to sustainable projects overall.

3. Tank backfill and final cover

Applicable permitting authority requirements should be referred to for determining acceptable backfill materials and other requirements for final cover of tanks. In the past, onsite wastewater systems have often been considered short-term wastewater service options with "out-of-sight-out-of-mind" attitudes for management, and few if any long-term sustainability considerations applied to their installation and care. Tanks were commonly covered over, with current and future owners having little to no knowledge of their location, nor any ability to check their condition without digging up and disturbing significant areas of properties.

As detailed in "Supplemental Information" provided under Chap. 6 on this book's website, routine checks and periodic servicing are needed for even the most basic types of onsite systems, all of which include a primary settling/septic tank. Tanks should be installed with watertight access risers and lids that reach just above the grade of the final soil cover over and around the tank so that lids can be easily removed as needed for checking sludge/scum levels, condition of inlet/outlet fittings, effluent filters, and so on. Figure 5.22(a) shows a long length of durable corrugated PVC access riser that can be cut to the proper length if the designer is not able to accurately specify riser lengths for each tank opening to bring risers to just above grade. If cutting riser lengths at job sites, installers should, however, take care to use proper cutting and smoothing tools (such as routers) to have flush connections between tank adapters or collars, risers, and lids. Manufacturers of the riser material can usually be very helpful with that type of information and recommendations.

Sand or easily compactable soil (e.g., sandy loam) free of any rocks or debris is commonly used for backfilling concrete tanks. For FRP and polyethylene tanks, pea gravel (or other manufacturer-approved material) should be used because the material fills in well behind itself along the curved bottoms of sidewalls of these tanks. Pea gravel or other approved material should be used at least up to the top surface of the tank body for polyethylene and FRP tanks. The depth and type of cover material needed for the specific type of tank is typically specified by the tank manufacturer and depends on the structural design of the tank.

Designers should determine whether tanks will be subjected to any traffic or vehicular loads, and if so, manufacturer-specified cover depths and materials used accordingly for the expected weight loading. Concrete slabs of varying depths can be poured over tanks in addition to certain minimal cover depths for wheel loading over tanks, along with traffic bearing access risers and lids. Typical minimal depths of backfill cover for fiberglass tanks are 18 to 24 in (45 to 60 cm) for *nontraffic* conditions. Pea gravel and/or sand should be used for at least the first 12 in (30 cm) above the top surface of polyethylene or FRP tank ribbing, with select soil or sand backfill free of rocks/coarse fragments or debris used for the remaining depth of cover to the surface (Fig. 5.23).

Green bolt-down and gasketed fiberglass lids are shown in Fig. 5.24. The bolts used need a specific size of hex wrench to remove, and so are at least somewhat resistant to tampering by children or other unauthorized persons. The lids are durable enough for foot traffic, typical residential-sized mowers, and in general light duty traffic (nonvehicular). Heavier traffic duty lids can be installed where

(a)

(b)

Figure 5.22 (a) The photo is a length of sturdy PVC riser shipped to a project site for cutting risers to meet proper elevations at the job site. (b) The photo shows risers adhered onto the plastic tank adapters. For piping and/or conduit penetrating risers, holes are cut into the risers using hole-saws of the proper diameter to fit the outside pipe diameter, and grommets inserted into the holes to seal around the pipe once it's inserted. Lubricants may be needed for cutting riser material, and manufacturers or their technical literature on the material should be consulted.

FIGURE 5.23   The FRP tank with risers and lids installed was backfilled up to the top surface of the tank with pea gravel, and then with sand for the next 12 in (30 cm) or so. Sandy loam was used for backfilling the remaining depth up to the access lids.

FIGURE 5.24   Proper burial depth and final grading to allow ready access for checking and servicing the tank are shown here. Note that the tank lid can be removed without soil, leaves and organic surface debris falling into the tank. Once grass is planted around these green fiberglass tank lids, they will hardly be visible to visitors at this park.

needed. Gasketed bolt-down lids of this type affixed to access risers with watertight connections to tanks or tank adapters prevent surface water from entering the tank, and prevent odors from escaping the tank.

4. *Comments on piping installed to and between tanks and other system components*

Piping into and out of tanks is typically installed after placement of the tanks or treatment units, though piping alignment and presence of adequate drainage/fall should be field-verified before setting tanks and treatment units. It's important to identify any conflicts, such as water and/or electric line crossings, setbacks from buildings/ structures and other site features, and so on, before installing any system components.

Piping layouts for sites prepared by the engineer/designer of the system should use angles in the alignments that correspond to available pipe fittings (e.g., 45° and 90° bends, and 22.5° in geographic areas where larger supplies of pipe-fittings are readily available). That's especially important for larger-diameter (>2-in to 3-in, or 50- to 80-mm) PVC piping that allows less flexibility with alignments and for trenches in rock cut with rock-trenchers, since they are much narrower than those dug with a digging bucket. Figure 5.25 shows a rock trencher of a size commonly used for decentralized systems.

FIGURE 5.25 This 8-in-wide (20-cm) rock trencher cuts to a maximum depth of approximately 40 in (1 m), which is sufficiently wide to accept a 4-in (100-mm) pipe as long as the trench is cut relatively straight (though some slight curvature of alignment is tolerated by 4-in (100-mm) PVC Schedule 40 pipe).

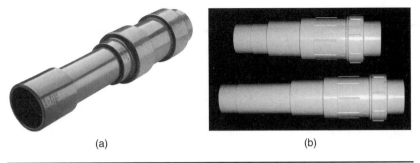

(a)   (b)

FIGURE 5.26   PVC or CPVC couplings (*a*) and (*b*) are commonly used for making watertight repairs to PVC pipe, but can also be used for new installations. (*Photo courtesy for (a) of Spears Manufacturing Company.*)

Figure 5.26 shows two types of "telescoping" couplings used for Schedule 40 or 80 PVC pipe, where there may be tight spaces and difficulty with making a watertight connection in a straight run of pipe. To install the coupling shown in Fig. 5.26*a*, collapse the expansion joint, solvent cement the ends, and expand the telescoping portion halfway, rotating the fitting around the pipe at both ends to make sure applied PVC cement covers the whole joint. The expansion coupling (Fig. 5.21*b*) is glued onto one exposed end of the pipe, and then expanded by pulling on the telescoping end (bottom image). It's then attached to the other exposed end of the PVC pipe with a standard PVC coupling. Coupling (*a*) tends to be the more expensive of these, and is more commonly used for situations where expansion and/or contraction of pipe are expected. Figure 5.27 illustrates offsetting pipe between tanks allowing for insertion of rigid fittings.

Where bell and spigot piping are used for projects, bell-ends of pipes should be directed *against* the direction of flow, that is, for pressure piping, the bell-end of the pipe lengths on the pump side of the line, and for gravity flow, on the uphill end of the pipe. In that way, wastewater runs over the blunt end of the pipe inserted into the bell end, rather than against it, thereby reducing opportunity for either buildup of solids in gravity lines or needless friction/energy losses in effluent pressure lines.

5.6   Further Comments on Watertight Tanks and Piping

Decentralized wastewater systems are commonly located in relatively remote rural or semirural areas where homes and buildings may be served by rural electric power utilities unable to quickly respond to power outages in many cases. Relatively minor storm events (whether ice, wind, or rain) may cause power outages lasting for days in parts of those utility grids. Even with backup generators,

FIGURE 5.27 The short length of piping and rigid Schedule 40 PVC fittings between the two tanks could be installed without the use of any flexible couplings here due to offsetting the tank elevations by a small amount (left tank draining to right). Tanks can also be offset horizontally slightly if they need to be installed at the same elevation to allow for the same rigid coupling and pipe installation.

it's important during those periods to (1) limit power needs associated with basic wastewater system functioning (as discussed further in subsequent chapters for various treatment processes), and (2) attempt to maximize reserve wastewater storage capacities without discharging inadequately treated waste to the environment.

Figure 5.28 compares the reserve storage capabilities for two types of tanks—fiberglass and precast concrete. As mentioned earlier, precast concrete lids can be "floated" and unseat when water levels inside the tank rise above the top surface of the tank. For the relatively common situation illustrated in the figure (septic tank pretreatment followed by low pressure dosed subsurface dispersal as discussed in Chaps. 9 and 10), the use of a fiberglass tank with watertight access risers would result in significantly more, and possibly multiple additional days of reserve storage capacity.

Assume both types of septic tanks (FRP and concrete) are 1500 gal, each with two 24-in (0.61-m) diameter access risers. Each of the two types of pump tanks has one 24-in (0.61-m) diameter riser and one 30-in (0.76-m) riser (with the latter over the pump vault), and tank inlets and outlets are watertight along with piping. While allowing water levels to rise into the sewer line is considered a "worst case scenario,"

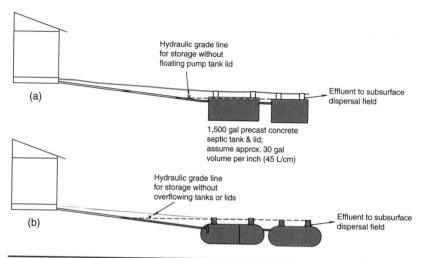

FIGURE 5.28 Comparisons of reserve storage capacities for (a) concrete and (b) FRP.

it illustrates an important point relative to the limitations of certain materials and methods of construction. The "hydraulic grade line" shown for each type of tank is considered the maximum possible level that would not result in floating the concrete lid, or wastewater rising above a bolt-down riser lid.

For the precast tanks, assuming a 3-in (7.6-cm) elevation drop between the inlet of the first tank and the inlet to the second/pump tank (which would be a minimum, and assumes minimal distance between the tanks), 3 in (7.6 cm) of tank depth would be lost for storage in the first tank to avoid the risk of floating the lid on the second tank. The shaded portion of the tanks shows the maximum possible storage volume in the tanks and into the sewer line, assuming that the inlet/outlet ports and sewer line are all watertight. Assuming a typical 30 gal/in (44.7 L/cm) of tank height for a concrete septic tank that size, that would remove 90 gal (340.7 L) from potential reserve storage capacity for those two tanks.

For the FRP tanks, the same elevation drop (3 in or 7.6 cm) is assumed between the first tank inlet and the inlet to the second/ pump tank. For that situation however, with access risers having watertight connections to tanks and watertight inlets/outlets, the wastewater level can rise to the bottoms of the lids in the second/ pump tank without risk of wastewater exiting the tanks or piping. The full depth of both tanks is available for reserve storage capacity in the event needed.

Assuming an average riser burial depth/length of 18 in (0.46 m) above the top surface of the tanks, three 24-in (0.61-m) diameter risers and one 30-in (0.76-m) riser, an *additional 153 gal (579 L)* of reserve wastewater storage is provided *just in the access risers*. That assumes a 15-in (38-cm) depth into the first two risers and 18-in (0.46-m) depth in the two risers on the second tank. Depending on the length and slope of piping to the house, that would add even more volume, up to the elevation of the second tank's access lids.

While these added reserve capacities are hopefully seldom needed, it's important on behalf of long-term system use and environmental and public health sustainability considerations to plan on one or more of those events occurring over the life of the system.

5.7   Wastewater Sludge/Septage or "Residuals" Management

Septage pumped from septic tanks, sludge needing to be pumped/removed periodically from secondary or advanced treatment units, wastes from marine vessels, and residuals/"biosolids" produced from composting toilets must all be properly handled and treated, and are all materials regulated in the United States under federal law. U.S. regulation of wastewater residuals is covered under the federal rule 40 CFR Part 503. Composting and lime stabilization followed by land application, are common methods of handling septage in the United States, along with treatment at wastewater treatment facilities able to handle septage and sludge. These activities are all regulated under this federal rule in the United States.

Many secondary or advanced decentralized treatment units that produce sludge needing to be periodically removed are designed and plumbed so that the sludge can either drain by gravity or be pumped back to the primary treatment tank(s). Chapter 6 discusses sludge production for various treatment processes as one of several important sustainability considerations. Excessive sludge production can significantly add to operational costs, and in the aggregate for multiple systems may exceed local sludge treatment and handling capabilities.

For geographic locations having suitable climates and sufficient land areas, cocomposting of waste sludge and septage with waste wood products or other sources of "bulking" material and carbon sources are an excellent way of diverting those wastes from landfills, and producing a good soil amendment for yards and gardening. For wastewater residuals composting operations, some of the key treatment considerations for achieving a product that can safely be reused for gardens or lawns are

- Treatment time
- Treatment temperature
- Moisture content during processing
- Carbon to nitrogen ratio in the composting mixture

Enough time is needed in composting operations for the various physical and biological processes to occur. The composting materials must reach a high enough temperature for a sufficient amount of time for pathogen destruction to occur. Moisture content must be sufficient for biological processes to occur, while not being overly moist, which would result in anaerobic, anoxic, or "septic" conditions. The carbon to nitrogen ratio of the organic materials being composted must be suitable for sustaining the bacterial populations responsible for much of the treatment occurring. Controlling and monitoring composting processes are necessary to make sure that enough overall treatment has occurred for the material to be safe for public exposure.

The results of a research project demonstrating the use of waste wood chips to filter raw septage pumped from septic tanks can be downloaded from this book's website under Chap. 5, "Supplemental Information." The filtrate was tested, with suspended solids, levels sufficiently reduced that the filtrate could be further treated using a variety of methods described in Chap. 6, including natural processes such as wetlands. The filtered septage residuals and wood chips were composted using the windrow composting method.

In the U.S. domestic wastewater sludge composting and other wastewater residuals treatment operations must all be managed so that the final products produced from those operations meet the requirements of the federal 40 CFR Part 503 rules. For operations directed toward beneficial reuse of those waste materials, there are specific testing requirements for the end products to ensure that they are safe for public exposure (e.g., soil amendments for lawns and gardens). The frequency of testing and other specific requirements tend to vary by size and type of facility, and end use of the product. However, there are no specific volumes of that regulated material that are exempt from regulation under U.S. federal rules, such as the residuals produced from residential composting toilets.

Composting toilets present an interesting case for wastes regulated in the U.S. under Part 503 rules. The federal rules do not exempt the waste from those units, and therefore at least a certain amount of testing is required prior to applying the composted residuals to gardens or lawns. Because of the importance of effectively controlling the various physical reactions and processes involved with composting, and the variations in use and operation of composting toilets in homes, it seems reasonable that the waste produced from these units was not exempted from the federal rules. It is not difficult to imagine scenarios in which inadequate processing of the waste material could occur prior to applying it to gardens or lawns. Examples might include household gatherings or parties, during which a much greater volume of waste is entering and overloading a device that has only so much processing or storage capacity. It's therefore important for those owning and using composting toilets to responsibly manage their operation and residuals produced, and to carefully read manufacturers' recommendations and prescribed limits on usage. While composting of sludge and septage can contribute significantly to more sustainable long-term waste management practices, there are certain public health and environmental considerations that must be addressed.

References

1. V. D'Amato, Principal Investigator, "Factors Affecting the Performance of Primary Treatment in Decentralized Wastewater Systems," Final Report, Water Environment Research Foundation, Alexandria, VA, 2008.
2. EPA "Onsite Wastewater Treatment Systems Special Issues Fact Sheet 2, High-Organic-Strength-Wastewaters (Including Garbage Grinders)," U.S. EPA, Cincinnati, OH, 2002.
3. T. R. Bounds, "Design and Performance of Septic Tanks." *Site Characterization and Design of On-Site Septic Systems*, ASTM STP 1324, M. S. Bedinger, J. S. Fleming, and A. I. Johnson, Eds., American Society of Testing and Materials, New Orleans, LA, 1997.
4. Austin Water Utility, Special Services Division, Industrial Waste Control/Pretreatment Program, City of Austin, TX.

CHAPTER 6

Secondary and Advanced Treatment Methods

Where a physical site evaluation shows that desired levels of treatment cannot be naturally provided by the soil following primary treatment, some method of further treatment is needed prior to final effluent soil/land dispersal. U.S. states have varying requirements for site conditions needing added treatment, along with the level of treatment needed for the specific geophysical conditions. In geopolitical locations where no such regulations apply, technical judgments must be made regarding the level of treatment needed to prevent adverse effects from any limiting design parameters (LDPs). In locations where regulations specify minimum levels of treatment for certain conditions, consideration of other factors may be needed to achieve truly sustainable systems meeting the types of criteria discussed in Chap. 2.

Methods of treatment discussed in this chapter include (1) those providing mainly BOD and TSS reduction (secondary treatment), (2) those providing secondary treatment along with a significant amount of nitrogen reduction, and (3) disinfection processes for pathogen reduction. Systems in the latter two categories are commonly referred to as "advanced" onsite treatment systems (suggesting that treatment beyond secondary is provided). Decentralized wastewater industry research and development has tended to focus on those three categories of treatment, and is generally consistent with levels of treatment needed to address the principal LDPs for domestic wastewaters applied to various site conditions. That is, secondary treatment parameters (BOD, TSS, and NH_3/NH_4^+ reduction), total nitrogen reduction, and pathogen reduction.

While secondary treatment parameters (BOD and TSS) don't tend to be LDPs as such for most sites, the reduction of those wastewater constituents to certain levels has important operational and system longevity implications for a variety of treatment and dispersal

137

methods commonly used for onsite systems. As discussed previously, BOD and/or TSS reduction are needed for most pathogen and nitrogen reduction processes, with pathogens and/or nitrogen found to be LDPs for many sites. Significant reduction of BOD and TSS can also contribute to substantially longer subsurface dispersal field service lives, and minimize clogging problems for subsurface dispersal systems as discussed further in Chap. 9.

6.1 Secondary and Advanced Treatment Processes Commonly Used for Decentralized Systems

Several basic physical/chemical and biological processes are employed by a wide range of proprietary and nonproprietary treatment systems developed during the past few decades. Manufacturers of proprietary treatment systems tend to incorporate at least minor variations with treatment processes in the design of their units that distinguish their product from those of their competitors. Those features may or may not actually enhance performance, depending on the process and specific modification. The focus of this chapter will be on comparing basic treatment processes, particularly as related to sustainability considerations.

While treatment processes continue to be developed, most decentralized wastewater systems today use one of the following, or combinations of these:

- Single-pass biofiltration—free-draining (e.g., sand filters, peat biofilters, or other synthetic filter media); these systems rely on naturally aerobic conditions with the effluent draining through an unsaturated media of some type, without the need to mechanically oxygenate the treatment media to maintain aerobic conditions; this process is also referred to as fixed film or attached growth treatment, using an unsaturated media.

- Recirculating biofiltration—free-draining (e.g., recirculating sand/gravel filters, recirculating textile media filters, trickling filters, etc.).

- Submerged/saturated biofiltration (e.g., upflow filters, and the predominant process associated with subsurface flow wetlands).

- Partially submerged attached growth, or aerated submerged attached growth (e.g., rotating biological contactors "RBCs," or continuously submerged attached growth media in aerated chambers).

- Suspended growth—continuous flow (e.g., aerated tank units which process the wastewater at essentially the rate wastewater flows from the facilities served).

- Suspended growth—batch reactors (e.g., sequencing batch reactors, or SBRs).

A relatively large body of technical literature has been compiled covering all of the above treatment processes, as applied to both centralized and decentralized treatment systems. "Resources and Helpful Links" on this book's website references some of those works, including several engineering educational texts, design guidance manuals, and technical publications. Combinations of the above treatment processes are incorporated into some proprietary/manufactured treatment systems. Combinations of nonproprietary processes can also be used to meet advanced treatment levels. Each basic process tends to have its advantages and limitations, with certain combinations used to overcome those limitations as discussed further in this chapter. Conceptual illustrations of each basic process category are presented below in Figs. 6.1 to 6.6.

Specific treatment units/systems using these processes may have variations from the configuration shown, based on specifically engineered products and systems features. Some understanding of these basic processes, and their differences, can provide helpful insights into why some will tend to be more appropriate and sustainable on a long-term basis than others for specific site and project conditions.

Single-pass unsaturated/aerobic biofilters: Single-pass biofilter treatment systems may be dosed by gravity, periodically over the day with a pump using a timer, or on a "demand" basis actuated by floats in the pumping compartment. Single-pass biofilters can also be dosed using nonelectric dosing devices discussed further in Chap. 9, including dosing siphons and Flouts™. The merits of these various dosing methods are discussed in Chaps. 7, 9, and 10 in much more detail. In contrast to aerated tank units (Figs. 6.4 to 6.6), this category of treatment system is only dosed a few minutes per day, rather than 20 to 24 hours/day as with suspended growth processes entraining oxygen into the fluid for bacteria through mechanical aeration or air diffusers. These packed media

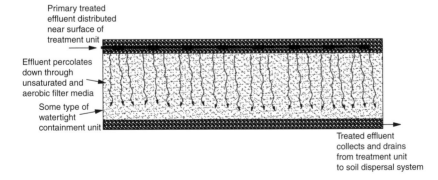

FIGURE 6.1 Side view of single-pass unsaturated (aerobic) biofilter.

filters need only to be dosed periodically to maintain moist conditions needed for bacterial populations to survive (with frequency depending on climate and media type), since they are naturally aerobic.

In Fig. 6.1 primary treated effluent is applied to the treatment media surface through a perforated piping distribution network. Effluent percolates through the media by gravity, providing filtration and biological treatment processes. As long as the media is unsaturated, conditions remain aerobic naturally. Effluent collected in piping at the base of the treatment unit exits as shown. This type of system is discussed in detail in Chap. 7. Some type of dosing (pumped or nonelectric method) achieves better distribution and treatment.

Peat biofilter units marketed in the United States tend to use significantly higher loading rates than buried single-pass intermittent sand filters. However, some studies suggest that loading rates higher than about 2 to 2.5 gal/day · ft² (80 to 100 mm · day⁻¹) result in some adverse effects on performance, including surface clogging and higher effluent ammonia levels.[1] The type of peat used has been found to have significant effect on performance however, and more coarse and fibrous peat is recommended.[2] A filtration media depth of at least or about 24 in (0.6 m) is recommended for sand filters, and is comparable to recommended depths for peat biofilters.

Unlike sand filters, peat media is organic and tends to break down over time, with faster rates of decomposition reported for warmer climates. On average, peat in modular treatment units would be expected to need replacement once every 7 to 8 years. Routine annual checks of effluent quality are needed to see if decomposition is occurring. Peat decomposition is indicated by the presence of visible particulate matter in the effluent.

Recirculating unsaturated/aerobic biofilters: Like single-pass media filters, recirculating media filters are typically dosed intermittently throughout the day based on a recirculation rate established by the engineer/designer of the system. The recirculation rate used depends on the treatment media and type of treatment system, desired effluent quality and waste stream characteristics. Ratios most commonly range from about 3:1 to 5:1 (ratio of recirculation/return rate to system design flow), or possibly higher, depending on the filter media, waste strength and treatment needs. Pumps actuated with timers are typically used to achieve the proper recirculation rate. These systems tend to use more energy than single-pass biofilters for treating the same amount of flow, but significantly less than suspended growth processes that require aeration occurring most if not all of the day.

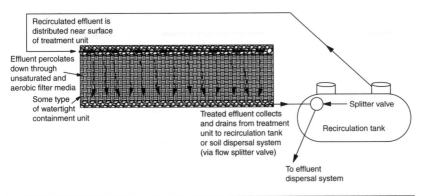

Recirculated effluent is distributed near surface of treatment unit

Effluent percolates down through unsaturated and aerobic filter media

Some type of watertight containment unit

Treated effluent collects and drains from treatment unit to recirculation tank or soil dispersal system (via flow splitter valve)

Splitter valve

Recirculation tank

To effluent dispersal system

FIGURE 6.2 Side view of flow diagram for a recirculating unsaturated (aerobic) biofilter.

For the recirculating biofilter shown in Fig. 6.2 effluent is applied to the surface of the treatment media through a piping network. A certain amount of filtration happens along with mostly biological treatment as effluent percolates through the media by gravity. A common variation of the above is use of a sump at the base of the treatment unit for collecting and recirculating the effluent, rather than a separate tank (e.g., some trickling filter units such as the one shown in Figs. 6.11 to 6.13). Media used for this type of system tends to be coarser than for single-pass filters. Sand/gravel or synthetic media may be used, including high surface area plastic (as in trickling filters) polyester/textile, foam, and so forth.

Recirculating sand/gravel filters (RSF/RGFs) may be used to overcome some of the limitations associated with single-pass sand filters, including (1) occupying less area for the filter, (2) increased total nitrogen removal capabilities, and (3) potentially lower construction costs due to much smaller filter size and lesser amount of select filter media needed. The same physical aspects of recirculating media filters (i.e., greater porosity) that enable recirculation and greater nutrient removal processes along with occupying a smaller treatment unit footprint, also, however, tend to reduce their pathogen reduction capabilities. Single-pass sand filters tend to rely on a significant amount of physical filtration occurring as effluent migrates through the treatment media, whereas recirculating media filters rely more on the biological treatment processes occurring as effluent comes into contact with the aerobic and anaerobic bacterial populations established on the surface of the media (fixed film process). RSF/RGFs have also been found on average to perform better when left open to the atmosphere (rather than in enclosed containers), and so are more commonly used for commercial scale or clustered residential systems with the treatment unit located in a non-public-access location. RSF/RGFs are discussed further in Chap. 7 along with buried intermittent sand filters.

Upflow anoxic (saturated) biofilters: Upflow filters are also sometimes referred to as "anaerobic upflow filters." However, this implies the absence of oxygen entirely, which is not the case if oxygen is biochemically available in the form of nitrate-nitrogen (NO_3). It is the absence of free oxygen in this type of system that enables their successful use in conjunction with effective nitrification treatment processes, for achieving higher levels of total nitrogen removal. As described in Chap. 2, an anoxic environment with some source of carbon is needed following (or in recirculation mode with) an aerobic treatment process that effectively nitrifies the wastewater. Chapter 8 describes in detail a combined subsurface flow wetland and trickling filter treatment system that uses this approach for total nitrogen reduction to relatively low levels.

Figure 6.3 shows primary treated effluent entering the base of the upflow biofilter unit and distributed across the bottom through perforated piping. This conceptual illustration shows multilayered horizontally placed filter media, although some variations of upflow filters have vertical baffles, with effluent directed through the media both horizontally and vertically. That type of horizontal flow configuration is similar to the functioning of subsurface flow wetlands. In all cases though, upflow filter media is saturated. Some units may also have valves allowing for backflushing and sludge removal, and/or recycle of effluent.

Upflow filters have been increasingly used in conjunction with secondary treatment methods for total nitrogen reduction where that's needed, and would not usually be employed as a stand-alone method of treatment. Subsurface wetlands also have limited nitrification ability, although they are capable of providing up to about 25 to 30 percent total nitrogen removal without the use of other processes, along with low levels of BOD and TSS. Wetlands use horizontal flow through the submerged media, which offers much more surface area for oxygen transfer along the air/water interface than upflow filters, as well as a certain amount of nutrient removal through a combination of vegetative uptake and a limited

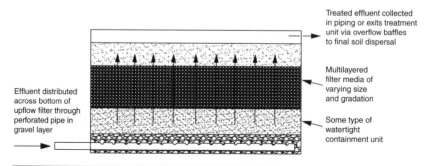

FIGURE 6.3   Side view of upflow anoxic (saturated) biofilter.

amount of nitrification/denitrification. Wetlands are discussed further in Chap. 8, and are included in Tables 6.1 to 6.3 comparing stand-alone treatment processes.

Alternating submerged and unsubmerged attached growth treatment: Rotating biological contactors (RBCs) have alternating submerged and aerobic conditions which enables the growth of multiple layers of microbial species needed for certain key biological treatment processes. Bacterial populations growing on the rotating media include those needed for nitrogen removal. Large decentralized systems studied in the state of Massachusetts that employed RBCs along with some type of anoxic unit treatment process were found to perform better than most other types of systems studied for producing low levels of total nitrogen.[3] Effective nitrification is needed for subsequent denitrification and total nitrogen reduction using nitrification/denitrification processes.

In Fig. 6.4 the treatment unit illustrated is an RBC, with the media shown in the left compartment rotating around the horizontal shaft. For the configuration shown, the amount of time treatment may occur depends on the rate that flow enters the unit. Variations of RBCs include multiple "stages" in the process, with multiple sets of media disks either in the same or multiple compartments. This allows for the growth of predominantly different bacteria in subsequent stages as treatment proceeds, allowing processes such as nitrogen removal to better occur. These bacteria die and sink to the base of the treatment unit ("biomass"), requiring periodic pumping/removal. When properly designed, built, and operated, aerobic attached growth (or "fixed film") processes illustrated in Figs. 6.1, 6.2, and 6.4 have been found to reliably and

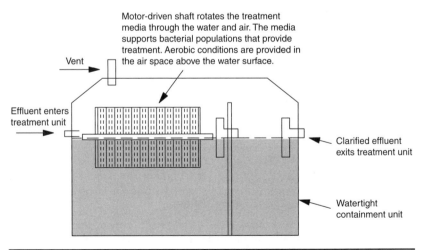

FIGURE 6.4   Side view of an alternating submerged and unsubmerged attached growth treatment process.

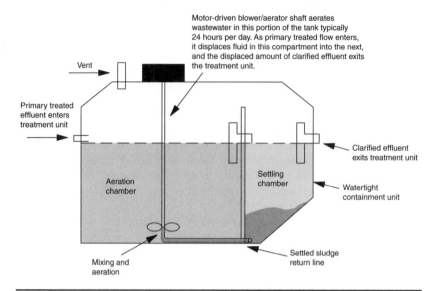

Motor-driven blower/aerator shaft aerates wastewater in this portion of the tank typically 24 hours per day. As primary treated flow enters, it displaces fluid in this compartment into the next, and the displaced amount of clarified effluent exits the treatment unit.

Vent

Primary treated effluent enters treatment unit

Aeration chamber

Settling chamber

Clarified effluent exits treatment unit

Watertight containment unit

Mixing and aeration

Settled sludge return line

FIGURE **6.5**   Side view of suspended growth—continuous flow (aerated tank unit). Variations on these units include conical settling chambers inside of single-compartment tanks, with aeration occurring outside the settling cone, with sludge settling out the bottom of the conical clarification zone. For that type of unit, effluent exits the tank from the top of the cone.

consistently treat wastewater to secondary effluent levels, including nitrification (conversion of NH_4 to NO_3).

Continuous flow suspended growth treatment (aerated tank units): Continuous flow aerated tank units like the one illustrated in Fig. 6.5 aerate essentially 24 hours a day and 7 days a week. This is necessary to maintain aerobic bacterial populations in the aeration chamber responsible for the treatment processes. Re-startup of the system, if it's allowed to go "dormant" or inactive, may take up to several weeks. Some manufacturers include a submerged attached growth component (submerged portion of the media shown in Fig. 6.4) in their suspended growth treatment units that helps overcome some of the performance limitations observed with this process, especially for smaller-scale decentralized wastewater systems. The National Sanitation Foundation (NSF) in the United States uses laboratory testing protocol (the NSF 40 Standard) to evaluate manufactured treatment units for their ability to meet secondary treatment levels. Aerated tank units are routinely tested under that program, along with other types of treatment units.

Sequencing batch suspended growth treatment (sequencing batch reactors, or SBRs): A series of illustrations (Fig. 6.6a to 6.6e) show a single process in its multiple stages of operation. That process is

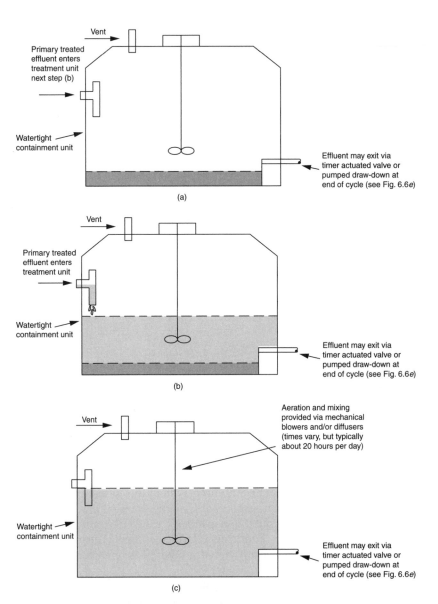

FIGURE 6.6   (*a*) Side view of SBR. Beginning of five-step operating cycle. Only sludge/settled solids are shown in the tank here. This step follows Fig. 6.6*e*. (*b*) SBR fills with primary treated wastewater. (*c*) SBR aerates the wastewater for some number of hours during the day (as much as 20 to 21 hours). (*d*) Settling period to clarify aerated/agitated wastewater in tank. The sludge blanket at the bottom of the tank re-forms. (*e*) Settled effluent exits the SBR tank.

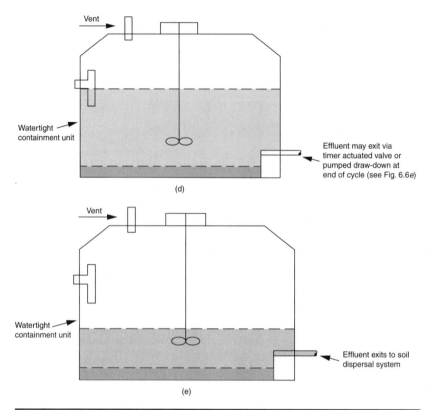

FIGURE 6.6 *(Continued)*

similar to the one shown in Fig. 6.5, except that a single compartment goes through sequential operational steps, rather than using multiple compartments or a baffling system to segregate those steps in the process. For suspended growth treatment processes, some means of settling out the suspended bacteria performing the treatment is needed for producing a clarified effluent. The process illustrated in Fig. 6.6 is a suspended growth "batch reactor" process, or a sequencing batch reactor (SBR).

The five steps shown are for an SBR that operates in discrete "batches," with treatment typically occurring in two tanks configured for different steps in the process so that flow isn't interrupted. Or, a flow equalization tank can be used to dose the SBR during the "fill" cycle. The "intermittent decanted extended aeration" (IDEA) process is similar to SBRs, in that both aeration and settling occur in the same tank, though there are just three steps in the process (aeration, settling, and decanting). Other variations also exist.

A variety of detailed technical design resources for different methods of treatment are referenced on this book's website. Process design guidance for both attached growth/fixed film and suspended growth treatment processes can be found in *Small and Decentralized Wastewater Management Systems*[4] and *Wastewater Engineering: Treatment and Reuse*.[5] Manufacturers of recirculating media filters should be consulted regarding the surface area and appropriate design assumptions for their specific media and treatment units, in combination with basic design approaches presented in the above or other technical resources for the specific type of system.

Membrane bioreactors (MBRs) represent yet another variation on the activated sludge/suspended growth treatment process, and have been used for larger-scale decentralized wastewater systems as well as for centralized (municipal scale) systems. MBRs use a membrane (immersed in the aeration tank or in a separate membrane tank) for the liquid/suspended solids separation (clarification) process. The use of the membrane helps overcome limitations with other types of activated sludge systems associated with less effective settling and clarification of the effluent. It allows higher concentrations of bacteria treating the wastewater, due to its not being limited by normal settling phenomena, through the use of membrane solid/liquid separation instead.

MBRs utilize varying types of membranes that filter out molecules of different sizes and chemical characteristics. Four membrane processes employed by MBRs include microfiltration, ultrafiltration, nanofiltration and reverse osmosis, with each using different types of membranes.[4] MBRs are significantly more operationally intensive than most other treatment methods typically used for decentralized treatment systems, and would tend to be used only where very low levels of certain wastewater constituents were consistently needed (e.g., total nitrogen). MBR systems continue to be developed and have become increasingly used for wastewater treatment projects requiring very low effluent levels for certain pollutants. However the technology would usually not be cost-effective for smaller- to mid-sized decentralized wastewater projects as compared with other options unless land area available for other methods is severely limited, or waste characteristics and needed effluent quality require that type of process. Energy consumption, sludge wasting and maintenance levels are very high for MBRs as compared with commonly used fixed film processes. Other technologies with significantly lower operational costs, in combination with certain methods of final soil dispersal and treatment, would be expected to produce effluent quality comparable or superior to MBR effluent.

For stronger waste streams such as restaurant/commercial kitchens, and especially those that may contain higher concentrations of fats, oils, and grease, some activated sludge/suspended growth processes may have advantages over fixed film/attached growth processes. A suspended growth process may still need to be coupled with another process to achieve sufficient overall treatment. However, providing at least secondary treatment in a tank environment would tend to avoid some of the potential clogging problems that might occur with oils/greases directly entering a packed media filter system, or impacts to essential bacterial population growth on fixed film surfaces.

Suspended growth/tank reactor treatment systems may be more vulnerable to operational upsets from variations in domestic waste strength and flows as compared with many attached growth/ fixed film and filtration treatment processes. The bacterial "habitat," survivability and overall "buffering" capacities of these two categories of treatment processes tend to be very different. It is, however, important to consider climate and seasonal temperature variations where the system will be operating to ensure that wastewater processing temperatures are in keeping with those needed for targeted levels of treatment. In colder climates more insulation over and/or around the system is often needed for bacterial populations on treatment media used for fixed film systems, as well as for suspended growth systems in cold conditions. For decentralized wastewater systems, which are much more subject to variations and fluctuations in use and flows than larger centralized systems, the ability of the system to continue reliably treating wastewater during all periods of use is a critical element of their long-term sustainability.

6.2   Comparison of Commonly Used Decentralized Treatment Systems

Table 6.1 summarizes the advantages and limitations associated with some of the most commonly used predispersal treatment methods for decentralized systems, and which use one of the six basic process categories described above. The table places particular emphasis on sustainability considerations, and is generally organized by the basic treatment process used by each of the systems.

Advantages and limitations associated with each treatment method assume properly functioning systems, treating typical domestic strength wastewaters. Wastes from restaurants and other commercial facilities producing higher strength wastewaters may require special planning considerations, with some of those discussed in previous chapters.

The way in which treatment occurs in each treatment technology, and certain basic elements of their operation, offer insights into why some technologies tend to be more sustainable than others. For example, single-pass or recirculating packed media filters, or "fixed film"/ "attached growth" treatment processes in which effluent flows by gravity through an unsaturated and naturally aerobic treatment media have an inherent energy usage advantage over suspended growth treatment systems into which air has to be constantly entrained for oxygen transfer and aerobic conditions to occur. Likewise, fixed film/attached

Treatment Technology and Process Category	Advantages	Limitations
Single-Pass Aerobic Biofilters		
1. Intermittent sand filters (detailed in Chap. 7) NOTE: It is assumed here that the treatment units is watertight, and not "bottomless." Bottomless sand filters will be discussed in Chap. 10. The use of bottomless filters for secondary treatment may result in high levels of nitrate-nitrogen migrating to groundwater or surface water supplies.	• Capable of reliably producing a high quality of secondary wastewater effluent, with very high levels of nitrification • Reduces total nitrogen levels by approximately 25–30% on average • Consistently low organic and suspended solids content of the effluent safely allows for reduction of land area requirements for subsurface disposal systems in certain soil types • Provides significant pathogen reduction (average fecal coliform levels of 200 col/100 mL or less) • Effluent is relatively odorless (faint soil odor) • Considerable performance data is available over 25–30 years and in many locations and conditions • Nonproprietary system, so may use locally available materials of construction and specialized design • Should require routine maintenance only once annually	• Greater space requirements (footprint) than some other secondary treatment options • Higher level of design detail needed to ensure proper operation and performance, as compared with some proprietary packaged treatment units • Local sources of filter media of proper gradation needed • Infiltration of rainwater into filter during wet weather periods increases effluent loading to dispersal field (though proper grading of soil cover over filter can minimize this) • More vulnerable to very cold weather conditions as compared with some other systems, and potential for accompanying lesser performance during those periods • More costly to service/restore if media becomes clogged with oils/greases, due to need to replace impacted portion of media, as compared with some other biofiltration systems

TABLE 6.1 Secondary Treatment Process Comparisons

Treatment Technology and Process Category	Advantages	Limitations
Single-Pass Aerobic Biofilters		
	• Significantly less energy usage as compared with other advanced treatment processes • No waste sludge production from sand filter treatment process • Long useful service lives if properly designed, constructed, and maintained • Suitable for single family residences or larger flows • Can continue to provide high levels of secondary treatment during periods of sporadic use, or following periods of nonuse (such as vacation homes and seasonal business) • Can use area over filter for light foot traffic and yard activities	
2. Peat biofilters NOTE: It is assumed here that the treatment units is watertight, and not "bottomless." The use of bottomless filters for secondary treatment may result in high levels of nitrate-nitrogen migrating to groundwater or surface water supplies.	• Capable of reliably producing a high quality of secondary wastewater effluent, with relatively high levels of nitrification • As long as peat is in good condition, consistently low organic and suspended solids content of the effluent safely allows for reduction of land area requirements for subsurface disposal systems in certain soil types • Effluent is relatively odorless (mild musty odor)	• Average service life of peat in units ranges from 4 to 12 years due to breakdown of organic peat, with filter media needing replacement at that point. Some peats tend to break down more quickly in warmer climates • Greater space requirements (footprint) than some other secondary treatment options • Proprietary treatment unit (requires local availability or shipment to site)

- Should require routine checks and/or maintenance only once annually
- Suitable for single family residences or larger flows
- Can continue to provide high levels of secondary treatment during periods of sporadic use, or following periods of nonuse (such as vacation homes and seasonal business)
- Significantly less energy usage as compared with other advanced treatment processes
- No waste sludge production from peat filter treatment process
- Modular units requiring less skilled labor for proper installation as compared with other nonpackaged treatment options

- Requires availability of suitable type of peat for periodic replacement of filter media
- More costly to repair/restore if media becomes clogged with solids or oils/greases, due to need to replace impacted portion of media, as compared with some other biofiltration systems

TABLE 6.1 Secondary Treatment Process Comparisons (*Continued*)

Treatment Technology and Process Category	Advantages	Limitations
Recirculating Aerobic Biofilters		
3. Recirculating sand/ gravel filters	• Capable of reliably producing a high quality of secondary wastewater effluent, with very high levels of nitrification • Can provide 50–60% total nitrogen removal • Consistently low organic and suspended solids content of the effluent safely allows for reduction of land area requirements for subsurface disposal systems in certain soil types • Less susceptible to reduced performance as compared with some other secondary treatment options during periods of sporadic use, or following periods of nonuse (such as vacation homes and seasonal business) • Provides significant pathogen reduction (2–3 log reduction for fecal coliform) • Effluent is relatively odorless • Significantly less energy usage as compared with suspended growth treatment processes • No waste sludge production from sand/gravel filter treatment process • Considerable performance data is available for varying locations and conditions • Nonproprietary system • Should require routine maintenance only once to twice annually	• Greater space requirements (footprint) than some other secondary treatment options • Higher level of design detail needed to ensure proper operation and performance, as compared with some proprietary packaged treatment units • Local sources of filter media of proper gradation needed • Infiltration of rainwater into filter during wet weather periods increases effluent loading to dispersal field for non-enclosed units • Most suitable for nonpublic use areas (best suited for clusters of homes or public/ commercial applications) • Cannot use area over filter for residential yard activities or foot traffic • More vulnerable to lesser performance during periods of cold weather as compared with most other systems • Require control panels with timer capabilities to best control recirculation rates • More costly to repair/restore if media becomes clogged with solids or oils/greases, due to need to replace or clean impacted portion of media

4. Recirculating textile filters (e.g., AdvanTex® treatment units)	• Capable of reliably producing a high quality of secondary wastewater effluent, with very high levels of nitrification • Can provide 50–75% total nitrogen removal if effluent is recirculated to an anoxic zone or further treatment process with adequate carbon source • Less susceptible to reduced performance as compared with some other secondary treatment options during periods of sporadic use, or following periods of nonuse (such as vacation homes and seasonal business) • Consistently low organic and suspended solids content of the effluent safely allows for reduction of land area requirements for subsurface disposal systems in certain soil types • No rainwater infiltration/collection into treatment unit during wet weather periods • Very low space requirements (small footprint) • Effluent is relatively odorless • Significantly less energy usage as compared with suspended growth treatment processes • Significantly less waste sludge production as compared with suspended growth treatment systems	• Requires a recirculation tank along with the media filter unit • Less performance data is available as compared with some other secondary or advanced treatment options • Proprietary treatment units • Requires inspection/maintenance more frequently than some other treatment options, and typically requires manufacturer-approved service contracts • Depending on local availability of variety of proprietary secondary treatment units, may cost more than other secondary treatment units certified under the NSF 40 Standard • Require control panels with timer capabilities to control recirculation rates (AdvanTex controls are equipped with automatic timers and recirculation ratio adjustment features) • Typically requires a recirculation tank along with the media filter unit (unless AdvanTex unit recirculates back to septic/primary tank)
	• Long useful service lives if properly designed, constructed and maintained	

TABLE 6.1 Secondary Treatment Process Comparisons (Continued)

Treatment Technology and Process Category	Advantages	Limitations
Recirculating Aerobic Biofilters		
	• Should have long useful service lives if properly designed and installed • Suitable for single family residences or larger flows • Modular type treatment units can be added if design flows increase • Units with hanging textile sheets are relatively easy to service if there are filter clogging or excess oils/grease buildup	
5. Recirculating trickling filters	• Capable of reliably producing a high quality of secondary wastewater effluent, with high levels of nitrification • Can provide 50–75% total nitrogen removal if effluent is recirculated to an anoxic zone or further treatment process with adequate carbon source • No rainwater infiltration/collection into treatment unit during wet weather periods • Very low space requirements (small footprint) • Should have long useful service lives if properly designed and installed • Suitable for single family residences or larger flows • Modular type treatment units can be added if design flows increase • Less energy usage as compared with suspended growth treatment processes	• Sporadic use or periods of nonuse can adversely affect system performance and maintenance requirements • Typically proprietary treatment units • Requires inspection/maintenance more frequently than some other treatment options, and typically requires manufacturer-approved service contracts • Less performance data is available as compared with other secondary or advanced treatment options for single residence/small-scale applications • Costs can be higher than some other proprietary secondary treatment units • More electromechanical components than some other secondary treatment options, with possibly more maintenance/repair needs

	• Less waste sludge production as compared with suspended growth treatment systems • Improvements to some proprietary trickling filter unit media packaging have made it easier to clean or replace media if there is clogging or excess oils/grease buildup • Less susceptible to reduced performance as compared with some other secondary treatment options during periods of sporadic use, or following periods of nonuse (such as vacation homes and seasonal business) • For many packaged units, effluent is recirculated to the treatment media from a sump at the base of the unit (not separate tank)	• Requires control panels with timer capabilities to control recirculation rates • May require longer start-up period after extended periods of nonuse as compared with some other fixed film/attached growth treatment processes, depending on media type and surface area
6. Rotating biological contactor NOTE: Media rotates in and out of fluid, with alternately submerged and aerobic attached growth conditions	• Capable of reliably producing a high quality of secondary wastewater effluent, with high levels of nitrification • Can provide 50–75% total nitrogen removal if effluent is recirculated to an anoxic zone or further treatment process with adequate carbon source • Less susceptible to reduced performance as compared with suspended growth treatment options during periods of sporadic use, or following periods of nonuse (such as vacation homes and seasonal business)	• Typically proprietary treatment units • Requires inspection/maintenance more frequently than some other treatment options, and typically requires manufacturer-approved service contracts • Less performance data is available as compared with other secondary or advanced treatment options for single residence/small-scale applications • Costs can be higher than some other proprietary secondary treatment units

TABLE 6.1 Secondary Treatment Process Comparisons (*Continued*)

Treatment Technology and Process Category	Advantages	Limitations
Recirculating Aerobic Biofilters		
	• No rainwater infiltration/collection into treatment unit during wet weather periods • Very low space requirements (small footprint) • Should have long useful service lives if properly designed and installed • Suitable for single family residences or larger flows • Modular type treatment units can be added if design flows increase • Less energy usage as compared with suspended growth treatment processes • Less waste sludge production as compared with suspended growth treatment systems • Typically requires separate clarifier for treated effluent	• More electromechanical components than some other secondary treatment options, with possibly more maintenance/repair needs • May require longer start-up period after extended periods of nonuse as compared with some other fixed film/attached growth treatment processes
Submerged Biofilters		
7. Subsurface flow wetlands	• Capable of reliably producing relatively low levels of BOD_5 and TSS, though nitrification is typically limited unless size is increased significantly to allow for greater oxygen transfer • Considerable performance data is available for varying locations and conditions • Less susceptible to reduced performance as compared with some other secondary	• Greater space requirements (footprint) than most other secondary treatment options • Limited nitrification capabilities contribute to effluent odors • Higher level of design detail needed to ensure proper operation and performance, as compared with some proprietary packaged treatment units

treatment options during periods of sporadic use, or following periods of nonuse (such as vacation homes and seasonal business)	• Local source of suitable rock media of proper size is needed
• Nonproprietary system	• Rainwater collected in wetland during wet weather periods increases effluent loading to dispersal field
• Should require routine maintenance only once annually, or seasonally to remove excess vegetation	• Cannot use area over wetland for residential yard activities or foot traffic
• Relatively long useful service lives if properly designed and constructed	• BOD and TSS reduction are limited to levels lower than some other secondary treatment methods due to dissolved and/or suspended solids and organic matter from decomposing plant debris
• Requires essentially no electric power consumption for operation	• More costly to repair/restore if media becomes clogged with solids or oils/greases, due to need to replace or clean impacted portion of media
• No waste sludge production from subsurface flow wetlands	• Reduced performance during extended cold weather periods
• Can be used for single family or larger-flow systems	• Must be maintained to remove invasive and woody vegetation
• Offers certain amount of habitat for local fauna	
• Wetland plant respiration contributes positively to overall ecology	
• Used effectively in conjunction with other secondary treatment processes for achieving total nitrogen reduction, though occupies more space than upflow filters	

TABLE 6.1 Secondary Treatment Process Comparisons (*Continued*)

Treatment Technology and Process Category	Advantages	Limitations
Activated Sludge (Suspended Growth) Processes		
8. Suspended growth combined with attached growth/fixed film processes	• Capable of producing a high quality of secondary wastewater effluent, although nitrification may be limited depending on system design and operation • Can typically provide 30–50% total nitrogen removal • Somewhat less susceptible to reduced performance as compared with some other secondary treatment options during periods of sporadic use, or following periods of nonuse (such as vacation homes and seasonal business) • No rainwater infiltration/collection into treatment unit during wet weather periods • Very low space requirements (small footprint) • Should have long useful service lives if properly designed and installed • Suitable for single family residences or larger flows • Modular type treatment units can be added if design flows increase • Attached growth component helps to limited extend to reduce adverse performance effects from seasonal or variable usage patterns	• Periods of nonuse can adversely affect system performance and maintenance requirements • Typically significantly lower nitrification capabilities as compared with packed media filter (fixed film) processes • Proprietary treatment units • Requires inspection/maintenance more frequently than some other treatment options, and typically requires manufacturer-approved service contracts • Less performance data is available as compared with some other secondary or advanced treatment options • Costs sometimes higher than other secondary treatment units certified under the NSF 40 Standard • More electromechanical components than some other secondary treatment options, with more operation and maintenance and component servicing/replacements needed over time • Significantly greater sludge production than fixed film/attached growth systems, and accompanying higher sludge pumping needs • Significantly higher energy usage than fixed film/attached growth systems

| 9. Aerated tank treatment units | • Many different treatment units available from different manufacturers which have been certified under the NSF 40-Standard
• Many units available which have lower initial costs than most other secondary treatment options
• Very low space requirements (small footprint)
• No rainwater infiltration/collection into treatment unit during wet weather periods
• Can be used for single family residences or larger flows
• Modular type treatment units can be added if design flows increase | • Suspended growth treatment process is more susceptible to variations in flow than other secondary treatment options, and sporadic use or periods of nonuse (such as with vacation homes) can adversely affect system performance and maintenance requirements
• Typically significantly lower nitrification capabilities as compared with packed media filter (fixed film) processes
• Proprietary treatment units
• Requires inspection/maintenance more frequently than some other treatment options, and typically requires manufacturer-approved service contracts
• More electromechanical components than some other secondary treatment options, with more operation and maintenance and component servicing/replacements needed over time
• Significantly greater sludge production than fixed film/attached growth systems, and accompanying higher sludge pumping needs
• Significantly higher energy usage than fixed film/attached growth systems |

TABLE 6.1 Secondary Treatment Process Comparisons (*Continued*)

159

Treatment Technology and Process Category	Advantages	Limitations
Activated Sludge (Suspended Growth) Processes		
10. Sequencing batch reactors (SBRs) NOTE: SBRs operating in intermittent mode (dosed on a timed basis) would be expected to perform better and more reliably than continuous-flow units.	• Several different treatment units available from different manufacturers which have been certified under the NSF 40-Standard for smaller-scale systems • If properly designed and managed, can reliably produce a high-quality secondary effluent • If operated properly, can provide 40–70% (or possibly higher) total nitrogen reduction (depending on design and operation) • Very low space requirements (small footprint) • No rainwater infiltration/collection into treatment unit during wet weather periods • Can be used for single family residences or larger flows • Modular type treatment units can be added if design flows increase • Eliminates the need for separate secondary clarifiers • Return-activated-sludge pumps not required • Mode of operation (and timing of each step in process) can be adjusted to enhance treatment	• Suspended growth treatment process is more susceptible to variations in flow than other secondary treatment options, and sporadic use or periods of nonuse (such as with vacation homes) can adversely affect system performance and maintenance requirements • Proprietary treatment units • Requires inspection/maintenance more frequently than some other treatment options, and typically requires manufacturer-approved service contracts • More electromechanical components than some other secondary treatment options, with more operation and maintenance and component servicing/replacements needed over time • Significantly greater sludge production than fixed film/attached growth systems, and accompanying higher sludge pumping needs • Significantly higher energy usage than fixed film/attached growth systems

TABLE 6.1 Secondary Treatment Process Comparisons (*Continued*)

growth treatment processes tend to produce less waste sludge needing removal and hauling from the site as compared with properly managed suspended growth treatment processes (e.g., aerated tank units and SBRs).[5]

Figures 6.7 through 6.10 illustrate one type of "packaged"/proprietary treatment unit that uses the fixed film/attached growth treatment process. The AdvanTex recirculating packed media filter units use plastic treatment media and function very similarly to recirculating sand/gravel filters (discussed in Chap. 7). However, they occupy a much smaller footprint and are enclosed, which offers odor control and protection from weather. The units are typically dosed at rates between about 15 and 25 gal/day · ft (0.6 to 0.8 m⁻¹/day), which is about 3 to 5 times the areal loading rate used for a recirculating sand/gravel filter.

In Fig. 6.7, effluent is distributed over sheets of synthetic textile media packed into the unit, and drains by gravity to the base where it returns to the recirculation tank. The circular orifice shields shown here (Fig. 6.7b) snapped onto the piping over the spray orifices (also used for sand and gravel filters) help better distribute effluent over

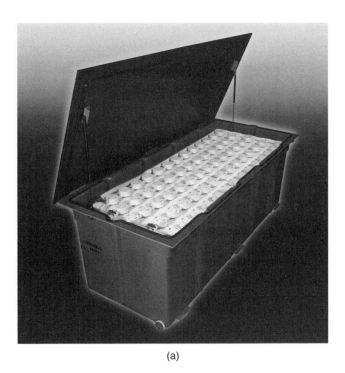

(a)

FIGURE **6.7**   (a) and (b) The interior of an AdvanTex treatment unit (manufactured by Orenco Systems, Inc.) is shown here. (*Photo courtesy of Orenco Systems, Inc.*)

(b)

FIGURE 6.7 *(Continued)*

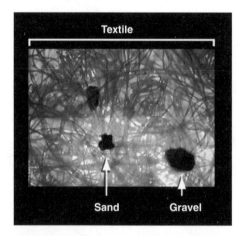

FIGURE 6.8 This image shows the higher surface area plastic textile filter media as compared with sand and gravel media used for those types of fixed film/packed media filter systems. (*Photo courtesy of Orenco Systems, Inc.*)

the media surface. If the media sheets require cleaning due to excessive buildup, they are pulled, sprayed off, and replaced. The 3 × 7.5 ft (0.91 × 2.3 m) unit is fiberglass, with antiflotation flanges located along the outside bottom edge for anchoring either with concrete or onto a tank "saddle."

Figure 6.9 Two AdvanTex units serving this large residence and riser lids for the tanks are barely visible amidst the cactus in this landscaped area near the home. Because the units are enclosed and easily landscaped, they can be located very near the structures served. These lids are colored brown to blend in with the desert landscape. (*Photo courtesy of Premier Environmental.*)

Figure 6.10 Constructed in 2003, L.A.'s Audubon Nature Center is entirely "off the grid," and was the first building in the United States to earn a platinum rating under the Green Building Council's LEED program. Wastewater from the center is treated onsite by three AdvanTex units. Pumps are solar powered, and the effluent is disinfected in a fourth unit and reused for toilet flushing and irrigation. The peak design flow for this system is 1200 gal/day (4542 L/day). (*Photo courtesy of Orenco Systems, Inc.*)

Seasonal usage, such as with vacation homes, resorts, and schools, has been found to greatly reduce performance especially for certain types of systems. Process features such as added flow equalization and timed dosing of treatment units can significantly improve performance, and reduce adverse performance impacts from sporadic flow patterns common to many decentralized wastewater systems.[3] Packed media/fixed film processes with high surface areas for biomat formation and bacterial populations to sustain themselves during "feast/famine" cycles may be better able to maintain good treatment levels during periods of sporadic use, although flow equalization capacity or methods should still be used.

Table 6.2 provides averages and ranges of treatment performance that would be expected from those same systems for key wastewater constituents. The effluent quality values in the table are for "stand-alone" treatment units of that type, not operating in conjunction with another unit process other than primary treatment. Averages and ranges given for total nitrogen removal for recirculating media filters may include configurations with the treatment unit functioning in combination with the septic/primary settling tank (in recirculation mode). Performance reported for the parameters included in the table for the different technologies are derived from a number of studies and data sources cited in works referenced at the end of the chapter.

It is assumed that effluent quality reported in Table 6.2 is for properly designed, operated, and managed systems. Averages and ranges given are typical of systems treating primary treated domestic wastewater comparable to those given in the table "Reported Raw Wastewater and Septic Tank Effluent Levels for Key Constituents–WERF Study Results," found on this book's website. The effluent quality levels presented in Table 6.2 don't include uncharacteristic outliers or "excursions" associated with abnormal operational conditions observed in data. Effects of temperature on performance will tend to vary by system type, along with a variety of other operating conditions. The values presented in the table pertain mainly to small- to mid-sized decentralized systems, although some of the technologies tend to be used mainly for larger-scale systems (e.g., RBCs). Where lesser amounts of independent data are available for those technology types for small-scale systems, it was necessary to include performance results for larger-sized decentralized systems of that type.

6.3   Factors Affecting Treatment Performance

A variety of physical and chemical wastewater characteristics can greatly affect wastewater treatment processes in various ways, and should be evaluated as a part of process selection and design. All of the treatment types discussed in this chapter rely on natural biological and physical processes, which in turn rely on a variety of factors

	Biochemical Oxygen Demand (BOD$_5$, mg/L)		Total Suspended Solids (TSS, mg/L)		Total Kjeldahl Nitrogen (TKN, mg/L)		Total Nitrogen Removal (% Removed)		Fecal Coliform[a] (Colonies per 100 mL)	
	Average (Typical)	Range	Average (Typical)	Range	Average (Typical)	Range	Average (Typical)	Range	Average (Typical)	Range
1. Intermittent sand filters,[b] loaded at 1.0–1.25 gpd/ft^2 (40–50 mm/day^{-1})	<5	2–10	5	2–10	<2	1–5	25%	20–40%	<200	10–400
2. Peat biofilters, intermittently dosed; loaded at 2.0–3.0 gpd/ft^2 (80–160 mm/day^{-1})	<10	5–15	<10	5–15	<10	5–10	30%	20–40%	NA[c]	NA
3. Recirculating sand/gravel filters, loaded at 4–5 gpd/ft^2 (160–200 mm/day^{-1})	<10	2–15	<10	5–15	<5	2–10	55%	50–60%	NA	NA

TABLE 6.2 Typical Performance Averages and Ranges for Secondary and Advanced Treatment Methods Commonly Used for Decentralized Systems

165

	Biochemical Oxygen Demand (BOD$_5$, mg/L)		Total Suspended Solids (TSS, mg/L)		Total Kjeldahl Nitrogen (TKN, mg/L)		Total Nitrogen Removal (% Removed)		Fecal Coliform[a] (Colonies per 100 mL)	
	Average (Typical)	Range	Average (Typical)	Range	Average (Typical)	Range	Average (Typical)	Range	Average (Typical)	Range
4. Recirculating textile filters[d,f] (e.g., AdvanTex), loaded at 15–25 gpd/ft^2 (600–1,000 mm/day^{-1})	11.3	5–20	7.9	5–25	<10	2–10	60%	50–70%	NA	NA
5. Recirculating trickling filters[d,e,g]	23.2	5–30	22.1	5–30	<20	5–20	55%	40–70%	NA	NA
6. Rotating biological contactors[g]	<20	5–30	<20	5–30	<20	5–20	55%	40–70%	NA	NA

Method										
7. Subsurface flow wetlands (sized for secondary levels of treatment, based on local climatic conditions)	<20	5–30	<20	5–30	<30	20–45	25%	20–30%	NA	NA
8. Aerated tank units with submerged attached growth component[h]	<20	10–30	<20	10–35	<30	5–45	40%	30–50%	NA	NA
9. Aerated tank units[h] (continuous flow)	Averages from studies vary greatly	10–65	Averages from studies vary greatly	15–65	<30	10–45	ND	10–25%	NA	NA

Table 6.2 Typical Performance Averages and Ranges for Secondary and Advanced Treatment Methods Commonly Used for Decentralized Systems (*Continued*)

	Biochemical Oxygen Demand (BOD₅, mg/L)		Total Suspended Solids (TSS, mg/L)		Total Kjeldahl Nitrogen (TKN, mg/L)		Total Nitrogen Removal (% Removed)		Fecal Coliform[a] (Colonies per 100 mL)	
	Average (Typical)	Range	Average (Typical)	Range	Average (Typical)	Range	Average (Typical)	Range	Average (Typical)	Range
10. Sequencing batch reactors[i] (conventional – 5-step)	<10	5–15	<15	5–30	<20	10–30	55%	40–70%	NA	NA

[a]Statistical indicator organism for levels of pathogens from human sources likely to be present in wastewater effluent.
[b]Values reported are for buried single-pass intermittent sand filters designed and constructed as described in Chap. 7.
[c]NA in this column indicates the technology would not be expected to produce acceptably low levels of pathogens without an added disinfection process.
[d]BOD and TSS values are based on a large study of residential scale treatment units (*Variability and Reliability of Test Center and Field Data: Definition of Proven Technology From a Regulatory Viewpoint, 2005*).
[e]BOD and TSS averages shown here are for residential scale trickling filter units. BOD and TSS averages for larger-scale systems would be expected to be comparable to RBC's.
[f]Assumes treatment system configured for recirculation of effluent with primary treated effluent for enhanced nitrogen reduction.
[g]Trickling filters and rotating biological contactors are most often used for mid to larger-sized decentralized systems. Reported values are for systems without additional unit processes that would enhance total nitrogen removal.
[h]Performance may be improved by use of flow equalization and timed/intermittent dosing of treatment unit.
[i]Conventional SBRs are typically used for mid to larger-scale decentralized systems. Effluent nitrogen data from field studies were found to be highly variable, especially for smaller systems, and very dependent on system operation.
ND = No significant data; Aerated tank units have not typically been used for systems needing nitrogen removal, and therefore not appreciable amount of data gathered for total nitrogen removal.

TABLE 6.2 Typical Performance Averages and Ranges for Secondary and Advanced Treatment Methods Commonly Used for Decentralized Systems (*Continued*)

for achieving acceptable and targeted levels of treatment, including time, temperature, pH, alkalinity, moisture conditions, and food/nutrient levels for the various bacterial populations needed for treatment. Those factors, in combination with each other, will affect different processes to varying extents. Some processes are inherently more vulnerable to variations in performance than others due to changes in operating conditions. Small decentralized/onsite systems are particularly vulnerable to this, due largely to the lesser amount of liquid storage capacity, flow equalization, and attenuation of the effects from changes with the various factors. Design features such as ample primary settling capacity (and mixing of influent wastewaters having varying levels of pollutants), and timed dosing of processes can help greatly to attenuate the effects of variable wastewater quality and flows. Timed dosing of smaller treatment systems tends to simulate what occurs with larger-municipal-scale systems that often have large flow equalization tanks/basins. These design features are especially important for some treatment methods that have shown a greater susceptibility to performance "excursions," including high effluent levels of BOD, TSS, and ammonia/ammonium.

For example, a study in southwestern Illinois of 23 aerated tank unit systems showed effluent ammonium levels ranging from 0.09 to 66.5 mg/L, with an average of 19.3 mg/L. Nitrate levels from those systems only averaged 4.85 mg/L, showing very poor levels of nitrification based on the high ammonium levels.[6] In contrast, studies of intermittent sand filters consistently show very high levels of nitrification, and very low effluent levels of BOD, TSS, and ammonium. Several studies on intermittent sand filters conducted between 1982 and 1998 in the United States showed effluent TKN levels averaging between 0.5 and 5.9 mg/L.[7] Monitoring data from two sand filters built in accordance with the methods and materials described in Chap. 7 for intermittent sand filters showed average effluent ammonium levels of 0.6 mg/L, and average effluent TKN levels of 1.8 mg/L over a period of more than 2 years. Of 48 ammonium samples collected, concentrations exceeded 1.4 mg/L only once, when a service provider for one of the two systems mistakenly pumped the entire contents of one of the systems' filter dosing tanks through the filter at one time, resulting in a very high ammonium level of 24.2 mg/L for one sampling event. A week later effluent ammonium levels had returned to normal, as measured at 0.45 mg/L.[8] If that single sampling event is excluded from the data, effluent ammonium levels averaged 0.1 mg/L.

Because intermittent sand filters rapidly drain following each dosing event, it typically takes very little time to restore performance if uncharacteristically high flows or organic loading occur, as evidenced from the above extreme operational error and rapid return to normal conditions. It is of course important to use effluent filters, and avoid allowing significant amounts of fats/oils/greases to enter the system so as not to clog the filter. Because aerated tank units operate on a flow-through "continuously stirred reactor" basis, it takes significantly more time for the treatment unit to return to better operating conditions if an uncharacteristically large volume/slug of waste enters the treatment unit. Use of sufficient primary settling capacity, along with flow equalization and timed dosing of effluent into the treatment unit may lessen performance fluctuations observed for aerated tank units.

Physical/chemical factors such as temperature, pH, and alkalinity can significantly affect treatment performance. Lower temperatures adversely impact nitrification processes that are critical to total nitrogen removal through nitrification/denitrification. Much longer treatment detention times may be needed for achieving comparable levels of nitrification at cold temperatures. The optimal pH for nitrification is considered to be between 7.2 and 9.0, with rapid reductions occurring as pH drops below about 7.0.

Alkalinity levels in wastewater are important because of buffering effects against downward changes in pH, and as an indication of whether enough bicarbonate (HCO_3^-) is present for nitrification processes to occur. Measurements of alkalinity across secondary treatment processes can, therefore, be an indication of whether nitrification is occurring. Septic tank effluent may have alkalinity ranging from less than 50 to 500 mg/L (measured as $CaCO_3$) or higher, depending on local water supplies and geochemical conditions. About 7.1 mg of alkalinity are removed for each milligram of ammonium nitrogen converted to nitrite and nitrate (or 7.1 lb of alkalinity for each pound of ammonium nitrogen). For each milligram of nitrate nitrogen in turn converted to nitrogen gas (removed from the wastewater), approximately 3.6 mg of alkalinity (as $CaCO_3$) are formed.[4] A basic rule of thumb is that alkalinity should be at least about 8 times the concentration of ammonium for nitrification to occur. Lime, baking soda or soda ash can be added to wastewater to increase alkalinity levels.[9]

To determine alkalinity needs for nitrification, it's necessary to determine the desired reduction in ammonia/ammonium nitrogen across the treatment process, and convert concentrations to mass equivalents. For example, if a 33-mg/L reduction in ammonia is desired (e.g., reducing NH_4^+/NH_3 concentration from 35 to 2 mg/L), with a wastewater design flow of 1000 gal/day (= 0.001 million gal/day, or mgd), or 3785 L/day, the calculation for required alkalinity would be as follows in U.S. Customary units. [1 mg/L = 8.34 pounds per million gallons]

$$0.001 \text{ mgd} \times 8.34 \times 33 \text{ mg/L} = 0.28 \text{ lb/day of } NH_3\text{-N}$$
needing conversion to nitrate.

Since about 7.1 lb of alkalinity are needed for each pound of NH_3-N,

$$7.1 \times 0.28 = 1.99 \text{ lb of alkalinity needed}$$

Converting this to mg/L,

X mg/L alkalinity = (2 lb/day)/(8.34 × 0.001 mgd)
= 240 mg/L alkalinity needed

The calculation is more straightforward in SI units.

33 mg/L × 3785 L/day flow = 124,905 mg/day,
or about 125 g NH_3-N reduction needed

125 g × 7.1 = 888 g/day alkalinity needed

X mg/L alkalinity = (888 g/day)/(3785 L/day) = 0.235 g/L or 235 mg/L alkalinity needed (*difference between 235 and 240 mg/L is due to rounding of numbers*).

As mentioned in Chap. 5, water softeners should as a rule not have their backwash drain lines plumbed into the onsite wastewater system. While there are varying opinions as to whether or not this adversely affects treatment performance for systems, it is known that higher sodium concentrations (associated with brine/salt in backwash waters) affect bacterial populations. A testing program conducted in the State of Virginia on the performance of AdvanTex treatment units found that draining water softener backwash into the system increased average total nitrogen levels in the effluent by about 30 percent (from 15.3 to 19.9 mg/L), and increased ammonium-nitrogen levels by about 250 percent. The study found average *Escherichia coli* levels, an indicator for human pathogens in the effluent, were increased by 147 percent when water softener backwash was drained to the systems.[10]

The more sustainable decentralized systems will tend to be those capable of maintaining acceptable levels of treatment as operating conditions change. This is particularly so for very small- to mid-sized systems for which flow attenuation and moderation don't occur naturally as a result of wastewater coming in from numerous sources and on varied schedules. For smaller systems, incorporating enough flow equalization capabilities, either through treatment tank capacity or through methods such as timed dosing, is critical to ensuring sound and sustained levels of treatment on a long-term basis.

6.4 Combinations of Treatment Processes for Achieving Higher Levels of Nitrogen Reduction

All of the technologies listed in Tables 6.1 and 6.2 are capable of treating wastewater to secondary treatment levels, despite greater vulnerability to upsets and variable performance for some of those treatment methods. Operated within certain well-controlled guidelines, and including features like flow equalization and timed dosing to treatment reactors, even the processes that tend to perform less reliably can be managed to effectively reduce BOD and TSS to secondary treatment levels.

As discussed in Chaps. 2 and 3, nitrogen is a wastewater constituent of particular concern environmentally along with pathogens, and an LDP for many sites. As shown in Table 6.2, some of the processes do not appreciably reduce total nitrogen levels by themselves, and even those with significant nitrogen removal capabilities may not produce effluent with low enough levels of nitrogen for sensitive receiving waters. Where land/soil characteristics are such that natural soil treatment processes and vegetative uptake are not capable of protecting groundwaters or surface waters, some additional unit process may be needed to reduce pollutant concentrations to acceptable levels.

Certain processes in combination can provide significantly better removal of key wastewater constituents such as total nitrogen, where needed based on geophysical site conditions and effluent quality requirements. Treatment for nitrogen removal through nitrification/denitrification processes requires certain combinations of conditions in a particular sequence. One of the more challenging steps in the process for many treatment technologies is effectively accomplishing nitrification (conversion of NH_3-N to NO_3-N). Naturally aerobic fixed film/attached growth treatment processes (with downward flow of effluent through the media by gravity) have a significant advantage over suspended growth tank reactors, because of the natural availability of oxygen in the treatment media. That also has the advantage of requiring less energy usage as compared with suspended growth systems where blowers/aerators or diffusers are needed to entrain fluids with air. The TKN performance levels shown in Table 6.2 for the various technologies are an indication of their ability to better nitrify wastewater.

Following nitrification, the next key step in the nitrification/denitrification process is conversion of NO_3 to N_2 gas, which requires "anoxic" conditions (absence of free/available oxygen). It also requires an available source of organic carbon, which is why some of the recirculating fixed film processes are often configured to recirculate effluent back to the septic tank (which has anoxic conditions). The total nitrogen removal levels shown for technologies 3, 4, 5, and 6 in Table 6.2 assume the systems may be configured for recirculation to the septic tank.

To achieve higher levels of total nitrogen removal (<15 to 20 mg/L), it is often more effective to combine processes that offer the biochemical conditions needed for both nitrification and denitrification. Because this can add significantly to overall system costs, it's typically most cost-effective to cluster homes and businesses together onto a single treatment system providing higher levels of total nitrogen removal, using, for example, an effluent collection system for which primary treatment would be provided in the septic tanks serving each building. Economies of scale for initial treatment system costs and long-term operational expenses tend to make this approach significantly more cost-effective over time.

Two technologies used effectively in conjunction with secondary treatment processes for achieving high levels of total nitrogen removal for decentralized systems are anoxic upflow filters (Fig. 6.3) and subsurface flow wetlands. Attached growth treatment processes used in combination with upflow filters have proven very effective for reliably achieving average effluent total nitrogen levels on the order of 10 mg/L. A trickling filter treatment system used in combination with a subsurface flow wetland is described in detail in Chap. 8, and was found to produce very low effluent total nitrogen levels.

Figure 6.11 shows the media being dosed inside a Bioclere™ proprietary/packaged trickling filter system. Figures 6.12 and 6.13 show the same type of treatment system serving a residential condominium complex in the northeastern United States. That system includes an anoxic zone for denitrification, further filtration using a proprietary

Figure 6.11   This photo shows the interior of a Bioclere trickling filter unit in operation (cover removed). The biomass growth on the surface of the dosed media is a chocolate-brown color, which is typical of a healthy bacterial population for trickling filters providing secondary treatment. When these units are used for nitrification following secondary treatment, the biomass tends to have more of a reddish copper color (indicating much greater presence of nitrifying bacteria). (*Photo courtesy of Aquapoint Inc.*)

FIGURE 6.12   This Bioclere treatment system serves a residential condo project in upstate Massachusetts, and is designed for a flow of 10,000 gpd (37.9 m³/day). The system is designed to produce an effluent quality not to exceed 5 mg/L BOD, TSS, and total nitrogen, and includes an anoxic zone for denitrification and UV disinfection. (*Photo courtesy of Aquapoint Inc.*)

FIGURE 6.13   The operations building shown here for the Bioclere treatment system houses the disc filter, UV disinfection unit, control panels, and chemical feed equipment. Alkalinity is added for nitrification, and carbon is dosed for denitrification. The primary treatment and flow equalization tanks, secondary treatment and nitrification Bioclere units (access lids shown at surface), and anoxic reactor are all below grade. (*Photo courtesy of Aquapoint Inc.*)

cloth filter disc for added TSS reduction and ultraviolet (UV) disinfection. The operations building shown in Fig. 6.13 is designed to blend in aesthetically with the condo development. Landscaping around the system allows it to be relatively close to residences and buildings served. They are enclosed, which provides odor and vector control and protects the treatment unit and processes from cold, wet, or otherwise adverse weather conditions. Much of this system is below the ground surface, so for sites with extensive rock, it may be more feasible to have shallower burial depths for the unit treatment processes and increase the height of landscaping.

Detailed technical information about possible system configurations that may effectively reduce nitrogen to very low levels can be found several of the technical works referenced in "Resources and Helpful Links" on this book's website. As more technologies providing advanced levels of treatment are researched, developed, and lab and field tested, those technologies can be matched with the natural treatment capabilities of different soil and land types to provide sustainable levels of overall wastewater treatment.

6.5 Sustainability Considerations for Evaluating and Selecting Treatment Methods

After evaluating a project's needs, waste characteristics and flows, physical site conditions, and determining LDPs for the area to be used for final effluent disposition, if treatment beyond primary treatment is needed, a candidate list of systems should be developed that will meet those performance requirements. Table 6.3 presents a comparison of the treatment technologies included in Tables 6.1 and 6.2 using a relative rating scheme for several key sustainability factors. All of the systems are assumed to treat typical domestic wastewater, and for the effluent qualities given, not operating in conjunction with another unit treatment process beyond primary treatment. Sustainability factors in the matrix include

- Energy consumption for normal operations
- Ability to reliably produce average effluent quality cited in Table 6.2, including for seasonal, vacation, or other conditions of use
- Sludge production (with higher levels of sludge production requiring more frequent pumping and trucking of sludge to offsite locations capable of treating/handling that regulated material)
- Land area requirements for the treatment system
- Useful service life of major system components

- Effects of cold weather (significant duration) or cold climate on performance
- Susceptibility to increased flows from significant rainfall

All of the technologies in Table 6.3 are rated relative only to each other. Initial capital costs are not included in the table because those costs will be very dependent upon locally available products, materials of construction and labor resources, and other project-specific factors. The rating factors pertain to essentially all settings of operation for those particular methods of treatment, and assume each system would be treating wastewater having the same flows and waste characteristics.

With respect to treatment reliability ratings, not all of the systems are capable of treating wastewater to the same levels. The ratings assigned in Table 6.3 are based on the specific technology's ability to treat to the typical levels shown in Table 6.2 for those wastewater constituents. Judgments must be made for each project about what treatment levels need to be achieved, and which systems are capable of meeting those levels. It's also important to note that the ratings assigned to technologies for the various sustainability factors are not "absolute." That is, system's configurations and design details will play a large role in determining where the system might fit in this type of matrix. Variations in operation and maintenance practices (overall system management) have been found to have very great impacts on system performance.[3] The ratings in the matrix are intended to provide general guidance, and reflect trends observed for these systems by the decentralized wastewater industry. The ratings also tend to pertain mainly to small- to mid-sized decentralized systems. Larger systems tend to become much more complex with more "custom" designs and features. All treatment technologies may of course not be suitable for any particular project or site conditions.

The relative ratings shown are based on common system configurations for small- to mid-sized decentralized projects. Specific project design details may offer advantages that would improve relative standings. The table is intended as a planning tool for generally comparing treatment options that are potentially feasible for a given project.

The physical makeup and operation of the various basic treatment processes, and their ability to efficiently accomplish removal of targeted wastewater constituents (LDPs) using natural physical phenomena such as gravity, affects their degree of sustainability. Some processes, and combinations of processes, tend to be much more effective than others in producing effluent with sufficiently low levels of certain LDPs such as nitrogen. Project sites and the level of treatment needed to avoid adverse impacts differ. The appropriateness of using one method of treatment versus another will vary accordingly.

The following relative rating factors are used in Table 6.3.

	Treatment Performance Reliability	Energy Consumption	Sludge Production	Land Area Needed	Expected Useful Service Life	Effects of Cold Weather on Performance	Susceptibility to Increased Flows from Rainfall
Intermittent sand filter	Excellent	Good *	Excellent	Poor	Good	Good ‡	Satisfactory
Peat filter	Excellent	Good *,†	Excellent	Marginal	Poor	Excellent	Excellent
Recirculating sand/gravel filter	Good	Satisfactory	Excellent	Satisfactory	Good	Marginal	Marginal
Recirculating textile filter (e.g. AdvanTex)	Good	Satisfactory	Good	Good	Excellent	Satisfactory	Excellent
Trickling filter	Satisfactory	Satisfactory	Good	Excellent	Excellent	Satisfactory	Excellent
Rotating biological contactor	Satisfactory	Good	Good	Excellent	Satisfactory	Satisfactory	Excellent
Subsurface flow wetland ("rock reed filter")	Excellent	Excellent	Excellent	Poor	Good	Poor	Poor
Combined activated sludge and submerged fixed film	Good	Poor	Poor	Excellent	Good	Satisfactory	Excellent
Aerated tank unit (continuous flow)	Poor	Poor	Poor	Excellent	Good	Satisfactory	Excellent
Sequencing batch reactor (sequencing aerated tank process)	Satisfactory	Good	Poor	Excellent	Good	Satisfactory	Excellent

* Assumes pumped dosing. With increasing size of system above single family, energy usage for single-pass ISFs would approach that of recirculating media filters, due to area of filter and flow limitations per pump needed to serve a given area. Single-pass ISFs are assumed to be loaded at areal rates about 1/4 to 1/5 that of recirculating filters.
† Assumes peat biofilters using pumped dosing, and having a similar dosing (piping) configuration as single-pass ISFs.
‡ Assumes buried intermittent sand filter (see Chap. 7).

TABLE 6.3 Comparison of Sustainability Factors for Treatment Methods

6.5.1 Discussion of Sustainability Comparisons for Treatment Methods

Ratings for sustainability considerations assigned in Table 6.3 to each of the treatment technologies are discussed below, and organized by rating factors listed across the top row of the table.

1. *Buried intermittent sand filters*

a. Buried intermittent sand filters (ISFs) are capable of reliably producing low levels of BOD, TSS, TKN, and fecal coliform (pathogen indicator organisms). Of the technologies listed in Tables 6.1 to 6.3, ISFs are the only one capable of producing effluent with average fecal coliform levels below 200 colonies per 100 mL. ISFs are also found to perform well immediately following start-up. During periods of minimal or sporadic use, as long as downtime isn't excessively long, some amount of moisture tends to remain in the sand and maintain bacterial populations that contribute to treatment.

b. The only significant electric power expended for ISFs is for the effluent pump(s) dosing filters (where that's the dosing method used). Control panels use a very small amount of electricity. For typical single family residential systems, a 1/2-hp (0.37-kW) effluent pump operates on average less than 15 minutes/day to dose an ISF. Because filter dosing only occurs if sufficient wastewater enters the system, there is no energy demand during periods of nonuse. If site topography and system layout offer enough elevation drop ahead of the filter, single-pass sand filters may also be dosed using nonelectric dosing devices discussed in Chap. 9, such as siphons or Flouts. One project example discussed in Chap. 11 uses Flouts to dose a sand filter.

c. The only sludge produced and requiring removal for this type of system is in the primary settling/septic tank. Filter loading rates, media gradation, and other design elements should be such that premature clogging with biomat or suspended solids is avoided.

d. Land area needed for ISFs is somewhat greater than for most other technologies listed in the tables, and comparable to areas needed for subsurface flow wetlands treating to levels shown in Table 6.2. Loading rates to achieve treatment levels presented in that table would be limited to 1.0 to 1.25 gal/day · ft² (40 to 50 mm/day) for ISFs.

e. Properly designed, constructed, and maintained ISFs have been found to operate well for 35 to 40 years or longer. Particle and/or biomat buildup in the filter over time would be the principal reason for eventual failure of an ISF.

f. ISFs have been used successfully in cold climates, though in general they are more vulnerable to lesser performance in cold weather conditions as compared with some other systems that can be better insulated. For reliable treatment to levels shown in Table 6.2, a balance must be struck for ISFs wherein sufficient available air/oxygen exists through the final soil cover (filter allowed to "breathe"), while providing enough cover to prevent adverse effects from very cold temperatures.

g. ISFs are somewhat vulnerable to increased water loading (flows) from rainfall, though these effects can be minimized with proper grading of the final soil cover over the filter. A certain amount of increased flows to the soil dispersal system would be expected from the filter during periods of prolonged and/or intense rainfall.

2. *Peat biofilters*

a. Peat filters are capable of consistently producing effluent with the levels shown in Table 6.2, as long as the peat has not begun to decompose. When organic decomposition begins to occur with peat filter media, suspended solids, and other pollutant levels will begin to rise significantly in the effluent. The treatment performance shown for peat filters in Table 6.2 would be for intermittently dosed filters, using a coarse fibrous peat filter media.

b. Energy usage for single-pass intermittently dosed peat filters is comparable to that of ISFs.

c. The only sludge produced for peat filters is in the primary settling/septic tank. At such time when the peat media needs to be replaced, it would need to be hauled to a facility permitted for that type of material.

d. Land area requirements for peat filters loaded at 2 to 2.5 gal/day · ft^2 (80 to 100 mm/day) would be about one-half that of ISFs.

e. The useful service lives of peat biofilters without media replacement are limited to about 7 to 10 years on average, depending on climate and the specific type of peat used for the system. The cost and availability of peat for periodic replacement of the media should be considered when evaluating treatment alternatives.

f. Peat filters have been used successfully in cold climates (e.g., Canada), with better performance often reported for systems during cold weather periods. Modular units can be insulated for protection against excessively cold temperatures.

g. Enclosed/modular peat filters are not subject to increased flows during rainfall events as long as units are relatively watertight, with surface or groundwaters prevented from entering the unit.

3. *Recirculating sand/gravel filters*

a. Recirculating sand/gravel filters (RSF/RGFs) are capable of reliably producing effluent levels shown in Table 6.2. They can provide 50- to 60-percent total nitrogen removal, though cold weather conditions may significantly reduce nitrification capabilities. Reliable and consistent treatment to very good secondary treatment levels (low BOD and TSS levels) is well documented for RSF/RGFs. The use of coarser media results in much lower pathogen reduction as compared with ISFs.

b. RSF/RGFs use very modest amounts of electric power, using much less energy as compared with suspended growth treatment systems. Recirculation of the effluent causes them to use more energy than single-pass filter systems, though this tendency may decrease as system size (design flow) increases.

c. The only sludge produced for this type of system is in the primary settling/septic tank. A certain amount of biomass builds up within and at the base of the filter over time, though a well-designed system that's properly built and maintained should last for several decades.

d. Typically about one-fourth as much land area is needed for RSFs/RGFs as compared with ISFs, depending on wastewater characteristics and loading rates to the filter. Effluent quality reported in Table 6.2 would be typical for domestic wastewater loading rates of 4 to 5 gal/day · ft^2 (160 to 200 mm/day).

e. Properly designed, constructed, and maintained RSF/RGFs should have long useful service lives of up to 40 years or longer. Clogging of the filter over time would be the principal reason for eventual need for major servicing or replacement of the filter media.

f. Because of the need to maintain access to naturally available air/oxygen supply to the filter media, cold weather conditions can have an adverse impact on RSF/RGF performance. Increased ammonia levels (reduced nitrification) have been observed seasonally for RSFs in cold weather climates. Unlike ISFs, recirculating sand/gravel filters are typically not buried, and those found to perform the best are not enclosed in containment units.

g. RSF/RGFs are susceptible to increased water loading from rainfall events, though plastic covers are sometimes temporarily anchored over the filters to reduce those effects. It is critical to prevent surface waters or groundwaters from entering the filters.

4. *Recirculating textile media filters (e.g., AdvanTex treatment units)*

   a. The AdvanTex treatment unit, manufactured by Orenco Systems, Inc. in Oregon is a good example of a recirculating synthetic textile media filter. These systems function very similarly to RSFs/RGFs, but are covered rather than open, in watertight containment units. The larger units are equipped with fans for maintaining ample air/oxygen supply. A synthetic high surface area fibrous polyester plastic media is used for these treatment units (per Fig. 6.7*a* and 6.7*b*), with the media sheets packed into the containment unit. Similarly to RSFs/RGFs, effluent is dosed across the surface of the packed media, and drains by gravity to the base of the unit where it exits for either recirculation or to the effluent dispersal system (or followed by disinfection if needed). AdvanTex treatment units can reliably produce effluent of the quality shown in Table 6.2.

   b. Energy usage for AdvanTex treatment units is comparable to other recirculating media filters utilizing gravity flow of effluent through the treatment media, and with dosing of the effluent over the surface of the media. Depending on system configuration, it may be slightly less than for recirculating sand/gravel filters, due to shorter runs of piping (and accompanying lower friction energy losses).

   c. AdvanTex units rely on fixed film/attached growth treatment, which produces significantly less waste sludge than activated sludge (suspended growth) treatment processes. The units typically have a drain line leading back to the septic/primary tank for removing sludge that sloughs off the treatment media to the base of the unit over time.

   d. Areal loading rates to AdvanTex treatment units typically range from about 15 to 25 gpd/ft^2 (0.6 to 1.0 m/day). They therefore occupy about 1/3 to 1/5 of the space needed for RSFs/RGFs treating wastewater of comparable strength and effluent quality.

   e. Properly installed and maintained AdvanTex units should have very long useful service lives of 40 to 50 years or longer. If clogging occurs in the treatment media over time, the suspended sheets can be removed, cleaned, and replaced, unlike sand/gravel media filters. AdvanTex systems use fiberglass containment units for housing the filter media, which is not subject to corrosion over time. All components in the treatment units are constructed of some type of plastic, noncorrosive material (with the possible exception of some fan elements on the larger units).

f. Like other fixed film/attached growth treatment processes, textile media filters can be vulnerable to lesser performance during very cold weather conditions if not adequately insulated. AdvanTex units are enclosed, and can be insulated as needed for protection from severe cold weather conditions. The recirculation rate and fans used for the larger units, and gravity flow through the treatment media maintains aerobic conditions even for insulated and enclosed units.

g. AdvanTex modular treatment units are enclosed and not subject to increased flows during rainfall events as long as units are relatively watertight, with surface waters or groundwaters prevented from entering the unit.

5. *Recirculating trickling filters*

a. An example of a proprietary recirculating trickling filter treatment system is Bioclere (manufactured in the United States by Aquapoint, Inc., and by Ekofinn-Pol in Europe), which uses a high-surface-area plastic media in a watertight fiberglass containment unit. These units are relatively tall as compared with other fixed film technologies discussed earlier, with effluent dosed over the surface of media, and draining to a sump at the base where a pump continues to recirculate it over the media. Effluent quality shown in Table 6.2 can be reliably achieved with these systems.

b. The energy usage rating assigned to trickling filters in Table 6.3 applies primarily to smaller-scale decentralized systems (up to just a few thousand gallons per day, or 10,000 to 12,000 L/day). For smaller systems, since trickling filter pumps dose the media essentially all day (recirculating it from a sump in the base of the unit or an adjacent clarifier), they would be expected to use more energy than smaller-scale RSF/RGF or AdvanTex recirculating filters since those two dose media only intermittently through the day using a timer. With higher flows, energy usage for these systems would likely begin to be less overall as compared with RSFs/RGFs, due to the shorter pumping distance for dosing trickling filter media as compared with RSF/RGF media filters occupying greater areas and longer piping distances (friction losses over those distances). For the same flows the energy requirements for trickling filters would be expected to be significantly lower than that needed to effectively entrain air into activated sludge processing tanks. Very low power effluent sump pumps are used for packaged trickling filter units treating small- to mid-sized decentralized systems. As with the other fixed film media treatment processes described above, the use of gravity in unsaturated media for aerobic treatment processes is very energy efficient.

c. Similarly to AdvanTex units, accumulated sludge/biomass from trickling filter media sloughs off and collects in the sump at the base of the unit. Alternatively, a separate clarification unit may follow the trickling filter, with recirculation occurring with a pump located there. A sludge pump periodically removes the sludge, typically pumping it back to a primary settling/septic tank for decentralized systems. Like other attached growth/fixed film processes, sludge production is significantly lower than activated sludge processes.

d. Trickling filter units have land area requirements comparable to AdvanTex treatment systems, though likely somewhat less depending on system configuration (especially with AdvanTex units having separate recirculation tanks).

e. Properly installed and maintained packaged trickling filter treatment units such as Bioclere should have very long useful service lives. Those systems use fiberglass noncorrosive treatment enclosures and plastic treatment media. The units can be opened and serviced as needed. The media can be cleaned and/or replaced if needed.

f. Trickling filter systems can be vulnerable to lesser cold weather performance as with other fixed film treatment systems, but enclosed units can be adequately insulated/protected from very cold temperatures as needed. Many of these systems are in use in the northeastern United States and other cold-weather regions of the world.

g. Enclosed trickling filter treatment units are not subject to increased flows during rainfall events as long as they are relatively watertight, and surface or ground waters prevented from entering the treatment enclosure.

6. *Rotating biological contactors*

a. Rotating biological contactors (RBCs) are capable of reliably producing effluent of the quality cited in Table 6.2. Alternating submersion into and emergence from the fluid by the media likely contributes to better growth of anoxic bacteria layers on the media along with aerobic bacterial layers, which would reasonably offer at least a certain amount of enhanced nitrogen removal capabilities over fixed film processes without that anoxic element. As with other fixed film/attached growth processes, they are fairly susceptible to lesser performance in cold temperatures, though this can be controlled with insulated treatment enclosures.

b. Energy usage by RBCs tends to be somewhat lower than other types of recirculating fixed film processes covered in this chapter, due to the rotating drum mechanism used to cycle the treatment media through the liquid waste stream. With increased system size, this comparatively lower energy usage tends to be more pronounced.* Other types of recirculating media filters must lift and dose effluent against gravity, whereas RBCs rely on a slowly rotating drum onto which the treatment media is affixed.

c. RBCs produce sludge at rates comparable to trickling filters, and significantly lower than aerated tank reactor treatment systems (suspended growth processes).[7,11] Sludge sloughs off of the media and settles to the bottom of the unit for periodic removal. Secondary clarifiers are typically needed for RBCs due to biomass suspended in the wastewater from the action of the rotating media and continuous sloughing of biomass.

d. Land area/space requirements for RBCs are comparable to trickling filter treatment systems.

e. Properly installed and maintained RBCs can have long useful service lives, although their period of operation without major servicing is expected to be less than for fixed film systems having only noncorrosive components in the waste treatment unit (e.g., AdvanTex and Bioclere). The electromechanical device used to rotate the media shaft needs routine maintenance, including lubricating motor(s) and bearings and replacing seals as needed in accordance with manufacturers requirements.

f. RBCs are vulnerable to lesser performance in cold weather as with other fixed film treatment systems, but are enclosed and can be adequately insulated/protected from cold temperatures as needed.

g. RBCs are not subject to increased flows during rainfall events, since they should be relatively watertight, with surface waters or groundwaters prevented from entering the treatment enclosure.

---

*Email communications with Michael Hines, P.E., Southeast Environmental Engineering.

7. *Subsurface flow wetlands*

a. Subsurface flow wetlands, sometimes known as "rock reed filters" are capable of reliably reducing BOD and TSS to the levels shown in Table 6.2. Over time, subsurface wetlands tend to reach a limit with respect to BOD and TSS reductions due to the decomposition of plant litter from the vegetation, and reintroduction of that material into the waste flow. Subsurface wetlands are very limited in their ability to nitrify wastewater, as reflected in the higher TKN average shown in Table 6.2. Wetlands can be significantly up-sized to achieve greater nitrification (thereby increasing surface area over which oxygen transfer and vegetative uptake processes can occur). However, nutrient removal is much more effectively accomplished using wetlands in combination with other processes that efficiently nitrify. Subsurface flow wetlands are discussed in much more detail in Chap. 8.

b. Of the technologies included in the matrix, wetlands are the only one that does not rely on an electric power supply for either aeration or pumped dosing, or on some other method of dosing. Wetlands are typically gravity-fed, as detailed in Chap. 8.

c. Like single-pass and recirculating sand or gravel filters, the only sludge production associated with subsurface flow wetlands is in the primary treatment/septic tank. As with RSF/RGFs and ISF's, biomat and solids will tend to accumulate in the wetland over time. Proper media sizing and loading rates, along with adequate primary treatment are essential for avoiding problems with premature build-up of solids and biomat.

d. Along with intermittent sand filters, subsurface wetlands have the largest land area requirement for achieving effluent quality levels presented in Table 6.2. Subsurface wetlands may be further limiting in that the treatment area cannot be used for foot traffic or yard area, unlike buried intermittent sand filters situated so that the final soil cover and vegetation blend with the natural grade.

e. As compared with other treatment technologies in the matrix, and based on regulatory staff reports on wetlands operating over varying lengths of time, wetlands tend to have shorter useful service lives. Reasons for that depend greatly on specific designs and installations, but typically have to do with clogging and hydraulic channeling and short-circuiting observed to happen with subsurface wetlands over time. For wetlands treating primary treated effluent, a service life of at least about 15 to 20 years would be expected, with larger systems likely tending to last longer. That is likely due to the greater potential for preferential flow patterns in larger wetlands to change over time (e.g., due to changes in vegetation and rooting patterns), balancing out of some otherwise adverse effects. Use of multiple cells, with periods of resting for each cell, may help to extend service lives and improve long-term performance.

f. Wetlands are considered the most vulnerable of the technologies included in the matrix to adverse performance effects from cold weather conditions. While straw, mulch, and other materials are used to insulate wetlands from cold temperatures, plant growth and contributions to treatment in the wetland tend to be greatly reduced during prolonged and very cold weather periods.

g. Subsurface wetlands are considered to be the most susceptible to increased system flows from periods of intense of prolonged rainfall. Recirculating gravel filters are not vegetated, and can therefore be covered with plastic sheeting to help reduce the volume of rainfall entering the system. Subsurface wetlands are vegetated, thus preventing that measure. Greenhouse-type covers could be constructed over a wetland, but would add greatly to its costs.

8. *Aerated tank treatment units with submerged fixed film/ attached growth component*

a. A properly managed system of this type should be able to meet the treatment levels shown in Table 6.2. Inclusion of the submerged attached growth element for these systems reportedly improves their performance at least somewhat over conventional aerated tank units. Similarly to the fixed film processes described above, having a media onto which bacterial populations are affixed helps to better sustain them during periods of sporadic use or nonuse. It enables formation of layers of bacteria responsible for nitrification/denitrification processes. However, hydraulic retention times and operational constraints often present with small- to mid-sized decentralized systems tend to limit these favorable effects. Studies of mid- to large-scale decentralized systems have shown that, for example, nitrogen removal through nitrification/denitrification is usually much more effectively achieved through the use of multiple unit processes, with each offering a specifically designed biochemical environment for the targeted process. For systems available on the market today that are cost-effectively employed in decentralized settings, experience has shown that there are limits to the treatment efficiencies and capabilities that can be packaged into a single treatment unit.

b. Energy consumption for aerated tank treatment units, including those with a submerged attached growth component, tend to have the highest power requirements among the technologies included in the matrix. This is due to the need to continuously entrain or diffuse air into a tank of fluid to maintain aerobic conditions.

c. Sludge production is also the highest for aerated tank reactor systems, among those evaluated in the matrix. A study of sludge production rates for RBCs showed a substantial increase in sludge production rates when aeration was added to the same process.[11]

d. Of the technologies included in the matrix, packaged treatment units of this type use very little land area, especially for smaller enclosed units not requiring buffering distances for potential odors or other aesthetic reasons. Space requirements are comparable to trickling filter and RBC units.

e. Expected useful service lives for these systems will depend on the quality and types of materials used for the treatment unit and mechanical equipment; however, it would be expected to be comparable to packaged extended aeration plants without submerged attached growth elements. Studies of extended aeration package plants show useful service lives ranging from 15 to 30 years, with average lives typically around 20 years before major repairs or replacement. Concrete corrosion over time is a common reason for repair or replacement, along with replacement of blowers/aeration systems. Packaged units housed in fiberglass or other durable, sturdy and noncorrosive material should offer a longer average service period.

f. As with other biological treatment systems, cold weather conditions tend to slow and adversely impact suspended growth treatment processes, including those with an attached growth component. Enclosed or covered tank reactors can be insulated from cold weather, reducing those adverse effects.

g. Enclosed packaged or covered treatment units of this type are not subject to increased flows from rainfall.

9. *Aerated tank treatment units*

a. Studies on the performance of small-scale aerated tank units (ATUs) have shown highly variable, and in many cases poor performance. Reasons cited for poor performance have often been associated with management practices, along with design details. Flow equalization and/or timed dosing of the units to ensure adequate wastewater processing time in the units can improve performance. Also important are sludge removal on a frequent enough basis and servicing aeration equipment to make sure conditions remain sufficiently aerobic for treatment processes. It is clear however that this type of system tends to require significantly more ongoing operation and maintenance attention to ensure adequate performance as compared with several other treatment options.

b. Aerated tank treatment units have among the highest power requirements of those systems included in the matrix. Air must be entrained into the units continuously to maintain aerobic conditions.

c. Of those systems evaluated in the matrix, sludge production is among the highest for aerated tank reactor systems along with others using the suspended growth treatment process. The U.S. EPA recommends that sludge be removed from aerated tank units on average every 3 to 6 months, or otherwise as needed.[7]

d. Space requirements for this type of packaged treatment system are comparable to trickling filters and RBC units.

e. Expected useful service lives for these systems depend on the quality and types of materials used for the treatment unit housing and aeration equipment. Studies of extended aeration package plants show useful service lives ranging from 15 to 30 years, with average lives typically around 20 years before major repairs or replacement. Packaged units housed in fiberglass or other durable, sturdy and noncorrosive material should offer longer average service periods.

f. Enclosed packaged units or covered systems can be insulated from cold weather, reducing adverse effects on biological processes.

g. Enclosed packaged or covered treatment units of this type are not subject to increased flows from rainfall. However, treatment units that are open to the air would receive additional flows during periods of intense or prolonged rainfall events.

10. *Sequencing batch reactors*

a. Conventional five-step sequencing batch reactors (SBRs) are typically used for larger-scale decentralized and centralized wastewater systems, with three-step packaged SBR units marketed for smaller- (residential and small commercial) scale systems. Larger systems serving higher numbers of facilities tend to naturally equalize flows and waste stream characteristics, which tends to have a favorable effect on system performance. Variations in hydraulic retention time and waste characteristics, and the need for operational checks and adjustments to maintain targeted effluent quality tend to hamper the performance of smaller-scale SBRs having much less attention by trained operators. A conventional five-step SBR receiving mid- to larger-scale decentralized system domestic wastewater flows should be able to produce effluent of the quality shown in Table 6.2.

b. The absence of aeration occurring for a portion of the day in SBRs should cause them to on average have somewhat lower power requirements than activated sludge processes relying on reactor tank aeration 24 hours a day, 7 days a week. However, other electromechanical components are used in SBRs that consume power in addition to aerators (e.g., pumps or valves used to decant or discharge treated effluent).

c. Of the treatment processes discussed, SBRs along with other suspended growth treatment technologies tend to have the highest sludge production rates and removal needs.

d. Space needed for SBRs is comparable to aerated tank units, trickling filters, and RBCs.

e. Average periods of useful service for SBRs will depend on the quality and types of materials used for the treatment unit housing, aeration, and other electromechanical equipment. Packaged units housed in fiberglass or other durable, sturdy, and noncorrosive material can offer longer average service periods. Periods of service are expected to be comparable to aerated tank units, assuming mechanical components are maintained and serviced as needed.

f. Enclosed packaged treatment units or covered systems can be insulated from cold weather to reduce adverse effects on performance from this factor.

g. Enclosed SBR treatment units are not subject to increased flows from rainfall.

As seen in the matrix of sustainability ratings, there are tradeoffs of various types with each of these systems. Each tends to have its stronger and weaker points. For example, RSF/RGFs tend to perform more reliably on balance for secondary treatment processes and nitrification as compared with trickling filters and RBCs (two other fixed film processes). But trickling filters and RBCs tend to occupy less space, and particularly for larger-sized decentralized systems, use at least somewhat less electric power for achieving comparable levels of treatment. The specifics of each project and site conditions must be considered for making reasonable comparisons and judgments on the above and other sustainability factors when selecting treatment methods to be used.

Management practices play a very critical role in system performance, especially for those systems needing more frequent checks and servicing. The extreme variability of performance observed for smaller-scale packaged activated sludge (suspended growth) treatment systems underscores this point. It is recommended by the U.S. EPA that these units be checked a minimum of 2 to 6 times annually, with sludge removed "every 3 to 6 months as needed."[7] By contrast, properly designed and installed intermittent sand filters need routine servicing and checks only once annually. Several of the fixed film systems tend to be much less maintenance intensive than suspended growth processes having more constant reliance on electromechanical components and requiring more frequent process checks and adjustments. For small- to mid-sized decentralized wastewater systems not having the benefit of trained system operators available to check and service the system regularly, it's especially important to select technologies that can realistically be managed well enough over time to ensure their proper functioning.

The majority of the treatment technology types discussed in this chapter are sold by a number of manufacturers around the world as prepackaged units. Those include types 2, 4, 5, 6, 8, 9, and 10 in Tables 6.1 through 6.3. Because natural or synthetic aggregate of varying grades is used for nonproprietary single-pass and recirculating media filters and for subsurface flow wetlands, and because of the size of those treatment systems, those are constructed onsite. Therefore, the local availability of persons trained and/or experienced with their installation, or the presence of the design engineer or someone else able to oversee the details of construction should be considered when using those methods of treatment.

Manufacturers of proprietary packaged treatment units typically provide training to contractors for the installation of their units along with detailed instructions, and often field technical support for larger treatment systems. Because single-pass and recirculating sand/gravel filters and wetlands are all nonproprietary methods of treatment, such installation training and support may be less available. As a manufacturer of a wide variety of decentralized wastewater products and supplies, Orenco produces intermittent and recirculating sand filter "kits" that can be purchased and shipped to sites for installation, along with an installation video. Filter media would, however, need to be found and obtained locally, along with tanks in most cases. Even using materials kits and detailed instructions, some additional training and experience may be needed to properly install those systems.

Chapters 7 and 8 provide much more detailed information about single-pass and recirculating media filters and subsurface flow wetlands, along with illustrated steps in the construction of those systems. Their proper design and installation are critical to their effective use.

As mentioned before, sound long-term functioning of all wastewater treatment systems, regardless of simplicity of operation, requires some level of ongoing care. Periodic checks of systems are necessary to determine the need for servicing, including periodic sludge/septage removal and effluent filter cleaning. Recommendations for the frequency of periodic inspections, and routine service and maintenance intervals are provided on the website for this book, along with descriptions of basic maintenance procedures associated with all of the methods of treatment discussed in this chapter.

6.6 Disinfection Processes

6.6.1 Descriptions of Disinfection Methods

For soil/subsurface conditions calling for pathogen reduction prior to subsurface dispersal, or for surface application of wastewater where either there's public exposure to the effluent or the potential for runoff into receiving streams, some method of disinfection is needed. Because effluent quality characteristic of certain treatment processes and technologies may interfere with effective pathogen removal for one method of disinfection or another, the type of disinfection process used should be appropriately matched with the treatment process(es) used. Several sustainability considerations are important to the selection of a suitable disinfection method, with those discussed in this section.

There are three methods of disinfection most commonly used for decentralized wastewater systems. Those are

- Tablet (stack feed) chlorination
- Injection chlorination
- Ultraviolet (UV) irradiation

A fourth type of disinfection is ozonation, but it is not widely used for decentralized wastewater systems. While ozone is more effective than chlorine for pathogen destruction and requires a shorter contact time, its capital and operating costs (including power usage) tend to be high relative to other disinfection methods. Relatively complicated equipment is needed, and ozone is very corrosive and reactive so more specialized materials and safety measures are needed. More information can be obtained about ozonation in the U.S. EPA's fact sheet on it that is referenced on this book's website under "Resources and Helpful Links" Chap. 6.

Below are basic descriptions of each of the three principal decentralized wastewater disinfection methods. Table 6.4 presents

Disinfection Method	Advantages	Potential Disadvantages
Stack feed tablet chlorinators	• Relatively inexpensive • Very simple operation • Easy to install • Easy to service/maintain • Provides chloride residual if that is needed (e.g., drip dispersal lines) • No electric power required	• If not available locally, shipment of chlorine tablets may be subject to certain shipping restrictions. • Approved calcium hypochlorite wastewater tablets must be used (not swimming pool tablets). • Effluent flowing through unit may be "wicked up" into tablets, causing them to expand, jam the stack of tablets, and not drop into place properly. • Flow through unit occurs at rate of wastewater flow, without controlled exposure time in dispenser for dissolution of tablet. • Unless very low levels of ammonia are maintained in the effluent being disinfected, chloride will tend to chemically combine with ammonia to a less reactive form, requiring significantly longer retention times for effective disinfection. • Under or overdosing of chloride may easily occur. • Unless subsequent dechlorination is provided, results in formation and release to environment of organic halides (known carcinogen). • Discharge of chlorinated effluent to sensitive marine ecologies such as coral reefs may have serious long-term damaging effects, and continues to be studied.

| UV irradiation | • No chloride residual, with accompanying environmental or health concerns
• Flow rate through UV units tend to be controlled, resulting in more effective and reliable disinfection | • More expensive to install and maintain than stack feed chlorinators
• Requires reliable source of electric power
• Requires trained/skilled installers
• More frequent and greater amount of maintenance as compared with stack-feed chlorinators
• Does not produce a chloride residual, if that is desired for drip dispersal
• Must have very low levels of TSS to effectively disinfect (to avoid "shadowing" effect) |

TABLE 6.4 Disinfection Methods Comparison

(a)   (b)

FIGURE 6.14   (a) and (b) The arrows show the stack feed tablet chlorinator plumbed into a gravity effluent line leading to a pump tank, from which effluent will be dosed to a subsurface soil dispersal field.

advantages and disadvantages for the two methods most commonly used for small- to mid-sized decentralized wastewater systems: stack feed chlorinators and UV disinfection methods.

Stack Feed Tablet Chlorinators

Stack feed chlorinators consist basically of an in-line stack of tablets through which wastewater flows, dissolving the lowest tablet(s) in the stack until space is made for the next tablet up in the stack to drop into place (Fig. 6.14). In Fig. 6.14 the disinfection unit shown is one of the simplest available on the market, and is very inexpensive. The length of pipe above the tablet housing will be cut down so that the screw-in plug cap is just above grade for replenishing tablets to the unit as needed. The cap-fitting is temporarily placed without solvent-welding (gluing) until final grading around the tank.

Chlorination by Liquid Injection

This method uses pumped injection of liquid sodium hypochlorite into the wastewater effluent. Because it requires relatively frequent operational checks, calibration, and servicing by trained personnel, it is typically used for larger-scale decentralized or centralized wastewater systems.

UV Irradiation

These units operate by wastewater effluent flowing through a quartz sleeve in which a UV lamp is mounted along its length (axis). As wastewater flows into and out of the sleeve, the wastewater is irradiated. Figures 6.15a to 6.15d show a typical disinfection unit used for residential-scale onsite wastewater systems.

(a)

(c)

(b)

(d)

FIGURE 6.15   (a) and (b) UV disinfection unit being removed from the tank for cleaning. The arrow shows the lamp and sleeve being withdrawn from the unit's housing. The lamp (mounted/secured with brackets) is being carefully cleaned in the photo (c) and the housing rinsed out in the photo (d). (*Photo courtesy of Orenco Systems, Inc.*)

Advantages and limitations noted in Table 6.4 for tablet chlorinators as compared with UV disinfection units illustrate some of the major differences in planning and design considerations needed for each.

6.6.2   Comparisons of Sustainability Considerations for Disinfection Methods

Effluent Quality Considerations for Achieving Effective Disinfection

Use of stack feed tablet chlorinators requires very low levels of ammonia in the effluent (good nitrification performance) to effectively and reliably disinfect, unless significantly longer retention times are provided in the reaction/receiving tank downstream of the chlorinator (for breakpoint chlorination to occur), with controlled rates of discharge from the system. As noted in Tables 6.1 and 6.2, suspended growth/activated sludge treatment processes tend to be much more challenged with nitrification than packed media filters such as ISFs or RGF/RSFs. Studies on the performance of aerated tank units have shown relatively poor performance for many systems with respect to nitrification.

UV disinfection units require very low levels of suspended solids to avoid a "shadowing" effect by suspended particles on microbial species to be irradiated. A higher quality effluent also tends to reduce UV tube-cleaning frequency for the units. UV should be used in combination with those systems that can reliably and consistently produce very low levels of suspended solids (e.g., ISFs, and RSF/RGFs, or other treatment methods specifically designed to ensure very low TSS levels in effluent).

Power Supply and Energy Considerations
UV units require a reliable source of electric power, while passive tablet chlorinators have no electric power requirements.

Operation and Maintenance Considerations
Stack feed chlorinators have lower overall operation and maintenance time and activity requirements than UV disinfection units, though not necessarily less frequent. Stack feed units can experience wicking of effluent up into the stack of tablets, with resultant expansion and sticking/jamming of the tablets in the feeder. They should be routinely checked to make sure that this hasn't occurred, and/or that tablets are still present (not all dissolved). UV disinfection units require regularly cleaning and checks, typically at least every 2 to 3 months. More detailed checks and maintenance needed for disinfection systems may be found on this book's website.

Potential Environmental Impacts from Disinfected Effluent
Chlorinated effluent has certain public and environmental health implications, including the release of carcinogenic organic halides (chloro-organics) into the environment. Dechlorination of effluent is definitely preferred, and is required in many regulatory jurisdictions prior to release of disinfected effluent.

UV disinfection does not have harmful by-products. Public/environmental health concerns associated with UV units would reasonably be focused more on ensuring satisfactory levels of pretreatment (TSS removal), routine maintenance practices and reliable power supplies to ensure effective disinfection.

6.7 Sustainability and Treatment Systems Management
Much of what drives the sustainability of decentralized wastewater treatment approaches has to do with their long-term care and management. That is particularly so for smaller- to mid-sized systems, for which economies of scale don't tend to afford nearly as much time and attention by skilled service providers to ensure continued proper functioning. While there continue to be technological innovations and increased availability of devices to treat wastewater for at least certain key pollutants, and which have increasingly smaller footprints,

the higher level of sophistication usually carries with it much higher operation and maintenance needs. The extent to which that is sustainable in the long term will vary by setting and circumstances.

It is much more feasible to have higher-maintenance systems located in more populated areas near urban centers where there is readier access to skilled and specially trained technicians, and a wide range of material resources. However, decentralized systems are used throughout the world in locations where there is an inherent shortage of the resources needed to manage more complex systems. It's therefore essential that the methods and materials selected for treatment and dispersal match the setting. Some natural methods of treatment having less intense management requirements are covered in detail in Chaps. 7 and 8. Those systems may be better suited for use where appropriate to the particular terrain and climate. Although treatment systems relying on natural processes tend to require greater land areas as seen in Table 6.3, where sufficient land area exists of suitable character, use of those methods offers significant long-term cost and sustainability benefits.

References

1. "Suitability of Peat Filters for On-Site Wastewater Treatment in the Gisborne Region," Taihoro Nukurangi, NIWA Client Report, 2006.
2. S. M. Geerts, B. McCarthy, R. Axler, J. Henneck, H. Christopherson, J. Crosby, and M. Guite, "Performance of Peat Filters in the Treatment of Domestic Wastewater in Minnesota," *Proceedings of the 9th National Symposium on Individual and Small Community Sewage Systems*, Fort Worth, TX, ASAE, 2001.
3. S. M. Parten, "Analysis of Existing Community-Sized Decentralized Wastewater Treatment Systems," Water Environment Research Foundation, Alexandria, VA, July 2008.
4. R. Crites and G. Tchobanoglous, *Small and Decentralized Wastewater Management Systems*, McGraw-Hill, New York, 1998.
5. Metcalf and Eddy, *Wastewater Engineering: Treatment and Reuse*, 4th ed., McGraw-Hill, New York, 2003.
6. S. V. Panno, W. R. Kelly, K. C. Hackley, and C. P. Weibel, "Chemical and Bacterial Quality of Aeration—Type Waste Water Treatment System Discharge," *Ground Water Monitoring and Remediation*, Spring 2007 issue, pp. 71–76, Wiley-Blackwell Publishing, Westerville, OH.
7. *Onsite Wastewater Treatment Systems Manual*, U.S. EPA, 2002.
8. "Alternative Wastewater Management Project, Phase II," Final Report, Prepared for the City of Austin, TX, Water Utility by Community Environmental Services, Inc., August 2005.
9. *Operation of Municipal Wastewater Treatment Plants, Manual of Practice 11*, Water Pollution Control Federation, Alexandria, VA, 1990.
10. "Virginia AdvanTex Testing Program," Final Report, Rural Engineering Services, Inc., Fayetteville, AR, 2005.
11. R. Y. Surampalli and E. R. Baumann, "Sludge Production in Rotating Biological Contactors with Supplemental Aeration and an Enlarged First Stage," *Bioresource Technology Journal*, vol. 54, no.3, pp. 297–304, Elsevier Science, 1995.
12. Variability and Reliability of Test Center and Field Data: Definition of Proven Technology from a Regulatory Viewpoint, (National Decentralized Water Resources Capacity Development Project (NDWRCDP) Research Project), New England Interstate Water Pollution Control Commission Lowell, MA, 2005.

CHAPTER 7

Sand Filter Construction

Two types of natural media filters are covered in this chapter, with detailed information provided on the construction of intermittent sand filters (ISFs). These are most commonly used to serve smaller projects ranging from single residences to clusters of homes and small commercial facilities. ISFs are typically used for lower wastewater design flows than recirculating sand/gravel filters (RSFs/RGFs), although the construction methods and materials used for the two types of filters are very similar.

7.1 Intermittent Sand Filters (Single Pass, Intermittently Dosed)

In cases where secondary or "advanced" treatment of septic tank effluent is needed for residential-scale onsite systems, intermittently dosed sand filters are considered one of the most sustainable options where suitable to the specific site and project conditions. Some reasons for that include

- These systems use approximately 5 to 10 percent of the electric power needed by activated sludge-based systems (e.g., aerated tank units, sequencing batch reactors, etc.) treating the same flows. The reason is that ISFs are dosed only intermittently during the day, and for only a very few minutes at a time. Small- to mid-sized aerated tank treatment units are aerated either continuously throughout the day, and all but a few hours per day for SBRs. ISFs can also be dosed using non-electric devices such as those described in Chap. 9, and as exemplified in one of the projects described in Chap. 11.

- Aerobic conditions are maintained naturally by gravity flow through unsaturated filter media. In contrast, suspended growth treatment processes maintain aerobic conditions in a treatment tank by having to aerate (using blowers, diffusers, etc.)

the wastewater in the tank most of the day if not continuously, thus using much more energy.

- Sand filters consistently produce a very high quality of effluent, and don't tend to have "excursions" of high levels of pollutants, in contrast to what is often observed for many types of small-scale wastewater treatment systems. Typical effluent quality consists of average BOD_5 and TSS levels of less than 10 mg/L (and usually approximately 5 mg/L). Effluent ammonium levels are typically on average less than 1 mg/L, and average total nitrogen reductions of 20 to 30 percent are commonly produced.

- Like other fixed film technologies, and in particular packed media filters, sand filter treatment systems are much less subject to operational vulnerabilities and upsets than systems using the suspended growth/activated sludge process for small-residential-scale projects. These systems have been found to perform well even under variable occupancy and seasonal use conditions.

- Properly designed and constructed ISFs have long useful service lives, with much lower routine maintenance needs as compared with other secondary/advanced treatment options, including significantly less waste sludge production than systems using suspended growth/activated sludge treatment processes.

- Sand filters can provide pathogen reduction naturally to levels acceptable for recreational swimming waters in many U.S. states (average fecal coliform levels of less than 200 colonies/ 100 mL). This can help avoid having to unnecessarily chlorinate effluent dosed into the environment.

- Sand filter effluent is consistently colorless and nearly odorless (except for a slightly musty scent of soil/sand). Properly designed, constructed, and operated sand filters consistently produce effluent with very low ammonia levels, and are thus well-suited for systems if surface application of the treated effluent is to be used. Higher ammonia levels in wastewater are responsible for significant odors, and suggest that conditions in the treatment unit are not sufficiently aerobic for good secondary treatment to occur. With high ammonia levels in residential wastewater, disinfection through chlorination is also not able to occur efficiently. (See Table 6.2 for treatment process comparisons relative to effluent ammonia levels.) Low suspended solids levels in sand filter effluent also allow for effective UV disinfection.

Basic elements of an ISF are shown in Fig. 7.1. Following primary treatment in a properly designed septic tank as discussed in Chap. 5,

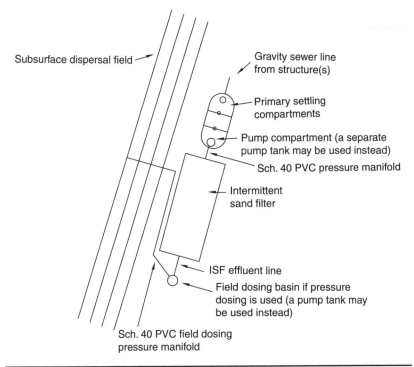

Subsurface dispersal field

Gravity sewer line from structure(s)

Primary settling compartments

Pump compartment (a separate pump tank may be used instead)

Sch. 40 PVC pressure manifold

Intermittent sand filter

ISF effluent line

Field dosing basin if pressure dosing is used (a pump tank may be used instead)

Sch. 40 PVC field dosing pressure manifold

FIGURE 7.1 Layout of a typical residential intermittent sand filter with subsurface low pressure dosing of treated effluent.

wastewater is dosed intermittently into a piping/distribution system placed in the upper layers of the sand filter. Treatment occurs as the wastewater trickles downward by gravity through approximately 2 ft (0.6m) of sand filter media (see Fig. 7.2). Conditions in the filter are naturally aerobic as long as the system is functioning properly and not subjected to saturated conditions (such as flooding or clogging).

Sand filters can be either open, as with the recirculating filter shown in Fig. 7.33, or buried. They can also be located in enclosures such as concrete tanks or troughs, but there must be sufficient availability of oxygen through either ventilation or forced air for them to remain aerobic and function properly. Buried sand filters are the most commonly used type in the U.S. for residences and small-scale decentralized wastewater systems. That is the type of intermittent sand filter detailed in this chapter.

For sand filters as with many other types of secondary or advanced onsite treatment systems, it is very important to use an effluent filter ahead of the media filter unit. Basic information about effluent filters was provided in Chap. 5, and additional technical information can be found in *Small and Decentralized Wastewater Management Systems.*[1]

Sand filter dosing cycles may be on a "demand" basis (using floats in the pump chamber set to actuate the pump at certain levels), or on a timed basis using a timer function in the pump's control panel. Timed dosing and demand dosing are discussed further in Chap. 10. Essentially the same approach can be used for pump and pipe sizing/design for dosing sand filters as is presented in Chap. 10 for low-pressure-dosed effluent dispersal fields. The residual pressure head (or "squirt height"), however, for a sand filter would typically be at least 5 ft (1.5 m), and distribution lines placed along a level plane. For sand filters dosed using a pump as described in this chapter, distribution lines and orifices are typically spaced no more than about 30 in (76 cm) on center. Consideration should always be given to achieving good distribution, and avoiding channeling of effluent through sand filters. For sand filters dosed using less pressure head (such as with a Flouts™ dosing device discussed in Chap. 9 and used to dose a sand filter in one of the project examples in Chap. 11), it may be necessary to use closer spacing between lines and orifices.

For optimal performance, sand filters should be dosed using from 10 to 20 microdoses spread out during the day and night. In this way, the filter sand remains moist and the bacterial populations providing biological treatment are better sustained. Floats are less capable of dosing smaller amounts at a time as compared with control timers, so timers are preferred for accomplishing microdosing of sand filters. This is due to the less sensitive functioning and the inherent mechanical nature of floats in a tank where a certain amount of disturbance may be occurring in the water column, and the need to set floats for different functions at least a few vertical inches from each other to prevent operational problems.

Controls with a timer are set with a certain number of minutes/seconds "on" and "off." Controls can be set with timer overrides (activated with a separate float at a higher elevation on the float "tree") to decrease "off" periods when flows to the system are higher. This helps avoid unnecessary high water alarm conditions during those periods. Timer "on" periods are typically just a very few minutes each for an intermittent sand filter, if not less than a minute, with typical "off" settings ranging from 30 minutes to 2 hours. The engineer/designer of the system should calculate appropriate timer settings based on a variety of factors including pump drawdown rate in the particular tank, float settings, and total daily design flow and potential flow variations. Some of these considerations are discussed further in Chap. 10 for low-pressure-dosed effluent dispersal.

Filtered effluent from the sand filter is collected at the base of the treatment unit in a "French drain" type collection pipe, as shown in Fig. 7.2. The effluent pipe exits from the filter through a watertight boot fitting. The effluent flows by gravity to a point in the system where some method of final effluent disposition/dispersal will occur. Either gravity flow or pressurized subsurface dispersal might be used

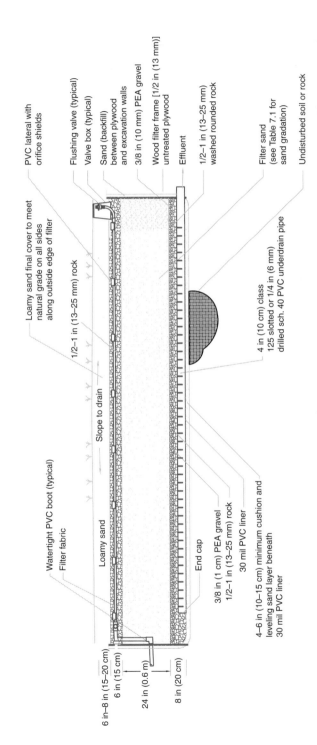

Figure 7.2 Sectional view—intermittent sand filter with gravity discharge.

PVC lateral with orifice shields

Flushing valve (typical)

Valve box (typical)

Sand (backfill) between plywood and excavation walls

3/8 in (10 mm) PEA gravel

Wood filter frame [1/2 in (13 mm)] untreated plywood

Effluent

1/2–1 in (13–25 mm) washed rounded rock

Filter sand (see Table 7.1 for sand gradation)

Undisturbed soil or rock

Loamy sand final cover to meet natural grade on all sides along outside edge of filter

1/2–1 in (13–25 mm) rock

Slope to drain

4 in (10 cm) class 125 slotted or 1/4 (6 mm) drilled sch. 40 PVC underdrain pipe

Watertight PVC boot (typical)

Filter fabric

Loamy sand

End cap

3/8 in (1 cm) PEA gravel

1/2–1 in (13–25 mm) rock

30 mil PVC liner

4–6 in (10–15 cm) minimum cushion and leveling sand layer beneath 30 mil PVC liner

6 in–8 in (15–20 cm)

6 in (15 cm)

24 in (0.6 m)

8 in (20 cm)

Figure 7.3   Intermittent sand filter effluent.

for final effluent disposition, depending on site conditions and applicable regulatory requirements.

Figure 7.3 shows a photo of a beaker of sand filter effluent sampled from a working system. The clarity of this effluent is typical of that produced by sand filters. This intermittent sand filter system was monitored for one year (total of 15 samples), with average BOD_5, TSS and fecal coliform levels of 2.6 mg/L, 3.5 mg/L, and 137 colonies/ 100 mL, respectively. The system was designed and constructed in a manner consistent with the methods and materials described in this chapter, and subsurface low pressure dosing was used for final effluent disposition. Applicable regulations should be referred to when considering suitable options for the final disposition of the treated effluent from ISFs or any other method of predispersal treatment.

Some reasons why sand filters may tend to be used less than packaged treatment units in some parts of the world include

- More skilled labor is needed to properly install a sand filter. Some other treatment methods tend to be "black box" treatment units assembled by the manufacturer.

- Many treatment units used for residential-scale systems are proprietary systems, and have local distributors to market them. Sand filters are nonproprietary, and thus not marketed like manufactured treatment units. Engineers/designers and installers of sand filters and other nonproprietary systems wouldn't therefore receive any type of commission on their installations/sales, as often occurs with proprietary treatment units, and particularly for installers.

- Sand filter treatment systems require more space than some other secondary/advanced treatment options.

- Some geographic areas don't have ready access to suitable filter media to use in intermittent sand filters. Increasingly, however, the availability and use of suitable recycled materials such as crushed glass having the proper size distribution will hopefully reduce this potential limitation. Note however that ground glass is subject to breakage and accumulation of fines during placement, so measures must be taken to minimize those effects, including possibly applying excess water to the material before placement of the final cover and putting the filter into service (fines should not be rinsed into dispersal trenches/beds or field dosing tanks).

7.2 Intermittent Sand Filter Construction

Below are a series of photos illustrating the process of constructing a residential-scale intermittent sand filter. While these systems can perform reliably and cost-effectively for many years, it's critical that they be designed and installed properly to prevent problems or premature failure. Manufacturers or their distributors typically provide training and technical support for proprietary treatment units. Because sand filters and other natural methods of treatment are nonproprietary and are without vendor support and promotional training, contractors often don't have readily available training on the installation of those systems. Many of the construction details below apply to the installation of various types of natural onsite treatment systems, including subsurface flow wetlands and recirculating sand/gravel filters, both of which must be installed in watertight/lined excavations or containment structures, or installed in soils with very low infiltration rates (tight clays). Local wastewater regulatory authorities may have sand filter design and installation requirements that vary in certain ways from the details described in this chapter, and should therefore be consulted before designing and installing this or any other types of systems.

The intermittent sand filter constructed for the project described below was sized for a design flow of 240 gal/day (908 L/day) and a three-bedroom home with water-saving fixtures. Descriptions of the various steps in the construction process are provided. The system was designed and installed for a site with shallow depths of soil above rock. The owner preferred not to have most of the depth of the filter above grade, but also wished to reduce the amount of rock removal needed for the filter to be entirely buried, or below grade. Therefore approximately one-third of the overall depth of the filter is above natural grade for this ISF. In cases where sand filters are to be installed on sites with more easily excavated soils, it tends to be more cost-effective to install the filter below grade as long as there is no shallow groundwater to complicate construction activities.

The filter for the project described below is 10 ft (3 m) wide by 24 ft (7.3 m) long, with a septic tank effluent design loading rate of 1.0 gal/day · ft² (4 cm/day), and a total depth of 42 to 48 in (1.1 to 1.2 m). The actual sand media filter depth is 24 in (61 cm). For the gradation of sand to be used for this filter, a loading rate of 1.0 to 1.25 gal/day · ft² (4 to 5 cm/day) is recommended, so this design is relatively conservative. Table 7.1 shows the gradation range that should be used for the above effluent loading rate when applying primary treated domestic wastewater effluent (and following an effluent filter).

The use of crushed recycled glass instead of sand has been demonstrated and used successfully in projects. In some geographic areas where there is not a local supply of suitable filter media, crushed recycled glass would be a good option to consider. The particle size distribution shown in Table 7.1 should also be used for ISFs having crushed glass as the treatment media.

The importance of using clean properly graded media cannot be overemphasized for these filters. Too many fines in the gradation lead

Sieve Size	Particle Size	Percent Passing
3/8 in (9.525 mm)	9.50 mm	100
#4	4.75 mm	95–100
#8	2.36 mm	80–100
#16	1.18 mm	50–85
#30	0.6 mm	15–60
#50	0.3 mm	3–10
#100	0.15 mm	0–2
#200	0.075 mm	<1

TABLE **7.1**   Intermittent Sand Filter Particle Size Distribution

to premature clogging, and too coarse a gradation will not provide high levels of treatment for single-pass intermittent sand filters. Care must also be taken not to use sands manufactured from softer rock such as limestone that dissolve over time in water. Sand composed of hard dolomite crystals can be acceptable for use in sand filters as long as it is well-washed and free of the fines that tend to be present with limestone-based material. Concrete sand produced in the United States (ASTM C-33) does not meet the gradation shown in Table 7.1 due to the higher amount of fines allowed in that sand specification, and would need to be washed free of most of the fines to be acceptable for sand filters. Silica quartz sand such as that produced in many volcanic regions of the world has been found to work very well for sand filters, as long as it is properly graded. Depending upon the local/regional geology for a given project, the engineer/designer of the system should determine the availability of an acceptable filter media early in the planning stages of the project if use of a sand filter is being considered.

The silica sand used for this filter (Fig. 7.4) was trucked to the site from a plant producing various gradations of filter media. This sand is about as coarse as should be used for single-pass sand filters. Ideally the design engineer or installer of the system would be at the plant supplying the sand at the time the delivery truck was loaded to verify the quality of sand sent to the job site. In many if not most cases the sand produced by aggregate plants tends to vary slightly through any given pile of material, depending on moisture conditions, settling of fines, and other factors. Therefore, coordination with the producing plant is important to ensure that the sand loaded onto the delivering truck is of the proper gradation.

FIGURE **7.4** Intermittent sand filter media.

FIGURE 7.5   Filter media being delivered to site by a 24-ton (21,773-kg) end-dump truck.

To prevent soil and grass from mixing with the sand, the filter media for this project was unloaded onto sheets of plastic moisture barrier until placed into the filter as shown in Figs. 7.5 and 7.6. While it's most cost-effective to have media delivered to sites in larger trucks and using minimal numbers of delivery trips, weight limits on roads should be checked to ensure that they are not exceeded, and that truck lengths aren't excessive for entering sites.

It's important to locate sand filters in as level an area as possible on sites, and where the filter can be protected from any other construction activities, site improvements, any vehicles (including heavy mowers), and the like. Where filters must be located along slopes, it's important to try to orient the longest dimension of the filter along the contours. The site used for this project was terraced, and the area used for the filter naturally sloped at about 5 percent.

Figure 7.7 shows the sand filter site excavated, leveled across the bottom, and cleared of all rocks and excavated material (smaller rocks raked). The presence of relatively shallow limestone rock was the reason why treatment beyond septic tank primary treatment had to be provided before subsurface dispersal of the treated effluent. The exterior footprint of the sand filter excavation was first cut using an 8-in-wide (20-cm) rock saw with a maximum depth cut of 40 in (100 cm), similar to the one shown in Fig. 5.25. A mini-excavator was used to remove the diggable material first, and then a backhoe-mounted hydraulic rock hammer used to break the harder areas of rock out for

FIGURE 7.6 Filter media unloaded onto plastic to prevent contamination with soil and grass.

FIGURE 7.7 The filter excavation is raked free of larger rock fragments before building the wood frame for the liner and placing cushion sand under the PVC liner.

removal by the excavator. Smaller remaining rocks were removed using shovels and a rake.

Because a wood frame will be constructed in the excavation for the placement of a liner for the sand filter as shown in subsequent photos, it's necessary to overexcavate laterally by approximately 1 ft (30 cm) on all four sides of the sand filter. It's also helpful to do any trenching necessary for inlet/outlet piping for the filter prior to final preparation of the filter excavation.

Figure 7.8 shows the sand filter box (10 × 24 ft, or 3 × 7.3 m) during framing. Approximately 1 ft (30 cm) of excavated space is left outside the frame for backfilling with select fill (fill that won't damage the sand filter PVC liner as the frame rots). The wood frame is 40 in (100 cm) tall, and constructed of 1/2-in (15-mm) plywood (untreated) cut down to 40-in (100 cm) width, with 2 × 4 in (50 × 100 mm) boards used for the framing. The corners are built to fit flush into each other. Vinyl-coated sinkers are used all facing outward, or down from the top on the outside of the frame, so that the PVC liner that will be placed inside the frame won't be damaged by the nails (#8 and #12 "vinyl sinkers" are used, and are intended to corrode). Galvanized nails should not be used. Note that the frame here is intended to decompose over time.

Because of the presence of hard limestone rock on this site that was difficult to excavate, it was decided that the sand filter would extend to approximately 18 in (45 cm) above grade on the downhill side when completed (including final cover). A natural rock retaining

FIGURE 7.8   Filter frame being built in panels and assembled.

wall would be placed around the filter to offer support over time when the wood frame begins to deteriorate. For filters installed below grade, the exterior frame supporting the watertight liner should extend to a few inches above natural grade, so that surface water or infiltrating moisture does not enter the filter.

Where sites have greater depths of more easily excavated soil, sand filters may be constructed such that the edges of the completed filter meet natural grade. Final grading of the soil cover should be such that the centerline of the filter is crowned by 2 to 3 in (5 to 8 cm) to drain rainfall and site runoff away from the filter. This will help prevent excessive added flows to the system during wet weather periods.

The sand filter frame is completed in Fig. 7.9, with felt roofing strips placed along seams and wherever there are nails (to protect the liner—not yet placed). The cross brace helps secure the structure.

Figure 7.10 shows the last of the rock debris being cleared from inside of the filter frame, and final leveling being done before placing a minimum of 4 to 6 in (10 to 15 cm) of cushion sand underneath the PVC liner. Concrete sand was used as cushion sand for this liner due to the more angular and sharp-edged quality of other local sands.

Figure 7.11 shows cushion sand being placed and raked level. Notice the two holes in the wood frame. The lower hole at the far end is cut just large enough to accommodate the 4 in underdrain pipe

Figure 7.9 Completed filter frame with roofing felt placed over nails and seams to protect the liner. A wood brace is used to stabilize the sides of the frame until backfilling and placement of media at least a portion of the way up into the frame.

FIGURE 7.10   The last of the rock debris being cleared here from within frame before the cushion sand is placed.

FIGURE 7.11   Cushion sand under liner being placed and leveled.

exiting the filter. The smaller hole (about 18 in or 45 cm above base of frame) is cut just large enough to accommodate a 2-in (50-mm) Schedule 40 pressure manifold that will connect to a piping network that will dose the sand filter. When the PVC liner is installed, PVC boots will be solvent-welded to the inside of the liner to receive the 2-in (50-mm) pipe entering and the 4-in (100-mm) Schedule 40 PVC pipe exiting the lined filter.

The photo in Fig. 7.12 shows the 30-mil PVC liner placed, and tacked down with small plywood pieces along outside edge of frame. The temporary wood cross brace was removed and replaced after the liner was in place. Note that in the corners extra material is left to ensure that no stress is placed on the liner when sand/aggregate is placed inside the liner. When installing the liner, the manufacturer's packing/unpacking information should be reviewed to determine the correct orientation for positioning the liner inside the filter frame. This will minimize wear on the surface of the liner and disturbance of the leveled sand bedding under the liner.

Extra liner material was placed over the bottom of the liner to create added protection from pea gravel to be placed at the bottom of the sand filter. Some pea gravel can have very sharp (razor-like) edges, and will tend to puncture liners. Inexpensive plastic moisture barrier (thin black PVC liner) folded over a couple of times was used on top of this liner along with excess 30-mil liner to help prevent punctures from placed aggregate. Only socks were worn by workers when working inside the liner for this system, since small rocks/gravel on the bottoms of shoes would pierce the liner.

FIGURE 7.12   The PVC liner (30 mil) is placed here.

Liner boot will be solvent-welded into place where hole was cut in frame.

Liner boot hole cut in frame.

FIGURE 7.13   PVC liner boots are being installed here.

Black liner boots are being installed at this point in the construction process. The gray liner will be cut where the inlet/outlet holes are located, and the black PVC boot will be solvent-welded (using PVC cement or "glue") into place (Fig. 7.13). Liner boots should be glued (solvent-welded) from the inside so that the media exerts pressure against the solvent-welded lap to help promote positive sealing (for subsurface flow wetlands, water and media exert pressure on this connection). Suitable sealant/solvent (per PVC boot and liner manufacturer) and pipe clamps are used around the boots and pipe on the outside of the filter box or excavation.

A slotted PVC filter underdrain is placed along the centerline of the liner, lengthwise as shown in Fig. 7.14. Drilled 3- to 4-in (75- to 100-mm) Schedule 40 pipe may also be used, with holes drilled clean to a minimum diameter of 1/4 in (6 mm), spaced every 6- to 8-in (15 to 20 cm), and oriented along both sides of the pipe between 3 and 4 o'clock and 8 and 9 o'clock. For larger intermittent sand filters, multiple underdrains may be needed, spaced at 6- to 8-ft (2- to 2.5-m) intervals, and connected at the outlet end of the filter so as to minimize numbers of pipes exiting the filters. Cushion sand beneath the liner may be very slightly sloped toward the underdrain and also toward the outlet of the sand filter, though not more than about 1/2 to 1 in (15 to 25 mm) per every 10 ft (3 m). Extra liner material strips were placed under the drain for this system to help prevent punctures in the underlying liner material when media was installed around the underdrain (Fig. 7.15).

FIGURE 7.14 Slotted underdrain is placed.

FIGURE 7.15 Larger clean rock placed around underdrain to help prevent clogging from pea gravel and biomat build up over time; Space is left along one side of the underdrain for workers to walk and place material, to avoid puncturing the liner.

Pea gravel is placed along the liner just past the underdrain to leave an area where workers can walk without excessive foot traffic on the sharp pea gravel delivered to this job site. Shoes were removed after placement of the liner, and only socks worn until the remainder of the 8-in (20-cm) layer of pea gravel was placed so as to not puncture the liner.

Pea gravel was first placed up to about 1 in (25 mm) above the liner surface and below the slots on the underdrain. Pea gravel would tend to clog the slots, so 1-in (25-mm) rounded rock is placed just around the underdrain up to about 1 in (25 mm) above the top of the underdrain, along the whole length of the filter (Fig. 7.15).

Figure 7.16 shows the remainder of the 8-in (20-cm) layer of pea gravel being raked level before placing the filter sand. A chalk line could be used instead of the marking paint shown here for setting levels along the liner walls (see Fig. 7.17). Certain paints may react chemically and damage PVC liners. Although key levels in the process were marked along the sides of the liner, a laser level was set up and used throughout the process for achieving level conditions throughout.

Approximately 18 yd^3 (13.8 m^3) of filter sand were needed for this sand filter, with the sand meeting the gradation criteria presented in Table 7.1. The particular sand used for this project flowed very easily into place (very noncohesive). For more angular sand that may not self-compact well, it is helpful to place the sand when slightly damp so that it tends to compact better and not leave pockets of uncompacted sand in the filter layer.

FIGURE 7.16   Pea gravel is raked level before placing the filter media (sand).

FIGURE 7.17   Filter media is placed, with sand backfilling outside the wood frame placed to the same level to balance pressures for structural stability.

Note in Fig. 7.17 the sand placed along the outside of the filter box. When cushion sand was placed below the liner, sand was also placed outside of and around the filter frame, to close any gaps between the wood frame and the ground surface, and to prevent rock from outside of the filter from entering the frame. As material is placed inside the liner, for structural purposes (and because the frame only has limited strength), sand is progressively backfilled outside the frame to the same depth as is placed inside.

Twenty-four inches (0.6 m) of filter sand are then placed and raked level. The marking level lines along the inside walls of the liner are used for general leveling purposes, but again, the laser level is used throughout the process to maintain level material placement (Fig. 7.18).

The 2-in (50-mm) Schedule 40 PVC pressure manifold will be tied into the pressure distribution lines to be placed within the washed rounded rock layer (Figs. 7.22 to 7.24). Some type of protective covering should be placed around the end of the turnup of the 2-in (50-mm) manifold pipe inside the filter before sand/aggregate are placed, to keep dust and aggregate from entering the pipe. A variety of things can be used including a nonglued capped sleeve (3- to 4-in, or 75- to 100-mm, pipe), or a rag taped around the manifold's end, with the tape not adhered to the pipe itself where solvent welding is to occur (as shown for the outlet side of the pipe here).

A total of 8 in of 1/2- to 1-in (15- to 25-mm) washed rounded river rock are placed above the filter sand (Figs. 7.19 and 7.20). The

FIGURE 7.18 Filter sand placed and leveled with a rake. Notice that a cleanout is located outside the far end of the filter where the drain pipe exits the filter. It will be cut to meet grade when final cover is placed.

FIGURE 7.19 Washed rounded river rock is placed above the filter sand in two stages (with effluent distribution piping placed after the first 4 in or 10 cm).

FIGURE 7.20 Larger rock is carefully placed on the filter sand in small piles spaced around the surface, and leveled so as not to embed into the sand, or cause unlevelness of the filter media.

distribution manifold piping and orifice shields are located in the middle of that 8-in (20-cm) layer, so 4 in of the 1/2- to 1-in (15- to 25-mm) rock are first placed, and then the manifold assembled and installed. The next 4 in (10 cm) layer of washed river rock are then carefully placed above and around the distribution piping.

The first 4 in (10 cm) of rock are placed and gently leveled so as not to be embedded into the filter sand, or not otherwise affect the levelness of the filter sand. Small piles/rows of the rounded rock are placed at intervals, and then spread from the top to surrounding areas to a depth of 4 in (see Fig. 7.20).

Figure 7.21 shows the distribution piping being assembled. There are plastic pop-outs at ends of the turned up pipe, used (and were shipped with these predrilled lateral lines) to keep dust and debris out of pipes during installation. If these are not supplied with pre-drilled pipe, the distribution pipes should be otherwise temporarily covered at the ends to prevent dirt/debris from entering and later clogging the sand filter distribution piping orifices. These lateral lines are drilled clean, free of PVC shavings, with 1/8-in (3-mm) diameter holes at 30-in (76-cm) intervals along the length of the laterals. The holes are drilled in a line, and the distribution piping assembled (and solvent-welded) so that they are lined up at 12 o'clock along the Schedule 40 pressure pipe. This spacing and configuration is used because this filter will be dosed using a high head effluent pump

Figure 7.21 Effluent distribution piping is place after the first 4 in (10 cm) of river rock above the filter media.

capable of 5 to 10 ft (1.5 to 3 m) of residual pressure head and good distribution with orifice shields to be placed over the orifice holes (see Fig. 7.22). For sand filters dosed using methods having lower available pressure (e.g., dosing siphons or Flouts), lesser spacing may be needed (though never less than about 15 to 18 in or 38 to 45 cm) to better achieve uniform distribution across the filter. It may also be necessary to direct orifices downward when there is not sufficient pressure head available to flush distribution lines.

During design, the pressurized distribution system is analyzed hydraulically to size the pump(s) used for dosing the filter. The Hazen-Williams formula is used for calculating head losses (friction energy losses) in the piping system delivering primary treated effluent to and through the sand filter distribution piping. Calculations must reflect the elevation difference between the pump and the sand filter distribution system (and residual squirt height), orifice spacing and size, and pipe size and length of lines. The pressure head and orifice size determine the flow from each of the orifices along the distribution piping when the pump is on. The pump selected should be capable of delivering effluent to the sand filter system with sufficient residual pressure head to achieve a squirt height of at least 5 ft (1.5 m) at the ends of the furthest lateral lines on the filter. Hydraulic analyses for sand filter systems are similar to those for low-pressure subsurface dispersal systems, as discussed in Chap. 10.

FIGURE 7.22 Plastic/PVC orifice shields manufactured specifically for this size PVC piping are snapped into place over each of the orifices, oriented here at 12 o'clock.

For this size of system, and because the sand filter is located fairly near the pump serving the filter with relatively mild slopes on the site, a 1/2-hp (0.37-kW) high head effluent pump was used. High head effluent pumps are capable of meeting fairly wide ranges of operating pressure heads over relatively low flow ranges. Low head effluent pumps operate over wider flow ranges and lower pressure heads.

After the distribution piping is installed, orifice shields are snapped into place above each of the drilled orifices (all positioned at 12 o'clock). The orifice shields keep the holes from getting clogged with aggregate, and help distribute the septic tank effluent over the filter. Even distribution of effluent over the filter aids in achieving high-quality treatment by avoiding channeling of effluent and over-loading of portions of the filter. Prior to placing the rest of the 1/2- to 1-in (15- to 25-mm) rounded stone, geotextile fabric, and final cover over the manifold piping, a squirt height test should be performed along with checks for any leaks in the piping joints. The pressure head entering the filter should be a minimum of 5 ft or 1.5 m) (the "squirt height") above the lateral lines with the orifice shields removed from the orifices on distribution lines next to the turnups. Orifice shields are then replaced following the test and prior to placement of the remaining 4 in (10 cm) of stone.

The turnups at ends of the lateral lines in Fig. 7.22 are used for periodic flushing/maintenance of the sand filter distribution lines.

A ball valve is installed in each of the turnups, and used to clean/flush the lateral lines at least once annually. Valve boxes are placed over these turnups prior to placing final cover over the filter.

A cleanout is located outside the filter where the slotted underdrain pipe connects to a solid Schedule 40 PVC 4-in (100-mm) pipe (the connection is inside the liner, since the filter unit is to be watertight). The 4-in (100-mm) drain pipe reduces down here to a 2-in (50-mm) pipe just past the cleanout, and then drains at 1/8 in/ft (10 mm/m) fall/slope to a dispersal field dosing tank for this system.

Figure 7.23 shows the filter ready for the last 4 in of 1/2- to 1-in (15- to 25-mm) rock. Geotextile fabric is then placed before the final cover. The distribution piping is 1 in (25 mm) diameter Schedule 40 PVC pressure pipe, 3/4 in (20 mm) diameter Schedule 40 PVC pipe could also have been used for this filter.

Figure 7.24 shows the sand filter with the remainder of the 8-in (20 cm) layer of 1/2 to 1 in (15 to 25 mm) rock placed. Notice that the rock is placed up to about 1 in (25 mm) below the top of the lined filter frame. Figure 7.25 shows turned up ends of Schedule 40 PVC sand filter distribution lines with ball valves on each, which are used to periodically flush and clean the lines of any buildup that may occur. Valve boxes will be placed over these before placing geotextile fabric and final sandy loam cover. On the left side of Fig. 7.24 are trenches that will be used for the final subsurface dispersal of the sand filter effluent for this system. The method of subsurface dispersal used for

Figure 7.23   Sand filter with distribution piping and orifice shields in place.

FIGURE 7.24 Sand filter ready for geotextile fabric and then final soil cover.

FIGURE 7.25 Turned up ends of distribution lines are shown here with ball valves, periodically opened with a special fitting screwed on to the end to flush the distribution lines.

Figure 7.26 Geotextile fabric is shown here placed and anchored with rocks over the filter, now ready for final soil cover.

this system is "low pressure dosing," and will be discussed in detail in Chap. 10. Sand filter effluent will drain from the filter to a pump tank, from which the field will be dosed.

After placement of the washed rounded river rock, geotextile fabric is used to cover the river rock prior to placement of final cover/backfill (see Fig. 7.26). This prevents small particles from entering and clogging the filter over time. Valve boxes were placed around the turnups/ball valves, with the surface of valve boxes at finished grade of final cover.

It is critical that sand filters be able to continue to "breathe" (maintain aerobic conditions) over time. It's therefore very important to select the proper type of material to use for the final cover over the filter. Note the large clots in the soil surrounding the filter in Fig. 7.27. "Sandy loam" was ordered for this project to use as final cover for the filter, but the loam had considerable clay in it, which would "choke" the filter over time (causing anaerobic or septic conditions). Therefore different final cover soil was ordered that is more like sand than loam. This stage of construction occurred at the end of a work week and delivery of the new loam couldn't be made until early the following week. Several inches of coarse sand were placed over the geotextile fabric to protect any shallow filter components (from digging rodents, deer hooves, etc.) until delivery of the proper cover soil could occur.

FIGURE 7.27 A few inches of coarse sand were placed over this filter until suitable final soil cover free of significant clays could be delivered to the site.

The cleanout pipe shown in the above photo was trimmed to grade before finishing placement of final cover.

Loamy sand was used to finish covering the filter to finished grade, as shown in Fig. 7.28. For this particular system, excavated rock was stacked into a natural retaining wall around the filter. The final cover was graded to crown slightly along the centerline to drain rainwater away from the filter, and be seeded with a native grass species. The wooden frame around the sand filter will deteriorate over time (the nail material is selected to deteriorate along with the wood), and thus exterior support is needed for this sand filter to prevent settling and erosion.

Other types of support/retaining structures can be used for raised sand filter treatment units, and may be more cost-effective and cause less site disturbance where there is shallow soil depth to rock or shallow groundwater. An example of this is shown in Fig. 7.29 for a sand filter unit under construction behind a decorative stone wall. When completed this sand filter was xeriscaped with shallow rooted native plants, along with the adjacent subsurface dispersal field area as shown from a different angle in Fig. 7.30 (arrow points to raised sand filter). That system is the second of five project examples discussed in Chap. 11.

After placing the final cover, the sand filter is vegetated with shallow rooted grasses. No woody plants or deep rooted plants should be used (a maximum rooting depth of 3 to 5 in [8 to 12 cm]

FIGURE 7.28 A natural stacked-rock retaining wall was built around this filter, and will be vegetated with native grasses.

FIGURE 7.29 Lined filter box for a completely above ground sand filter built inside this landscaped masonry wall.

is recommended). There should be approximately 6 to 10 in (15 to 25 cm) of final cover placed before vegetating, crowned along the centerline of filter to drain to the edges of the filter.

Ideally, any needed control systems should be robust, waterproof and weatherproof, use corrosion resistant materials of construction, and should be designed for the same expected service life

FIGURE 7.30 Landscaped wall around the completed sand filter treatment unit.

as the system controlled. In practice, however, this last criterion cannot usually be accomplished, so provisions for eventual control system component replacement should be considered.

Where control panels operating pumps that dose sand filters and other equipment cannot be mounted on shaded/protected exterior or interior walls of structures served due to distance or other factors, free-standing protective structures are recommended to protect panels from extreme temperatures and wet weather periods. Many control panels have weather-resistant housings, but added protective temperature insulation and avoiding excessive exposure to moisture will likely significantly prolong their useful service lives. Lightning strikes can also be a problem in some geographic regions such as the southwestern United States.

Figures 7.31 and 7.32 show an example of such an enclosure that was constructed by the contractor to house both a breaker disconnect box and the control panel for a small onsite wastewater system that was installed prior to the construction of the building to be served.

7.3   Recirculating Sand/Gravel Filters

Recirculating sand/gravel filters (RSFs) can reliably produce effluent with very low BOD and TSS levels, and reduce total nitrogen by 50 to 60 percent on average. RSFs are constructed similarly to intermittent (single-pass) filters with the following major differences:

1. The areal loading rate for recirculating sand/gravel filters (RSFs) is typically 3.0 to 5.0 gal/day · ft² (12 to 20 cm/day)

Figure 7.31   Control panel located in protective enclosure next to tanks (watertight bolt-down access riser lids shown for two tanks).

forward/design flow, and thus occupies significantly less area (lower loading rates may be needed to achieve desired performance levels in colder climates).

2. RSFs are not "buried" (or covered with final soil cover) as with the intermittent sand filter described above. In many cases they are covered to prevent excess rainwater, or to reduce the potential for odor issues (e.g., with plastic liner material, or with the filter built into a concrete containment structure with a lid and some method of allowing for ventilation/air flow).

3. Pea gravel (3/8 in or 10 mm, and washed) is used in the top 6 to 8 in (15 to 20 cm) of RSFs (rather than the more coarse 1/2 to 1 in, or 15 to 25 mm, stone used for ISFs).

4. RSFs require a recirculation tank with a working capacity typically of about one day's design flow, and some type of flow "splitter" valve to divert a certain portion of the recirculating

FIGURE 7.32   The arrow shows where a hex bolt connection is used for securing the panel access door to help prevent tampering, but allows ready access by authorized persons. In that way keys or combinations are not needed.

filter flow to final effluent soil dispersal or to discharge (see point 6 below).

5. The recirculation ratio used is based on the actual daily flow conditions, and is typically about 4:1, although it should be based on the quality of wastewater being treated and the desired effluent quality. Recirculation ratio is the daily volume of recycled effluent divided by design daily volume of the wastewater. Timers in the control panel are typically used to achieve the desired recycle rate, based on the flow from the pump for the given hydraulic conditions.

6. A "T" or "Splitter T" type float valve located in the recirculation tank or similar device is commonly used to split flow between effluent returning to the recirculation tank from the filter, and the final effluent dispersal system. Typically about 20 percent of the flow returning from the filter to the recirculation tank is directed to the final dispersal system by the splitter valve. Evaluation of and design for overflow and surge control at the recirculation tank should be included in the design plans.

Except in cold climates where some type of cover may be needed, RSFs are typically left open to the atmosphere to maintain ample

FIGURE 7.33 This California RSF treatment system consists of septic tank pretreatment followed by two recirculating sand filters and recirculation tanks. The primary treated effluent mixes with recirculating effluent, and is dosed several inches below the gravel surface, which provides vector and odor control for the treatment system. (*Photo courtesy of Orenco System, Inc.*)

access to air/oxygen for the filtration media, and aerobic conditions. For that reason they're most commonly used for systems with larger flows (e.g., multiple dwellings or commercial facilities). RSF treatment units are typically located in an area off-limits to the public. Some enclosed residential units have, however, been installed at residences in certain parts of the United States. Figure 7.33 shows an RSF system designed to serve a wastewater flow of 20,000 gal/day (75,700 L/day).

Sieve Size	Particle Size	Percent Passing
3/8 in (9.525 mm)	9.50 mm	100
#4	4.75 mm	70–100
#8	2.36 mm	5–78
#16	1.18 mm	0–4
#30	0.6 mm	0–2
#50	0.3 mm	0–1
#100	0.15 mm	0–0.2
#200	0.075 mm	0

TABLE 7.2 Recirculating Sand/Gravel Filter Particle Size Distribution

Suitable recirculation and loading rates must be determined for RSFs, depending on both influent wastewater characteristics and desired effluent quality. Chapter 7 of Resources and Helpful Links on this book's website provides several references for more detailed information on designing recirculating sand filters. Sources of material and equipment used for RSFs and ISFs are also provided there.

Table 7.2 provides typical media gradation for RSFs, with the media used needing to be clean/washed, and of a hard mineral, glass, or other suitable material as with intermittent sand filters.

Reference

1. R. Crites and G. Tchobanoglous, *Small and Decentralized Wastewater Management Systems*, McGraw-Hill, New York, 1998.

CHAPTER 8

Subsurface Flow Wetlands

Two types of subsurface flow wetlands are discussed in this chapter, each used for different treatment needs. Each type is considered a potentially sustainable option for the right set of site conditions and project needs, as discussed below. For small- to midsized decentralized systems covered in this book, subsurface flow wetlands tend to be used in preference to free-water surface wetlands for several reasons. Those include the likelihood of shorter distances between treatment units and the facilities served for smaller-scale systems, and the need to prevent direct public exposure to wastewater, and provide vector and odor control. Free-water surface (FWS) wetlands, for which wastewater is exposed to the air, tend to require greater separation distances and protective management practices. For a properly functioning subsurface flow wetland, the wastewater remains at least several inches below the surface of the treatment media, offering advantages for both vector and odor control as well as minimizing the opportunity for public exposure and related health risks.

8.1 Subsurface Flow Wetlands for Secondary Treatment

Subsurface flow wetlands have been used following primary treatment to provide removal of secondary treatment parameters (BOD and TSS reduction) along with a limited amount of total nitrogen reduction. Because of the relatively anoxic conditions in subsurface wetlands, nitrification tends to be very limited unless the system is sized large enough that it would most often be less cost-effective than other decentralized treatment options. With effective nitrification needed to accomplish higher levels of total nitrogen removal, wetlands are best used in combination with other processes that efficiently nitrify where low levels of effluent nitrogen are needed.

Use of subsurface wetlands following primary treatment might be considered a preferred and relatively sustainable option for residential to larger-scale decentralized wastewater treatment especially where:

- There's a need for secondary treatment (BOD and TSS reduction) along with a limited amount of total nitrogen reduction prior to the subsurface soil application of the effluent. This offers a longer service life to the dispersal trenches or bed by reducing biomat formation, as well as providing for some total nitrogen reduction for local water-quality protection.
- Limited or no power is available to a treatment site.
- There are no spaces or steep slope constraints (with steep slopes requiring substantial cut and fill to achieve recommended length to width ratios for wetland cells).
- The project isn't located where excessive or intense seasonal rains or other climate factors that would adversely affect performance are common occurrences.

There are other reasons why a subsurface flow wetland might be considered, including site aesthetics. Plant selection is, however, important for achieving treatment comparable to the effluent levels shown in Table 6.2. Ornamental plants are typically not deeply rooted enough to penetrate the full water depth of the wetland, which is typically about 18 to 24 in (46 to 61 cm). It's also important not to use woody plants, since the root system for those will become too extensive and possibly contribute to clogging problems over time, and be difficult to remove once established. Species such as reeds and cattails are often used for wetlands due to their deeper root penetration, nonwoodiness, resilience, ability to quickly propagate, and relatively high nutrient uptake potential as compared with other plant species.

Land area is always a consideration for the use of wetlands. Even in warmer climates, wetlands used for secondary treatment typically need at least as much space as a single-pass sand filter. Unlike buried sand filters, areas of sites occupied by wetlands are unavailable for other uses such as light foot traffic.

Figure 8.1 shows a conceptual illustration of a basic small-scale wetland treatment system used for achieving secondary treatment levels prior to final subsurface soil dispersal. Figure 8.2 shows the plan and profile view and basic elements of a typical subsurface flow wetland system, including critical elevations that should be set during design.

Wetland designs should be developed based on several critical factors including:

- Seasonal temperatures
- Historical rainfall, and monthly water balances based on that data, and including peak rainfall events

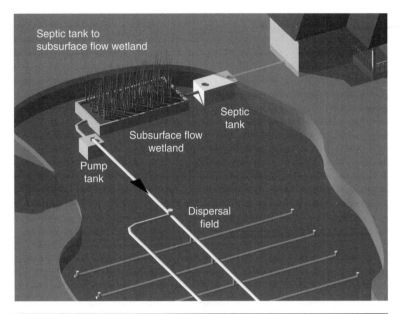

FIGURE 8.1   Conceptual illustration of a small wetland treatment system with low-pressure-dosed subsurface effluent dispersal.

- Influent waste characteristics
- Desired effluent quality

For subsurface flow wetlands leading to subsurface soil dispersal systems in geographic areas subject to periods of prolonged or intense periods of rain, some means of controlling the added rainfall flow to the wetland should be provided. That may be through either added storage capacity following the wetland (to the extent that's feasible or cost-effective), or by allowing the water level to rise in the wetland by way of either manual or automated controls. There are however some disadvantages and potential problems associated with the latter approach when the water surface rises above the media surface, particularly for small- to midsized wetlands located relatively close to the facilities served.

As noted above, several of the advantages of subsurface flow wetlands over FWS wetlands include vector and odor control, prevention of algae formation, and reliable TSS and BOD removal. If the water surface rises above the top of the treatment media, these types of benefits are eliminated, and there is the potential for substantial short-circuiting of treatment processes and adverse impacts to the subsurface dispersal field. For treatment to reliably occur in a subsurface wetland, wastewater must flow through the media with as little short-circuiting as possible.

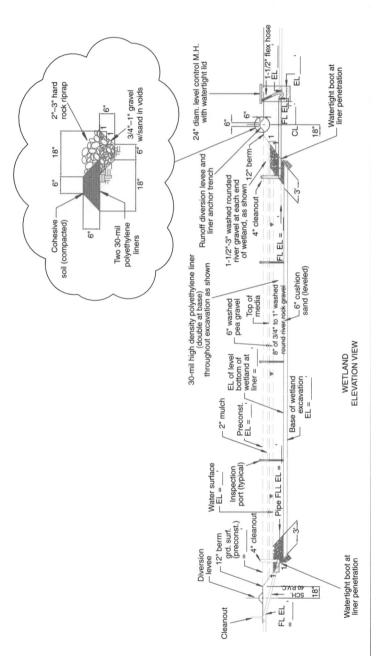

Figure 8.2 Side and plan views of a typical subsurface flow wetland.

236

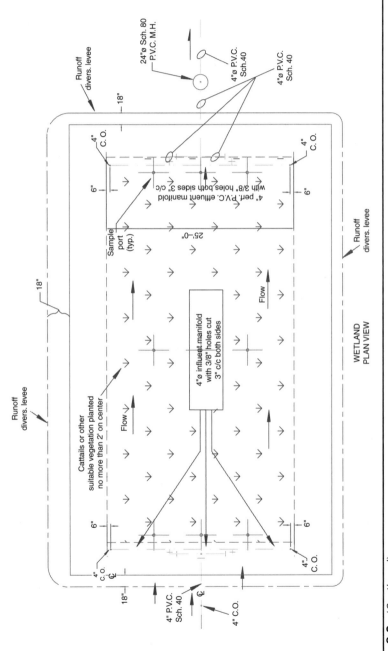

FIGURE 8.2 (*Continued*)

237

The performance of subsurface flow wetlands tracked over time for decentralized systems has been very mixed. Proper design details and management of the systems has proven to be critical to their successful use. Premature clogging, development of preferential flow patterns, and ultimate surfacing of effluent and failure have reportedly occurred for some systems even in drier climates that would presumably be well-suited for the use of wetlands. Surfacing of effluent in a subsurface wetland constitutes failure, since odors are present with wastewaters having significant ammonia/TKN levels, and since treatment would no longer be occurring through the subsurface media and root system. Mosquitoes and other "vectors" presenting public health risks would also be of concern for surfaced effluent.

Several models have been developed and verified during the past several decades for designing wetlands. Those design methodologies are detailed in some excellent technical resources now available, including the U.S. EPA and Manuals of Practice published by the Water Environment Federation. Several of those technical publications are listed in Resources and Helpful Links on this book's website.

8.1.1 Residential Wetland Example for Secondary Treatment

A residential wetland designed and built to serve a single family residence in Austin, Texas, was monitored over a period of about 2 years.[1] The system was designed to receive an average daily flow of 300 gal/day (1136 L/day), with a target BOD/TSS reduction of 90 percent based on reported typical residential septic tank effluent quality. The wetland was sized at 10 ft wide by 30 ft long (3 m × 9.2 m), a 3:1 length to width ratio, and an operating water depth of 24 in (0.6 m). The hydraulic retention time in the wetland was about 5.5 days, with an average areal loading rate of approximately 1.0 gal/day · ft^2 (40 mm/day). Manifold piping at the discharge end of the wetland drained to a pump tank, from which effluent was pumped to low-pressure-dosed trenches. Additional pathogen reduction was provided by placing 18 in (0.46 m) of filter sand at the base of shallow low-pressure-dosed trenches in the dispersal system. A family of four occupied the house throughout the monitoring period.

8.1.2 Residential Wetland Construction

Figures 8.3 through 8.9 show a series of photos detailing the system's basic features and construction. Certain elements in the construction of a subsurface flow wetland are very similar to a single pass or recirculating sand filter, and in particular the watertight containment structure for the treatment system. If clay soils with an infiltration rate of less than about 10^{-7} cm/s are not naturally present on the site, either a liner (clay or synthetic) must be installed, or some other type

Figure 8.3 The framing for this residential wetland is constructed in the same manner as the ISF in Chap. 7. This photo shows the cushion sand inside the frame leveled prior to placement of the 30-mil plastic liner. Conditions were very rocky on this site. So to avoid excessive rock removal, the wetland will extend significantly above grade, with a natural stone retaining wall placed around it.

of containment structure used. Figure 8.10 shows a cast-in-place concrete containment structure used for a different wetland system. For sites with very rocky conditions, it may be most cost-effective to excavate as deep as possible and then have the wetland extend above the ground surface, as shown in some of construction photos of the residential wetland below.

Figure 8.4 The HDPE liner here is temporarily anchored along the edges.

Figure 8.5   This photo shows the 30-mil HDPE liner installed, along with watertight boot inlet fittings, and inlet manifold piping. The manifold has 3/8-in (10-mm) drilled holes spaced every 3 in (8 cm) along the manifold. Extra liner material was used for "rub sheets" at the ends of the liner, with a second layer of HDPE liner material at the base of the wetland to prevent punctures from media placement. Outside edges of the liner will be rocked-in for anchoring.

A high-density polyethylene (HDPE) liner is used here because approximately 1 ft (0.3 m) of the liner above the media inside the wetland will be exposed to sunlight, which quickly degrades PVC liners. HDPE liners present certain challenges to projects, since inlet and outlet boot inserts must be heat fused/welded (not solvent welded as

Figure 8.6   Larger 3 to 6 in (80 to 150 mm) washed stone is used at the inlet and outlet structures, to ensure better hydraulic distribution of effluent across the head of the wetland, and better drainage of treated effluent from the tail of the wetland.

FIGURE 8.7 The wetland treatment media is placed and leveled in this photo. Washed rounded stone was found to only be locally available in the 1 to 1.5 in (25 to 38 mm) size range, although the design called for use of 3/4 to 1 in (20 to 25 mm) clean washed stone for the treatment media. While the larger stone would tend to have lesser clogging issues over time, its use here likely contributed to lesser overall treatment performance.

with PVC liners). As long as the proper size and rating/class of pipe are specified to the liner manufacturer, it is best if the manufacturer inserts the pipe into the inserts, and ships them inside the boots with the liner. Heat fabrication causes deformation of the HDPE, which

FIGURE 8.8 This wetland was planted with a native giant reed species (*Arundo donax*). Young plants are shown planted here about 2 ft on center (0.6 m) so that the tops of the roots are at the top of the treatment media. Pea gravel was then placed around them to hold them in place. Pea gravel was then carefully placed around them up to a depth of 6 in (15 cm) above the treatment media and water surface.

FIGURE 8.9 The wetland is completed here, with the stone work completed around the perimeter. The inlet to the wetland is in the near end, with a cleanout shown that can now be cut to finished grade. The *Arundo donax* (giant reed) plants are well established here. Watering of the plants was necessary to establish them since the system was started up during the summer.

FIGURE 8.10 This residential subsurface flow wetland was constructed in a concrete containment structure built on site. The length-to-width ratio of this wetland is much greater than recommended, though it illustrates the successful use of concrete for small-scale wetlands.

can make it very difficult to install the pipe later after it has cooled. The ends of the pipe can be capped with a rounded push-on cap to prevent damage to the liner from the pipe edges at each end during shipping of the liner. The manufacturer must of course be careful with the liner folding and shipping.

After placement of the liner, it should be tested for watertightness before placement of any rock media. Following placement of the first few inches of rock media, it should again be water-tested. Rock media should be very carefully placed into the liner to prevent punctures. Even with great care taken during media placement, this liner had to be sent back to the manufacturer for repair and shipped back for reinstallation, after initial installation and placement of about 6 to 8 in (15 to 20 cm) of rock media onto the liner. Placement of extra strips of liner material along the inside bottom of the main liner can help greatly to prevent punctures and leaks.

In Fig. 8.4 the outlet boot and piping are shown installed. The water level in the wetland will be controlled by a flexible heavy duty hose connected to the outlet pipe draining from the wetland and entering the concrete pump tank. The flexible hose will be connected to the 3-in (80-mm) PVC pipe in the pump tank using solvent welded fittings, and the mouth of the hose suspended from the interior sidewall of the pump tank at the water surface elevation with a stainless steel anchor bolt and plastic ties.

As with ISFs, sand backfilling behind/outside the liner frame should be placed at the same depth as rock media is placed into the lined wetland frame. The excavation for the wetland should allow for at least 6 in (15 cm) of sand between the sidewalls of the excavation and the wetland liner's wood frame.

For this wetland, a native stone exterior retaining wall was built along the perimeter of the frame to anchor the HDPE liner, with the completed exterior structure about 2 ft (0.6 m) above natural grade (Fig. 8.9). This also serves to prevent surface runoff from entering the wetland.

Six inches (15 cm) of pea gravel were placed on top of the leveled treatment media shown in Fig. 8.7. For this wetland, small piles of pea gravel were first mounded around the plant shoots and roots being planted at the top surface of the rock treatment media (see Fig. 8.8). The remainder of the 6 in (15 cm) of pea gravel was then carefully placed around the planted vegetation. Alternatively, pea gravel can be dug/moved aside to plant vegetation, and replaced around the plant root/shoot. This latter approach was used for the wetland discussed in Sec. 8.2. In either case, the tops of plant roots should be placed at the wetland's water surface, which is at the interface between the pea gravel and the larger treatment media. This requires digging down slightly into the treatment media for planting.

The use of chipped tires for subsurface flow wetland treatment media has been demonstrated, and offers an opportunity to use wetlands for treatment where suitable rock/gravel media may not be available or must be transported long distances to a job site.[2] One of the project examples in Chap. 11 uses chipped tires as the treatment media for a wetland. Recycling chipped tires for wastewater treatment processes is very favorable from a sustainability perspective, including such benefits as lower transport costs in most cases as compared with rock/gravel. While the metal wire protruding from chipped tires has been found to enhance phosphorus removal in wetlands for at least a limited period of time as compared with gravel media, the wire presents some additional wetland construction considerations. Some means of preventing punctures to watertight liners must be used, as well as care in handling the chipped tires on behalf of worker safety. Such measures might include at least 3 to 4 in (80 to 100 mm) of some type of "buffer media" beneath the tire chips. If at least a limited amount of gravel media is available locally, that might be used. Pieces of waste polystyrene (1 to 2 in or 25 to 50 mm) might also be used for this purpose, if available locally as a waste product needing to be recycled. The polystyrene would, however, need to be held in place over the liner using, for example, nylon netting. A product on the market called EZ Flow™ employs a similar concept with rolls of netted polystyrene used in place of aggregate for subsurface dispersal trenches.

Figure 8.10 shows a photo of a residential scale wetland constructed in a watertight cast-in-place concrete containment structure.

8.1.3   Residential Wetland Treatment Performance

Table 8.1 shows average monitoring results and ranges for influent and effluent from the residential wetland detailed above (Figs. 8.3 to 8.9). Numbers in parentheses show the numbers of samples for that parameter.

All samples for the wetland were collected and analyzed in accordance with *Standard Methods for the Examination of Water and Wastewater*,[3] with laboratory analyses performed by one of the City of Austin, Texas, Water Utility's laboratories. The subsurface dispersal field's performance was also monitored, and those results will be discussed in Chap. 10. To avoid the use of chlorination or UV disinfection, sand-lined low-pressure-dosed trenches were used following the wetland, with about 18 in (0.46 m) of concrete sand used at the base of the trenches.

While BOD and TSS levels were reduced sufficiently to meet local regulatory requirements, the system did not meet the 90% reduction targeted for those parameters. Very low levels of nitrification occurred in the wetland based on effluent nitrate/nitrite levels. Total nitrogen removal was 25 to 30 percent on average, but this was apparently due mainly to vegetative uptake of the nutrients. The wetland's performance illustrates the highly anoxic conditions characteristic of these systems. For site conditions and watersheds where nitrogen levels in effluent need to be limited, an added treatment process would be

Measured Parameter	Influent (Septic Tank Effluent)		Wetland Effluent	
	Average	Range	Average	Range
BOD_5 (mg/L)	187.3 (12)*	141–330	28.3 (19)	12–64
COD (mg/L)	407.8 (13)	185–594	116.4 (23)	46–202
TOC (mg/L)	105.2 (9)	89.4–168	34.2 (17)	13–145
TSS (mg/L)	30 (2)	13–46	17.2 (7)	4–62
NH_3-N (mg/L)	64.6 (13)	44–84.3	48.0 (22)	34–74.1
TKN (mg/L)	68.2 (13)	49.6–84.5	45.6 (23)	19.1–66.3
$(NO_3 + NO_2)$-N (mg/L)	0.1(13)	0.02–0.18	0.2 (23)	0.02–1.67
Fecal Coliform (Col./100 ml)	3.5×10^5 (10)	$6 \times 10^4 - 9.6 \times 10^5$	2.1×10^4 (15)	$7.3 \times 10^2 - 1.5 \times 10^5$
Fecal Streptococcus (Col./100 ml)	5.7×10^4 (2)	$6.0 \times 10^3 - 1.1 \times 10^5$	3.0×10^4 (4)	$1.2 \times 10^3 - 8.6 \times 10^4$
Chloride (mg/L)	84.3 (12)	68.1–130	99.4 (20)	70.2–200
Alkalinity (mg/L)	NA (1)	465	591 (4)	492–735

*Numbers in parentheses indicate numbers of samples taken for that parameter.

TABLE 8.1 Residential Subsurface Wetland Influent and Effluent Monitoring Data

needed in conjunction with the wetland to significantly reduce total nitrogen. The wetland system discussed in the following section was designed to accomplish that.

Some additional observations during the monitoring period for this wetland included formation of a layer of black suspended organic particulate matter at the bottom 3 to 6 in (8 to 15 cm) of the field dosing tank receiving the wetland effluent, and high nitrate levels in field monitoring samples collected. Care was taken to avoid this fairly well-defined "layer" when collecting samples from the pump tank with a bottom-fill bailer.

Based on the monitoring results and the above observations, several things might have been modified to improve overall performance:

- Increased wetland size, to as much as double the size used, along with a 2:1 rather than a 3:1 length to width ratio (2:1 is considered closer to optimal). The 3:1 ratio was used due to the small size of the wetland, and inadequate flow time and distance between the head and tail of the wetland.

- Reduce the water operating depth to about 18 in, to enhance oxygen transfer.

- Use of an effluent filter for the field dosing pump (to prevent clogging of field lines over time from any suspended matter exiting the wetland).

- Locate and use treatment media of 3/4 to 1 in (20 to 25 mm), instead of the 1 to 1.5 in (25 to 38 mm) found to be readily available locally.

- No lengths of half-pipe used over pressure-dosed field lateral lines, to increase evapotranspiration, and reduce the rate of downward effluent migration. Nitrate levels were high in the effluent samples collected from the field monitoring gravity lysimeters.

Models developed for wetlands may have limits to their applicability for smaller systems, such as those serving single residences. Much lesser data appears to have been gathered for smaller subsurface wetlands as compared with larger systems typically having more funding allocated to operations and monitoring activities.

8.2 Subsurface Wetlands Combined with Other Processes for Total Nitrogen Removal

As is also the case with upflow filters, the relatively anoxic conditions in subsurface flow wetlands make them well-suited for use in conjunction with an aerobic nitrification process to achieve low-effluent nitrogen levels through nitrification/denitrification processes. A project constructed at a City of Austin, Texas, treatment plant demonstrates this capability.[1] Areas to the west of Austin overlying the Edwards Aquifer, a rapidly recharged drinking water quality aquifer, are commonly very rocky and hilly with low natural soil treatment capabilities. A system consisting of a subsurface wetland functioning in series with a recirculating trickling filter demonstrated the ability to produce low levels of total nitrogen needed prior to land disposition of effluent in such sensitive watersheds.

Figure 8.11 is a conceptual illustration of this type of system used to serve a residential neighborhood. A gravity effluent collection system leads to a small mixing tank (not visible in the illustration), which drains to the head of the wetland. Wetland effluent then drains to the trickling filter, where media is dosed throughout the day. A recycle pump in the sump of the trickling filter operates on a timer and recirculates effluent back to the mixing tank at the head of the wetland. Treated effluent drains from the trickling filter to the final soil dispersal system as liquid is displaced in the treatment system (at the daily flow rate).

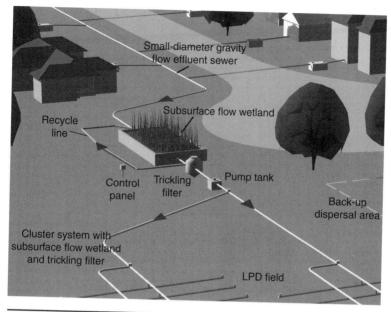

FIGURE 8.11 Conceptual illustration of a combination subsurface flow wetland and recirculating trickling filter system serving a clustered residential development. An effluent collection system serves the neighborhood, providing primary treatment at each residence.

Figure 8.12 shows a profile view of the demonstration system developed for Austin. Primary treated effluent enters the concrete mixing tank ahead of the wetland, passes through the wetland, and then drains by gravity into the trickling filter unit. The trickling filter is used primarily for nitrification, since the wetland treatment unit ahead of it produces relatively low BOD and TSS levels. The recirculation rate from the trickling filter back to the mixing tank ahead of the wetland was set at 3:1.

8.2.1 Wetland-Trickling Filter System Design

The recirculating wetland-trickling filter treatment system was designed for an average daily flow of 1000 gal/day (3785 L/day), and intended to demonstrate a process for achieving effluent total nitrogen levels of 5 to 6 mg/L, as well as very low levels of BOD and TSS.

A volumetric design model developed by Reed et al. (1995)[4] was used for this wetland design. An advantage of this type of model is that the design is based on average flow through the system, and this tends to compensate for water losses and gains from evapotranspiration and precipitation.[5] The model is based on input-output mass

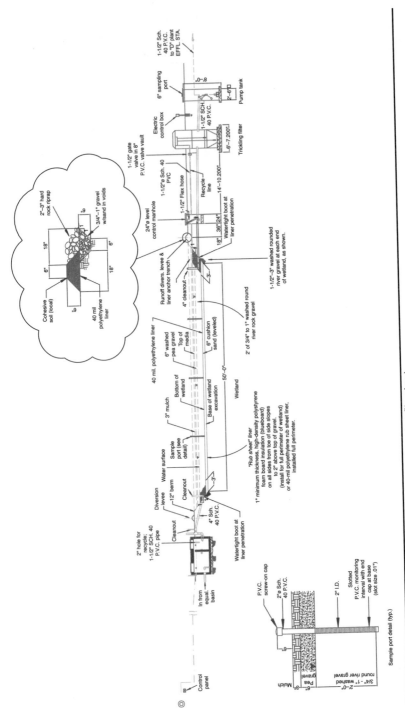

Figure 8.12 Recirculating wetland-trickling filter system (profile view).

248

balance relationships, and takes the general form of a first-order plug flow equation:

$$C_e/C_o = \exp(-K_T t) \quad \text{where } K_T = K_{20}(\theta)^{(Tw-20)}$$

and

$$t = LWyn/Q$$

The wetland treatment area is calculated from the following equation:

$$A_S = LW = Q_A[\ln(C_o/C_e)/(K_T yn)]$$

where C_e = wetland effluent concentration, mg/L
C_o = wetland influent concentration, mg/L
K_T = rate constant at temperature T, d^{-1}
T_w = average water temperature in wetland during period of concern, °C
θ = temperature coefficient at 20°C
A_s = treatment area (bottom area) of wetland, m^2
Q_A = average flow in the wetland,

$$= (Q_{In} + Q_{Out})/2, \text{m}^3/\text{d}$$

y = average depth of water in the wetland, m
n = porosity of the wetland, percent as a decimal
t = hydraulic residence time in the wetland, days

Quality data for the equalization basin was reviewed to determine expected levels of key parameters for the primary treated wastewater entering the system. Local historical weather data was used for temperature dependent design parameters needed for the design model.

Design parameters used for the wetland design included the following:

- $Q_A = 1000$ gpd = 134 ft^3/d = 3.79 m^3/d
- $\theta = 1.06$
- $K_{20} = 1.104$
- Influent BOD$_5$ (BOD$_o$) = 150 mg/L
- Influent TKN = 28 mg/L
- Winter air temperature = 38°F = 3.3°C
- Estimated water temperature = 50°F = 10°C
- $y = 2$ ft (0.61 m) of 3/4 to 1 in (20 to 25 mm) gravel, and $n \sim 0.38$
- A length-to-width ratio of 2:1 was used
- $t = 7$ days
- $T_w = 10$°C

Based on the above design parameters and treatment model, the wetland bottom area was calculated to be 1230 ft² (114 m²) to achieve an effluent with 5 mg/L total nitrogen, with a width of 25 ft (7.6 m) and length of 50 ft (15.3 m). A check was made for the available carbon sources for denitrification to occur in the wetland given that a C:N ratio of at least 5:1 is needed, and was found to be acceptable with a 7:1 ratio (521 lb/year, or 236 kg/year) calculated based on available carbon from BOD and plant residues. An available carbon of 30 percent was assumed for plant residues. Cattails were selected as the vegetation for the wetland for several reasons, including their ability to achieve a well-established root system the full depth of the wetland. Cattails are well-suited to the local climate and readily available from nurseries.

Primary treated wastewater from the municipal treatment plant equalization basin located near the demonstration system site was applied to the wetland. A sump pump was suspended via piping into the equalization basin, and operated with a timer in a nearby control panel to achieve the desired daily flow of 1000 gal/day (3785 L/day). Wastewater from the equalization basin was pumped into the first compartment of a 750-gal (2839-L) two-compartment tank. The second compartment of this tank served as a mixing chamber for primary wastewater and recycled wetland-trickling filter effluent. Effluent from this second chamber then drained by gravity to the inlet manifold piping in the wetland.

A 24-in (0.6-m) diameter manhole-type structure was used to control the water level in the wetland. Effluent drained by gravity from the wetland through an inlet pipe at the base of this structure that was connected to a flexible 1.5-in (40 mm) hose. This flexible hose was hooked to the sidewall of the manhole at a level such that the water surface in the wetland would be at the rock media and pea gravel interface. Wetland effluent drained out of this hose and exited through a pipe at the base of the structure, and flowed by gravity into the trickling filter.

An Ekofinn Bioclere modified trickling filter unit was used to nitrify effluent following the subsurface flow wetland. Although this is a proprietary treatment unit, nonproprietary trickling filter units can be constructed using locally available materials, and design methods available in wastewater engineering literature. For this application a Model 16/15 Bioclere unit was used based on system flows and performance needs. Wetland effluent drained by gravity to the clarifier in the Bioclere unit, and a single dosing pump automatically activated by a timer dosed the high-surface-area plastic filter media. Trickling filter effluent was recycled back to the head of the wetland (pumped from the base of the trickling filter unit to the mixing chamber ahead of the wetland) to accomplish denitrification. A pump test was conducted to measure the flow rate for the recycle line from the trickling filter unit to the concrete tank mixing chamber, for the purposes of setting the timer on the control panel. The pump timer was set to achieve a 3:1 recycle rate from the trickling filter to the mixing chamber and into the head of the wetland.

Effluent from the entire wetland-trickling filter treatment system drained from the trickling filter into a pump station from which the effluent was returned to the city's treatment plant for routine processing and discharge with the remainder of the plant flows. This final effluent pump was controlled on a demand basis by mercury float switches.

8.2.2 Wetland-Trickling Filter System Construction

This system was constructed at a site with relatively deep and easily excavated soils. It was therefore not necessary to construct a frame for the wetland, but rather to excavate a 1:1 side slope and use the natural soils. Six inches of sand bedding were placed under the liner, and lesser amounts along the sidewalls. Liner rub sheets were used along the sidewalls to protect the primary liner from any natural materials that might puncture the sides of the liner. However, in most ways the construction of this wetland was very similar to the one constructed for the residential system. Figures 8.13 through 8.20 show a series of photos illustrating key system elements and steps in the construction process.

8.2.3 Wetland-Trickling Filter System Performance

Tables 8.2 and 8.3 show results from monitoring the system over a 3-year period. During a portion of the monitoring period, the trickling filter was off-line due to control panel problems that were later resolved.

FIGURE 8.13 Soil at this site was found to be sandy loam and relatively free of rocks or other material that might damage the wetland liner. A 1:1 sidewall slope could be used since the soil was relatively stable, and the liner material was to be placed very soon following excavation (thus minimizing any opportunity for sidewall erosion).

FIGURE 8.14   The HDPE liner is shown installed here, with 10-ft strips of extra liner material ("rub sheets") placed at each end underneath the larger rock shown around the manifold piping. As with the residential wetland, about 12 in (0.3 m) of liner material was exposed to sunlight above the media, so a UV-resistant liner material was needed.

FIGURE 8.15   The wetland treatment media is placed and raked level here, with monitoring ports (12) and clean-outs showing at the surface. The liner is temporarily anchored down along the sides with gravel. An 18-in (0.46-m)-wide runoff diversion berm will be constructed to a height of about 6 in (15 cm) around the perimeter of the wetland to prevent site drainage from entering the wetland.

Figure 8.16 The level control structure for the wetland is shown here. The arrow points to the line marked in the plastic noncorrosive structure that corresponds to the operating water level in the wetland (at interface between pea gravel and treatment media). The 1-1/2 in (40 mm) hose connected to the inlet pipe is attached to a plastic chain, which is hooked to a stainless steel hook screwed into the wall of the structure. Just enough depth was provided at the bottom of the structure for collecting samples using a bottom-fill Lexan bailer (effluent exits through white pipe on left side of structure).

Figure 8.17 The recirculating trickling filter is shown here in the foreground. The arrow shows the pipe where effluent from the entire system will exit and drain to a pump basin with an access port for sample collection. The 1-1/2 in (40 mm) PVC pressurized recycle line from the trickling filter back to the head of the wetland is on the other side of the trickling filter (not shown), and leads to the 750 gal (2839 L) concrete mixing tank, where it mixes with primary treated effluent before draining into the wetland (see Fig. 8.19).

FIGURE 8.18 Cattails are being planted in the wetland here. The 6-in (15-cm) layer of pea gravel above the treatment media is dug out just enough to place the top of the root bulb at the interface between the pea gravel and coarser treatment media. That will also be the normal operating water surface for the wetland.

Mixing tank for effluent recycle and primary treated influent to system.

Bioclere recirculating trickling filter unit.

Concrete pump basin for sampling and return of effluent to city's treatment plant flow.

FIGURE 8.19 This photo shows all of the system components in place except for the level control structure which is just beyond the trickling filter unit and out of view. The control panel housing for the system is to the right of the trickling filter unit. A wood structure was built around the panels for added protection from wet weather, heat, and sunlight.

FIGURE 8.20 The subsurface flow wetland vegetation is well-established in this photo, and already in need of weeding to prevent woody and/or invasive plants from taking over. The level control structure can be seen here just to the right of the trickling filter at the far left.

Table 8.3 shows performance results for that period (trickling filter off-line), and Table 8.2 with the entire system operational.

The primary-treated influent to the wetland-trickling filter system had relatively low levels of BOD, TSS, and nitrogen species, and much lower than would be expected for septic tank effluent. However, comparisons of percentage removal for key treatment parameters show the importance of adding an aerobic treatment process capable of efficient nitrification.

Although the range of total nitrogen removal for the system when the trickling filter was off-line went up to about 95 percent, it's important to note that high levels of nitrogen removal occurred under those conditions only during early periods of operation having high plant growth rates. Levels dropped and leveled off substantially later in the project when again the trickling filter was off-line. If total nitrogen removal is tracked separately for those off-line (trickling filter off-line) periods, the effects of plant growth on the system's performance become evident. During the system's first year of operation when the cattails were newly planted and the trickling filter was off-line, total nitrogen removal averaged 88.3 percent. However, about 2 years later when the plants were fully developed, average total nitrogen removal under those same conditions dropped to 56.4 percent. Nitrogen removal performance tended to approach that of the residential system covered earlier in this chapter with the trickling filter off-line, although this wetland has about 25 percent greater hydraulic retention time and smaller average media size. Both of those factors together would be expected to contribute to better total nitrogen removal.

Measured Parameter	Influent (Primary Treated)		Wetland-Trickling Filter System Effluent		
	Average	Range	Average	Avg. % Removal	Range
BOD_5 (mg/L)	99.7	69–156	5.6	94.4	3.0–8.0
COD (mg/L)	229.2	161–300	39.4	82.8	13.4–170
TOC (mg/L)	60.4	36.7–114	17.6	70.9	5.6–56.2
TSS (mg/L)	46.0	23–74	2.5	94.6	0.5–12
NH_3-N (mg/L)	16.6	10.4–20.9	0.2	98.8	0.1–0.5
TKN (mg/L)	17.6	9.3–24.5	0.8	95.5	0.5–1.3
(NO_3+NO_2)-N (mg/L)	0.1	0.0–0.2	2.3	—	1.2–2.7
Total Nitrogen (mg/L)	17.7	9.3–24.6	3.1	82.5 (70.7–90%)	1.9–3.9
pH (standard units)	7.3	7.2–7.3	8.0	NA	8.0–8.1
Chloride (mg/L)	102.3	79.5–185	100.8	1.5	74.9–131
Alkalinity (mg/L)	230.4	205–270	190.1	17.5	138–223

TABLE 8.2 Wetland-Trickling Filter System with Trickling Filter Operating

Measured Parameter	Influent (Primary Treated)		Wetland Effluent		
	Average	Range	Average	% Removal	Range
BOD_5 (mg/L)	98.3	50–206	6.1	93.8	1.1–24.0
COD (mg/L)	218.6	19–384	31.6	85.5	10.2–135
TOC (mg/L)	56.2	25.5–144	20.4	63.7	3.8–247
TSS (mg/L)	41.1	4.8–98.0	1.7	95.9	0.5–7.6
NH_3-N (mg/L)	17.7	4.3–27.5	5.9	66.7	0.1–17.4
TKN (mg/L)	19.9	10.2–29.5	5.7	71.4	0.3–20.4
$(NO_3 + NO_2) - N$ (mg/L)	0.1	0.0–0.6	0.7	—	0.0–3.1
Total nitrogen (mg/L)	20.0	10.2–29.5	6.4	68.0* (15.1–95.1%)	1.2–20.4
pH (standard units)	7.1	6.7–7.5	7.6	NA	7.0–8.1
Chloride (mg/L)	88.2	71–170	86.0	2.5	65.0–104.0
Alkalinity (mg/L)	216.7	168–292	204.5	5.6	105–308

*See discussion in Sec. 8.2.3 earlier regarding monitoring periods and these results.

TABLE **8.3**   Wetland-Trickling Filter System with Trickling Filter Off-Line

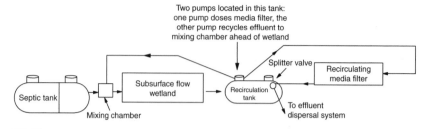

Two pumps located in this tank:
one pump doses media filter, the
other pump recycles effluent to
mixing chamber ahead of wetland

Splitter valve

Recirculating
media filter

Subsurface flow
wetland

Septic tank

Recirculation
tank

Mixing chamber

To effluent
dispersal system

Figure 8.21 Conceptual flow diagram of a combined subsurface flow wetland and recirculating media filter (unsaturated aerobic filter) system for providing high levels of total nitrogen removal.

During periods when the trickling filter was in operation, total nitrogen removal ranged from 70.7 to 90 percent, and averaged 82.5 percent. Average nitrification levels were much better with the trickling filter online. Other secondary treatment parameters including BOD and TSS levels were very comparable for periods when the trickling filter was on- and off-line. Overall the wetland's performance showed its ability to reliably reduce BOD and TSS to secondary treatment levels, but its nitrogen removal capabilities were limited unless operating in conjunction with a nitrification process.

Properly located, designed, and constructed subsurface flow wetlands in combination with nitrification-efficient processes can be expected to reliably produce very low effluent nitrogen levels where needed prior to final effluent disposition. Based on their demonstrated nitrification capabilities, two other treatment processes that would be expected to produce effluent with very low total nitrogen levels are recirculating gravel and synthetic textile media (e.g., AdvanTex®) filters in combination with wetlands.[6] The conceptual flow diagram in Fig. 8.21 shows that possible system configuration.

Chapter 9 provides general recommendations for levels of treatment needed for varying depths of soil of the types discussed in Chap. 3 for certain methods of effluent dispersal. Recommended effluent loading rates to the soil are also provided, based on method of dispersal, soil type, and level of predispersal treatment.

References

1. "Alternative Wastewater Management Project, Phase II," Final Report, Prepared for the City of Austin, TX, Water Utility by Community Environmental Services, Inc., August 2005.
2. A. Y. Richter and R. W. Weaver, "Phosphorus Reduction in Effluent from Subsurface Flow Constructed Wetlands Filled with Tire Chips," *Environmental Technology*, 24(12):1561–1567, Dec. 2003.

3. *Standard Methods for the Examination of Water and Wastewater, 20th Edition,* prepared and published jointly by the Water Environment Federation, the American Water Works Association, and the American Public Health Association, Washington, DC, 1998.

4. *Natural Systems for Waste Management and Treatment, 2nd Edition,* S. C. Reed, R. W. Crites and E. J. Middlebrooks, McGraw-Hill, Inc., United States, 1995.

5. *Natural Systems for Wastewater Treatment,* 2nd ed., WEF Manual of Practice FD-16, 2001).

6. M. E. Byers, K. E. Zoeller, J. D. Fletcher, "Recirculating Gravel Filter for Secondary Treatment of Septic Tank Effluent and Use of Time-Dosed Wetlands to Remove Nitrate from Gravel Filter Effluent," *On-Site Wastewater Treatment Proceedings of the 10th National Symposium for Individual and Small Community Sewage Systems, ASABE,* 2004.

CHAPTER 9

Effluent Dispersal Methods

T he method of effluent soil dispersal used for a decentralized wastewater system is critical to its overall performance, cost-effectiveness, and sustainability. The dispersal method should be matched with an appropriate level of predispersal treatment for proper functioning of the overall system. This will provide pollutant levels that are acceptably low for treated effluent combining with groundwater or surface water supplies. While predispersal treatment processes are often considered "the treatment system," they are actually only a portion of the full treatment system for those not directly discharging to surface waters. For wastewater systems relying on land/soil disposition of effluent where natural soil biochemical processes can contribute significantly to or be the principal means of pollutant attenuation, it's important to try to maximize opportunities for that to occur. Some dispersal methods have been found to facilitate those natural treatment processes better than others, at least for certain wastewater constituents of concern such as nitrogen and pathogen reduction. This chapter provides comparisons of several basic effluent dispersal methods that are commonly used, with again a focus on factors most affecting their long-term sustainability.

9.1 Basic Methods of Effluent Dispersal

Five basic methods of effluent dispersal are described and compared in this chapter. Those include

- Gravity flow subsurface dispersal beds and trenches
- Subsurface low-pressure dosing (or "low-pressure pipe" dispersal)
- Subsurface evapotranspiration (ET) beds
- Surface or subsurface drip dispersal/irrigation
- Surface irrigation/application of effluent

There are a number of design variations for the above five methods, with many proprietary products developed in efforts to address perceived limitations or disadvantages with some methods. However, most effluent dispersal methods used today would fall into one of those five basic types. Basic descriptions of each category of dispersal system are provided below, with emphasis on aspects that tend to affect sustainability.

While "leach pits" continue to be used in some parts of the world, they are not considered sustainable due to their inability to adequately protect groundwater supplies. Leach pits are relatively deep excavations where wastewater is allowed to pond and seep into the ground. Wastewater effluent applied to subsurface horizons that are well below plant root zones and aerobic soil conditions would not be expected to be adequately treated for nitrogen, pathogens, or other pollutants of concern. Leach pits are therefore not presented as a viable method of final wastewater disposition here.

9.1.1   Gravity Subsurface Beds and Trenches

This represents the oldest of the five dispersal categories, using no pumps or electric power or other method of dosing for the delivery of wastewater effluent to the soil. A conventional gravity trench cross section is illustrated in Fig. 9.1. This dispersal method uses "drain" rock at the base of the excavation, with perforated pipe bedded into that rock some distance above the bottom of the excavation. Geotextile fabric is placed above the rock to prevent soil backfill/cover placed over the rock from migrating into and clogging the rock layer/media. Gravity flow trenches and beds may be used for soil application of septic tank effluent or wastewater treated to higher levels.

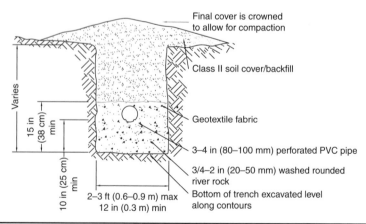

FIGURE 9.1   Conventional gravity trench detail (cross-sectional view, not to scale).

Common modifications to gravity drain fields and trenches are chambers or gravelless pipe products used instead of gravel in the trenches or beds. Some manufacturers of proprietary chamber products have claimed that gravel in trenches has a "masking," or "shadowing" effect on the soil, posing a barrier between downwardly migrating effluent and soil at the base of the excavation. Claims about these effects have been used as the basis for soil dispersal field sizing reductions allowed for chamber systems as compared with conventional gravel beds/trenches for a number of U.S. state regulatory programs. Research done by onsite systems experts at the University of Wisconsin-Madison and the University of Georgia-Athens, however, failed to show any evidence of that phenomenon.[1,2] Surface tension interactions between applied effluent and rock/gravel media in trenches and beds appear to negate any such tendencies for "shadowing" phenomena to occur.

Chambers may also reduce evapotranspiration that might more readily occur in rock/pipe trenches or beds, as long as backfill/cover soils used for conventional beds/trenches are low in clay content. If ponding of effluent along the infiltrative soil surface (base of chamber) occurs, saturated soil conditions may significantly reduce treatment capabilities, including nitrification/denitrification processes. Chamber systems are not recommended for use in fine-textured soils or clays (class IV textural category). However, use of chambers in more rapidly infiltrating soils raises concerns about too little retention time in the soil for treatment processes to occur, and possible groundwater impacts from nitrogen, pathogens, and/or other contaminants. The use of chambers may lessen natural ET rates and greater moisture retention (and time for treatment) from surface tension phenomena that occur with gravel/media-filled trenches. Rock or gravel placed in conventional or pressure-dosed trenches would also be expected to offer at least some added fixed film treatment, comparable to trickling filters.

Research comparing chamber systems with conventional beds/ trenches does, however, suggest that the presence of significant amounts of fines from gravel may reduce infiltration rates along bottoms of beds or trenches.[1] This underscores the need to use only well-washed rock or gravel free of fines, with sufficient hardness so that the aggregate doesn't crumble during installation or tend to dissolve when subjected to long-term effluent application.

For sites where it is difficult, due to distance from aggregate supplies, or terrain where it is difficult to move aggregate to the dispersal field site or place it, gravelless pipe products have been developed that may be used instead of gravel for beds or trenches. One of those products was mentioned in Chap. 8, EZ Flow™, which uses recycled polystyrene "nuggets" bound into rolls with netting around perforated distribution piping. That product is very light weight, and rolls can be placed on a pickup truck or even carried around sites. Use of

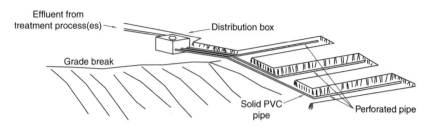

Figure 9.2 Conventional gravity trench layout along hillside using a distribution box for effluent delivery to trenches.

this type of material would also be favorable for minimizing infiltration rate reductions from heavy aggregate compacting soil in trenches/beds and fines in the aggregate. There may be concerns about useful service life and long-term performance of polystyrene used for this purpose. Those concerns would, however, need to be weighed against potential site disruption, transportation costs, and energy usage associated with using a much heavier natural aggregate product. Chipped tires have also been successfully demonstrated for use in beds and trenches instead of rock/gravel aggregate.[3]

As noted in Fig. 9.1, gravity flow trenches are constructed level along contours. Two types of structures commonly used to aid with distribution of effluent for gravity drain fields are drop boxes and distribution boxes. Figure 9.2 illustrates the use of a distribution box, typically constructed of plastic or concrete, and with each of the three outlets leading to lateral lines laid level along three different elevations on the hillside. Where settling may occur, as is the case for most sites, maintaining level conditions at distribution box outlets is often a problem. This may lead to overloading and failure of downhill lines with no effluent reaching upper lateral lines.

For adjacent trenches and drain/lateral lines, turned-up pipe fittings can be used to allow overflow from upper to lower trenches, as illustrated in Fig. 9.3a. Drop boxes are similar to using turned-up pipe fittings to overflow effluent to adjacent downhill lateral lines, and are illustrated in Fig. 9.3b. Drop boxes are typically made of plastic or concrete, with an inlet, two outlets to perforated drain lines (at the same elevation), and an outlet (at slightly higher elevation than trench outlets) that conveys overflow into the adjacent downhill trench or drop box. While distribution boxes have the advantage of distributing effluent to multiple trench lines, they are very vulnerable to problems from settling and unlevelness. Drop boxes and overflow/relief pipes tend to lessen this problem, but have a different disadvantage of not distributing effluent to adjacent trenches until a trench is filled to the overflow elevation. This tends to result in "creeping failure" for gravity flow dispersal fields. Individual trenches become used more than others, remain saturated and accumulate biomat clogging layers along the trench bottom and soil interface, and fail prematurely.

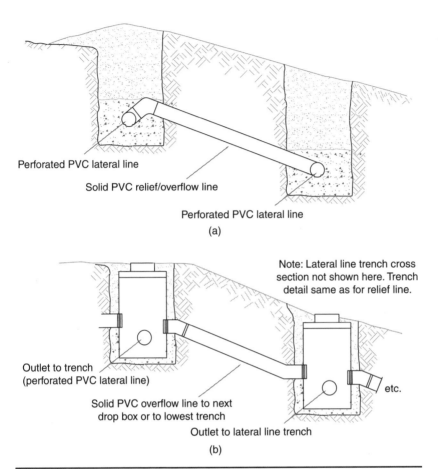

Perforated PVC lateral line

Solid PVC relief/overflow line

Perforated PVC lateral line

(a)

Note: Lateral line trench cross section not shown here. Trench detail same as for relief line.

Outlet to trench
(perforated PVC lateral line)

Solid PVC overflow line to next drop box or to lowest trench

Outlet to lateral line trench

etc.

(b)

FIGURE 9.3 (*a*) Relief/overflow lateral line side view (not to scale), (*b*) multiple drop boxes side view (not to scale).

With these types of problems for both types of components used to distribute effluent by gravity between multiple adjacent trenches, it is not surprising to find large sections of gravity fields with drain rock never having received any wastewater effluent at all, and other portions of fields completely saturated and failed. The second type of dispersal method discussed below was developed and is now commonly used to overcome these types of problems with gravity flow dispersal, and achieve relatively uniform effluent distribution.

9.1.2 Low Pressure Dosing

Low pressure dosing (LPD) systems are used to deliver septic tank effluent or higher quality effluent to subsurface dispersal fields, and can achieve relatively uniform distribution of effluent across the dispersal area when properly designed and installed. A pump (or multiple pumps) doses trenches intermittently, either on a "demand" basis with the water level and float switch activating the pump, or on a timed basis using timer settings in a control panel. Some nonelectric methods of dosing may also be used, and are described later in this chapter. The dosing and resting cycles associated with this dispersal method have been found to be very effective for achieving higher levels of soil treatment, particularly for nitrogen and pathogen reduction.[4-6] Dosing/resting cycles help to maintain aerobic conditions in the soil/trench interface, which is necessary for certain natural treatment processes. Uniform distribution of effluent across the dispersal area also greatly extends the useful service life of the field as compared with gravity flow systems, by avoiding oversaturating sections of the field and resulting failure.

In the 1970s and 1980s design approaches were developed for LPD systems, sometimes called low-pressure-pipe, or LPP dispersal, that have proven to be very successful and offer long useful service lives as long as adequate pretreatment is provided. A 1982 publication still routinely referred to, and that presents a sound stepwise design methodology, is *Design and Installation of Low-Pressure Pipe Waste Treatment Systems.*[7] The design approach detailed in that publication is best suited to smaller/residential-scale systems because it does not include certain information needed for sizing pumps and serving multiple fields with longer lengths of manifold piping between pump stations and the dosed field area. However, the basic design concepts and approach are clearly presented. Because it is considered a particularly sustainable method of final effluent dispersal, and because it is a nonproprietary method without benefit of vendor and manufacturer support for design and installation, Chap. 10 covers low-pressure-dosing systems design and installation in much further detail.

LPD/LPP trenches are typically shallower and narrower than conventional gravity trenches, allowing for more aerobic soil conditions, application of effluent into the plant root zone, and use of trench sidewalls along with bottoms of trenches for effluent infiltration. Plant root zones typically extend to about 20 to 24 in (50 to 60 cm) below the ground surface. LPD trenches should be aligned along contours and cut so that the bottom infiltrative surfaces of the trenches are still in this zone. Since a suitable (e.g., class II) final soil cover may need to be trucked to the site anyway, trenches may be cut as shallow as 12 to 15 in (30 to 38 cm) below natural grade, and soil cover placed over the entire field so that manifold/delivery piping still has enough soil cover. For sites having significant slopes, it is critical in this case to use sufficient erosion/sedimentation controls and some method of blanket-seeding that can quickly establish a vegetative cover over the field to prevent erosion of the newly placed soil. Chapter 10 covers these types of considerations for LPD field construction in much more detail.

Figure 9.4 shows a typical cross section of an LPD trench. Note that the distribution piping is embedded near the surface of the drain rock, so as to (1) provide a significant depth of rock for further treatment as effluent trickles across the surface of the rock where biomat will form

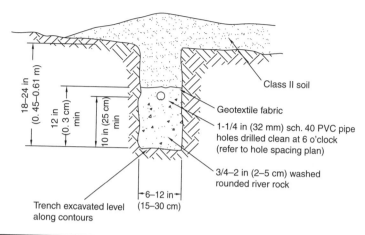

Figure 9.4 Typical LPD trench cross-sectional detail. A 5/32 in (4 mm) drilled orifice size is commonly used for LPD lateral lines. Trenches may be excavated either with a digging bucket or using a trencher or rock-saw. Details of LPD design and construction are discussed in Chap. 10.

over time, and (2) avoid submerging the distribution pipe if effluent drains slowly in some part or all of a trench due to either prolonged rainfall and saturated conditions, or biomat formation along the bottom of the trench. The same applies to conventional gravity dispersal trenches and beds. Figure 9.5 shows an LPD field ready for backfilling/placement of final soil cover.

Figure 9.5 LPD trenches are ready for placement of the final soil cover in this photo. The two CPVC valves at the upper left corner are used to set pressure head for the two field zones. CPVC was used here due to its corrosion resistance. Turnups of PVC lateral lines are installed at each end of the perforated lateral lines to allow flushing of lines if needed in the future. Geotextile fabric is placed above the rock/gravel and pipe to prevent soil cover from entering and clogging the rock or piping. Orifice holes are facing downward into the rock/gravel (6 o'clock) for this system.

Where slopes on a site offer enough elevation difference for dosing siphons to be used, LPD fields located downslope of buildings/facilities served can be dosed without the use of a pump. The use of dosing siphons and other nonelectric dosing devices, such as the Flout™, and tipping buckets are described later in this chapter.

Monitoring results for low-pressure-dosed trenches, even in soils considered marginal, have shown that very good nitrogen removal is possible. Studies performed for the City of Austin, TX, on LPD systems constructed in central Texas hill country limestone-based rocky soils showed average total nitrogen levels ranging from 2 to 4.0 mg/L over a period of about 2 years for four LPD systems. During dry weather periods, little to no percolate could be collected from the gravity lysimeters, suggesting that most of the percolate moisture was removed through ET processes and/or retained in the upper soil horizons. By contrast for the same hill country conditions, average total nitrogen measured for gravity lysimeters monitoring a conventional gravity flow drain field was 8.3 mg/L. All systems used only septic tank pretreatment prior to subsurface dispersal, and all systems served single family residences with typical (and nonseasonal) use patterns.[8]

Pathogen reduction appeared less reliable for LPD systems applying septic tank effluent in those soils based on fecal coliform monitoring in those same studies. Fecal coliform measurements for lysimeter samples ranged from 0 to 185,000 colonies/100 mL for four systems monitored. Averages for those systems during the 2-year period ranged from 475 to about 40,000 colonies/100 mL, indicating highly variable performance by site despite the apparent similarity of subsurface conditions. These soils tend to have very pronounced lateral bedding planes and have either naturally occurring preferential flow patterns or develop them over time with the dissolution of limestone. Without significantly more detailed long-term (and costly) study, it's difficult to evaluate what effects rainfall dilution, surface pollutants such as animal feces and other factors may have had on the monitored values observed for the various constituents. However, based on differences in total nitrogen levels observed for gravity flow and LPD systems, it appears that the various mechanisms relied on for soil treatment (including vegetative uptake and nitrification/denitrification) are in the aggregate improved with dosing and resting dispersal cycles.

A separate University of Texas at Austin study performed on systems located in hilly limestone-based ("caliche") soils to the west of Austin showed average total nitrogen of 5.2 mg/L from an LPD system (applying septic tank effluent). Monitoring results for a conventional gravity dispersal system (also applying septic tank effluent to trenches) located nearby in similar conditions showed average total nitrogen levels of 10.6 mg/L.[9] While only a limited number of samples were collected from those two systems over a period of months (three samples each), the conventional gravity flow system had maximum TKN and total nitrogen values of 17.3 and 20.4 mg/L, respectively, while the LPD system had maximum values of 4.5 and 9.0 mg/L for those same parameters, respectively. Results from other studies are consistent with these, showing that LPD systems with dosing/resting cycles, suitable application rates and relatively shallow trench depths can reliably provide high levels of total nitrogen reduction.[10,11]

Recommended application rates for different textural soil classes, dispersal methods, and levels of predispersal treatment are presented later in this chapter. Properly designed and installed LPD systems using dosing/resting cycles for alternating field zones, and suitable soil loading rates and pretreatment, should be expected to last at least 30 to 40 years. Much longer service lives would be expected where reliable methods of secondary or advanced predispersal treatment are provided. Relatively inexpensive components such as pumps, valves, or control panels would almost certainly need replacement periodically. However, the costliest part of the system—the distribution trenches with rock and pipe—should serve for a very long time with proper routine care.

Figure 9.6 shows a large LPD field located in a state park after the final grass cover was established (same system as shown before final cover in Fig. 10.18). Notice the faint stripes along the field contours. The darker green stripes throughout the field indicate that both moisture and nutrients are being uptaken by the grass over the LPD trenches. These trenches were as deep as about 24 in (0.6 m),

Figure 9.6 Large LPD field with final grass cover.

so evapotranspiration and nutrient uptake would be even greater for lines buried at shallower depths, with less distinguishable "striping" due to more spreading out of the moisture into root systems in the upper soil horizons.

Some of the common modifications discussed for gravity trenches and beds may also be used for LPD trenches, including gravelless pipe products (e.g., EZ Flow), chipped tires, and chambers (with pressure pipes suspended on the interior of chambers for LPD applications).

"Mound" dispersal systems can be considered a modification to LPD subsurface trench dispersal systems, and are constructed above natural grade using select fill material, although they may also be dosed using drip dispersal as described below. Where there's ample elevation difference between the pretreatment tank(s) and the mound, nonelectric dosing devices can also be used.

Mounds are used to overcome limiting site conditions, such as unsuitable soils, or shallow rock or groundwater. Mounds are essentially raised infiltration beds constructed using soils that offer much better treatment capabilities than the natural underlying soils. They rely on both evapotranspiration and infiltration, though for many systems and depending on climate, the infiltration rate through sandy soils used for mounds may well exceed ET rates. It's therefore important when designing mounds to remember that the areal loading rate used should match the rate at which the underlying natural soil is able to accept infiltrating moisture from the mound. Roughening or "scarification" of the natural ground surface throughout the base of the liner is also important to facilitate infiltration from the mound. This effluent treatment/dispersal method is not usually recommended for areas of sites with steeper slopes, due to slope stability and the potential for effluent "breakout" along their edges. Design recommendations for the construction of mound systems can be found on this book's website.

Sand-lined trenches are another modification to traditional LPD trenches, and are used to provide treatment beyond what would be expected from the natural subsurface conditions. Use of sand-lined trenches might be used for providing added pathogen reduction. However, monitoring has shown high nitrate levels may be expected unless predispersal treatment methods have reduced total nitrogen to acceptable levels prior to subsurface dosing of the effluent.[12] An example of the use of sand-lined trenches will be discussed in Chap. 10. Bottomless raised beds or gardens are a similar modification to LPD dispersal systems, and are comparable to mound systems, though enclosed around the perimeter using masonry, concrete, or some other type of retaining structure. An example of this approach will be discussed in Chap. 11 for one of the projects, with raised sand beds used instead of trenches.

9.1.3 Dosing Siphons and Other Nonelectric and Passive Dosing Methods

In areas with either limited power availability for wastewater treatment and dispersal components, where there are concerns about power supply reliability, or it's otherwise undesirable to use electric power for field dosing, some non-electro-mechanical devices have been used to accomplish intermittent dosing to both treatment and dispersal systems. The nonelectric methods discussed in this section rely exclusively on the "demand-dosing" approach, and so do not have some of the operational capabilities possible with timed dosing systems. Despite those potential limitations, passive methods that can reliably accomplish intermittent delivery of effluent to a treatment process or soil dispersal system have some advantages over continuous gravity flow systems for reasons discussed previously. Three nonelectric dosing methods that have been used primarily for smaller-scale onsite wastewater systems are dosing siphons, tipping buckets, and the Flout device, with siphons used most commonly of those devices. The rate of effluent delivery to the dispersal field for all three of these methods depends on the rate that wastewater enters the tank or chamber housing the device.

Dosing Siphons

The basic elements and functioning of dosing siphons are presented in Figs. 9.7 and 9.8. A good explanation of designing dispersal systems using dosing siphons is presented in an article by Eric Ball, P.E., and

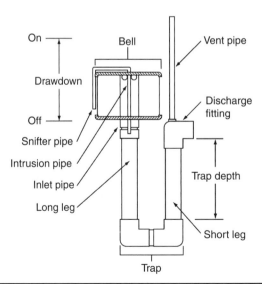

FIGURE 9.7 Basic parts of a dosing siphon. (*Image courtesy of Orenco Systems, Inc.*)

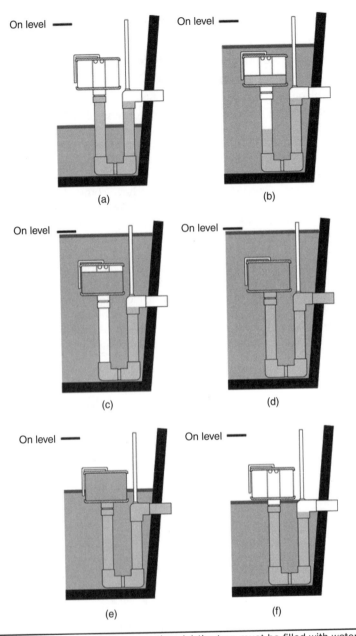

FIGURE 9.8 Dosing siphon operation: (*a*) the trap must be filled with water (primed) before raising liquid level above bottom of snifter pipe; (*b*) as the effluent level rises above the snifter pipe, effluent is pushed down in the long leg and air is released through the vent pipe; (*c*) just before dosing is triggered, the water level in the long leg is near the bottom of the trap, and the liquid level has almost reached the bottom of the vent pipe; (*d*) the dosing siphon is triggered when the liquid level reaches the bottom of the vent pipe, and effluent discharges from the tank; (*e*) the siphon continues to dose until the water level drops to the bottom of the bell; (*f*) air under the bell "breaks" the siphon; the snifter pipe ensures full recharge of air under the bell. (*Images courtesy of Orenco Systems, Inc.*)

is available for download under Resources and Helpful Links on the website for this book.

Dosing siphons require sufficient elevation drop between the dosing tank and the dosed system component to function, with siphons only discharging to lower elevations. When installed they must be primed to be able to operate (see Fig. 9.8*a*). As discussed in the above-referenced article, effluent transport lines leading to dosing siphons should have very smooth, constant alignments and transitions to avoid trapping air in the line, which will cause malfunctioning of the siphon.

Well-built siphons manufactured today are constructed of non-corrosive components with no moving parts, and have very few maintenance needs. They can, however, be subject to some operational problems such as loss of prime and lapsing into a trickling rather than a dosing mode. It's important that the effluent be filtered and sufficiently free of particulate matter to prevent those types of problems. An effluent of sufficient quality that biomat doesn't tend to build up significantly on components critical to a siphon's operation is preferred for continuous reliable functioning. Regardless of effluent quality dosed through a siphon, they should be regularly checked for proper functioning. Dosing counters are recommended to best monitor their operation.

Tipping Buckets, or Tipping Pans

Tipping buckets have sometimes been used as an onsite wastewater system dosing device, both for applying effluent to treatment processes (e.g., peat biofilter units) and to dispersal field lines. They are more commonly employed for metering flows and rain gauges. Figure 9.9 illustrates the use of a tipping bucket for intermittently dosing effluent into gravity dispersal field lines.

The size of bucket or pan used for this dosing method determines the volume of effluent discharged. The wedge-shaped bucket is mounted on a hinge or axle near the centerline of its balanced weight.

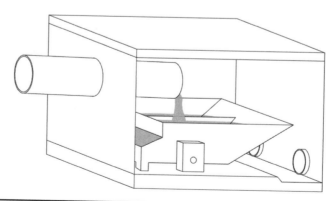

FIGURE 9.9 Conceptual illustration of a tipping bucket dosing manifolds to two dispersal field lines.

As effluent flows into the chamber and fills the bucket, the weight of the effluent tips the bucket forward, and its contents spill over into and out of the dosing chamber through the discharge pipe(s). Tipping buckets, hinges or axles, and other components associated with their operation in a wastewater environment should be constructed of noncorrosive materials.

The Flout System

The Flout system is another method of nonelectric/passive effluent dosing that relies on buoyancy and a relatively simple device for its operation. Figure 9.10 shows how a Flout operates, with a photo of a basic unit shown in Fig. 9.11. Unlike dosing siphons, this device does not need to be primed. A flexible coupling connecting the dosing tank's discharge piping to the Flout allows the buoyed Flout to rise from the bottom of the tank a certain amount as the liquid level rises in the tank. When the water level is high enough, liquid flows into the Flout body, causing it to lose buoyancy and sink back to the bottom of the tank. Liquid flows out of the tank through the piping until the level again recedes. The cycle then starts over again.

One of the project examples described in Chap. 11 uses Flouts for dosing both treatment and dispersal portions of a system.

9.1.4   Evapotranspiration Beds

Evapotranspiration (ET) beds are either synthetically lined dispersal beds, or dispersal beds constructed in clay soils having very, very low effluent infiltration rates. These systems effectively rely entirely on evaporation and transpiration as the means for handling the water component of wastewater. Lined ET beds have historically been used in geophysical locations with unsuitable subsurface conditions for soil treatment processes, including shallow depths to rock or groundwater. Wastewater effluent may be either applied to the bed by gravity or pumped.

ET beds have demonstrated very mixed performance due mainly to two key factors affecting their long-term use. Those are (1) climate and seasonal rainfall, and (2) the buildup of certain wastewater constituents including salts that can adversely impact vegetation needed to maintain aerobic conditions and contribute to ET processes. Even in parts of the southwestern United States with relatively low annual rainfall averages, there tend to be intense rainfall events or periods that saturate ET beds. Where ET beds have been used overlying rock and sensitive groundwater conditions, property owners sometimes report puncturing liners to allow beds to drain during or following periods of significant or extended rainfall. This type of system would be expected to have a significantly reduced useful service life as compared with dispersal methods that allow for infiltration of effluent as well as evaporation and transpiration processes. Since ET beds rely entirely on evaporation and transpiration, they must also be sized larger than systems that also rely on infiltration of effluent.

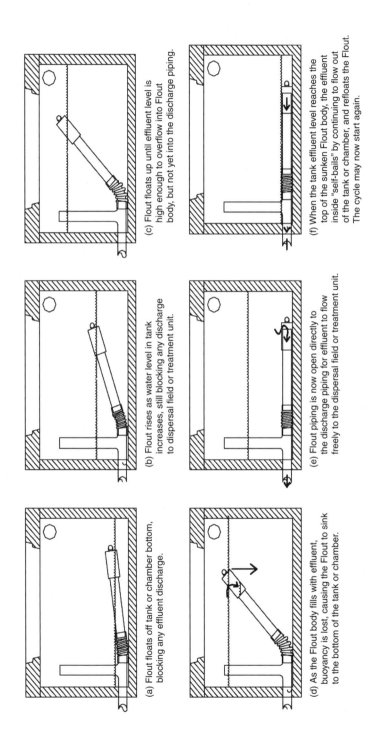

(a) Flout floats off tank or chamber bottom, blocking any effluent discharge.

(b) Flout rises as water level in tank increases, still blocking any discharge to dispersal field or treatment unit.

(c) Flout floats up until effluent level is high enough to overflow into Flout body, but not yet into the discharge piping.

(d) As the Flout body fills with effluent, buoyancy is lost, causing the Flout to sink to the bottom of the tank or chamber.

(e) Flout piping is now open directly to the discharge piping for effluent to flow freely to the dispersal field or treatment unit.

(f) When the tank effluent level reaches the top of the sunken Flout body, the effluent inside "self-bails" by continuing to flow out of the tank or chamber, and refloats the Flout. The cycle may now start again.

Figure 9.10 The Flout's basic operational cycle is shown here. (*Image courtesy of L.I.Z. Electric*).

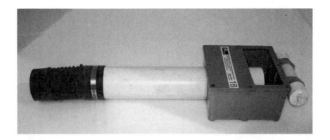

Figure 9.11 This photo shows a basic Flout assembly. Effluent enters through the end of the PVC pipe shown in the opening at the front of the device. (*Photo courtesy of L.I.Z. Electric*).

9.1.5 Subsurface Drip Dispersal

Drip irrigation/dispersal systems originally began to be used for agricultural applications, particularly in drier climates where water conservation and reuse are critical. Drip dispersal of wastewater effluent has become increasingly used for decentralized wastewater systems during the past two decades, particularly in geographic areas where wastewater reuse is desirable, and for sites having steep slopes and/or shallow soil depths. Surface application through drip emitter lines has also been used for decentralized wastewater systems, though to a much lesser extent than subsurface application. Like LPD systems, this method is also capable of delivering effluent uniformly across a dispersal field, and in smaller doses as compared with LPD distribution. While drip dispersal systems have been used for subsurface application of septic tank effluent, this is not recommended here due to the potential for excessive buildup of biomat along the soil/drip line interface observed in the field, and greater potential for clogging of emitters and/or excessive maintenance to prevent clogging.[13] Figures 9.17 and 9.18 illustrate those concerns.

There are two basic types of drip emitter tubing used: pressure compensating and nonpressure compensating. Pressure compensating emitters allow greater flexibility of installation along hillsides, so that lines do not need to be laid along contours.

Several key features distinguish pressure distribution using drip tubing from LPD effluent dispersal systems. Those include

- Thin-walled flexible tubing with very small emitter holes is used for drip dispersal, in contrast to Schedule 40 PVC pipe with larger (about twice the diameter) drilled holes used for LPD systems.

- Gravel trenches are typically not used for drip lines, which are either plowed directly into the soil using special vibratory plowing equipment, or placed after trenching.

- Drip line burial depth is usually 6 to 12 in (0.15 to 0.3 m).

- Although it is recommended that drip lines be placed along contour lines, there is more flexibility with pressure-compensating drip lines. LPD lines must be laid along contour lines.
- Drip lines must be flushed regularly to prevent clogging.
- Effluent must be filtered prior to dispersal through drip lines, due to the tiny emitter holes and potential for clogging.
- Recommended loading rates for drip dispersal are lower than for LPD dispersal, resulting in greater area requirements for drip for the same design flows. Some U.S. states, in recognition of this limitation, allow dispersal field size reductions for LPD/LPP dispersal where predispersal treatment to certain levels is provided, with drip dispersal given either no such "credit" or lesser amounts of sizing credit for the same levels of predispersal treatment. Concerns about long-term performance and soil loading rates used for effluent dispersal with drip systems have lead regulators in several states to call for more research on that topic.[14] Recommendations for soil loading rates are presented later in this chapter.

Figure 9.12*a* and *b* shows a section of pressure-compensating drip tubing.

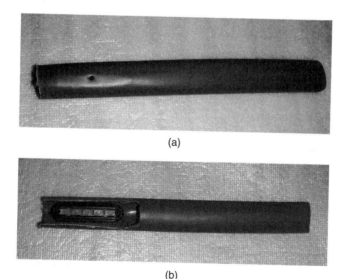

(a)

(b)

FIGURE 9.12 (*a*) Section of drip tubing showing effluent emitter hole. The tubing is less than 3/4 in (20 mm) wide, and the emitter hole diameter about 1/16 in (1.6 mm) (*b*) Other side of the same section of drip tubing showing the pressure compensating diaphragm where the emitter hole is located, and through which the effluent must pass to exit the tubing.

In addition to some type of predispersal treatment system, drip dispersal systems use the following basic components. Italicized components are those used for drip systems that would not be needed for most LPD systems, except under special site and design conditions.

- *Drip lines*
- Pumps
- *Filters*
- *Flow meter*
- Control system
- Supply line and manifold
- *Return manifold and line*
- *Flexible hose*
- *Air/vacuum valves*
- *Flush valve*
- *Specialty connectors and fittings*
- Valve boxes
- *Pressure regulators*
- Zone valves (may be needed for multiple dosing zones)
- Check valves (may be needed for multiple zones)

As seen from the list of components, pressure distribution of effluent using drip dispersal is significantly more complex than LPD dispersal, and requires a higher level of ongoing maintenance by persons having sufficient training and experience. Because of the shallow burial depth typically used for drip lines along with the relatively thin-walled tubing, factors such as cold weather, root intrusion that may clog emitters, and preventing damage to drip lines and effluent "breakout" from digging rodents may need to be considered for designs. Some manufacturers of drip tubing have incorporated herbicides and other elements into their product lines that help address these types of potential problems. Burial depths of up to 2 ft (0.6 m) may be needed for cold climates. In general, greater vulnerability to surface activities and animals should be considered if drip dispersal is to be used. Figures 9.13 and 9.14 show a photo of the effluent delivery manifold with drip dispersal lines placed and ready for final cover. Figures 9.15 and 9.16 show an air relief valve located at the end of a field zone, and a manual disc filter located ahead of the field zones (beyond treatment units), respectively.

Drip dispersal systems have tended to be used more in some U.S. states to meet required vertical separation distances between bottoms of distribution line excavations and rock or groundwater. The shallower burial depth of drip lines may offer advantages in that respect

FIGURE 9.13 This photo shows the Schedule 40 PVC manifolds connected to the drip lines. This site had shallow depth to rock, and drip dispersal was used following a secondary treatment system.

over LPD trenches, which tend to be dug at least 6 to 8 in (0.15 to 0.2 m) deeper. The shallower burial depth also helps to better optimize vegetative nutrient uptake and ET processes. Drip dispersal has been used on many school projects across the United States because the distribution method lends itself well to irrigating turf grass fields common to school grounds. Subsurface drip may present problems for some sports fields unless lines are buried deeper, due to the use of core or spike aerators for soccer, football, and other types of turf sports.

Initial costs for constructing drip dispersal systems as compared with LPD pressure distribution vary by site conditions and local industry practitioners. Long-term operation and maintenance costs, including power usage should, however, be lower for properly designed and installed LPD systems as compared with drip dispersal. Due to the use of much thinner-walled effluent distribution tubing for drip systems as compared with Schedule 40 PVC pipe for LPD systems, and the greater potential for clogging of emitters and biomat formation where emitter effluent is in contact with soils, drip dispersal would not be expected to have as long useful service lives as LPD systems receiving the same quality of effluent. Manufacturers of proprietary materials and equipment used for drip dispersal systems should be asked about product warranty periods, and for larger projects

Figure 9.14   The other side of the two drip dispersal zones in Fig. 9.13 is shown here, ready for final soil cover.

perhaps even warranties specific to those projects. Engineers/designers of decentralized systems advising their clients on long-term sustainability considerations can help property owners weigh initial costs with long-term system management and those costs, and expected useful service life and eventual replacement costs.

As mentioned previously, the application of septic tank effluent is not recommended here for subsurface drip dispersal. Figure 9.17 shows a photo of an excavated section of a drip dispersal system applying septic tank effluent. After primary septic tank treatment, the effluent here was filtered through a disc filter (similar to unit shown in Fig. 9.16) prior to distribution in the drip lines. Although the ground surface did not yet show signs of saturated soil conditions, when sections of this field were excavated the soil around drip line emitter holes was found to be very black with biomat formation, and saturated (notice the sheen around the emitter line). This drip dispersal field had been operational for only a few years prior to this investigation. Figure 9.18 shows a sample of the blackish-brown wet soil

FIGURE 9.15 This photo shows one of two air relief valves, located at the end of one of the two field zones for this system. A valve box will be installed over this valve before placing the final soil cover over the field zones.

FIGURE 9.16 A manual disc filter is shown here in a valve box. The filter is in the pressure manifold leading away from a secondary treatment unit. The filter unit consists of many plastic rings stacked together that comprise the cylindrical filter. This filter unit will be checked and cleaned every 3 months, at the same time the treatment unit is checked for service needs by a trained technician.

FIGURE 9.17 Saturated soils around drip emitters distributing primary treated effluent.

FIGURE 9.18 Saturated soil sample with biomat held against soil color chart.

held next to a soil color chart, with the sample darker than the very darkest color chart frame. Soil conditions were clearly not aerobic as needed for good long-term natural treatment processes to occur. For this reason, along with reducing the potential of clogging of emitter lines, the minimum quality of effluent recommended for dispersal with drip dispersal lines is secondary treated effluent using a treatment method capable of reliable performance.

Several good sources of detailed information are available on the planning/design, installation, operation, and long-term maintenance of drip dispersal systems, all of which are critical to their successful use. An excellent technical resource on drip systems is "Wastewater Subsurface Drip Distribution: Peer Reviewed Guidelines for Design, Operation and Maintenance," a report published in 2004 by the Electric Power Research Institute (EPRI) and cosponsored by the Tennessee Valley Authority (TVA). Manufacturers of proprietary drip tubing used for these systems provide technical assistance with designs and installation, including training. The Supplemental Information, Chap. 9 portion of this book's website provides a detailed technical design reference (EPRI publication), along with information for two U.S. manufacturers of drip tubing and other drip dispersal equipment: Geoflow, Inc. and Netafim USA.

An important consideration in comparing drip dispersal with LPD pressure distribution is the function of the rock/gravel in the LPD trench. Effluent storage capacity is provided in the significant pore space in trench gravel, offering aerobic conditions for treatment to occur along with a media surface much like a trickling filter. Alternating dosing/resting cycles where effluent is applied to the surface of rock media would be expected to offer both aerobic and anoxic conditions needed for at least some amount of total nitrogen reduction in the trench rock prior to soil infiltration along the base and possibly sides of the trenches. Effluent storage would also tend to prevent saturated surface conditions from occurring, particularly since LPD trenches tend to be deeper than drip dispersal lines.

Drip dispersal was originally developed as a method of irrigation, and to meet a water demand for vegetation. Drip systems are well-suited for irrigation in drier climates because they allow water to be dosed to the soil at lower rates consistent with agronomic needs. The method has been used increasingly in the south and southwest regions of the United States for effluent distribution, often with wastewater reuse and water conservation in mind. While drip dispersal can be a very good means of efficiently meeting the soil moisture demand from vegetation, wastewater production, and the need for reliable final disposition of effluent continues during those periods of time when there may already be enough soil moisture to meet plant moisture demands. In selecting the most sustainable system, a balance must therefore be struck between various potentially competing factors.

During periods of wet weather, and in climates where ET rates tend to be lower than rainfall rates for most if not all months, facilitating higher rates of water infiltration for an effluent dispersal system may be more important than supplying moisture to the soil at rates needed for plant uptake. In general, and as discussed in Chap. 3, design principles using the limiting design parameter (LDP) approach will contribute to the most sustainable systems overall. As long as the overall system effectively deals with pollutants, returning water to the hydrologic cycle at higher soil loading rates that don't exceed a soil's infiltrative capacity would tend to result in the most cost-effective and sustainable system.

Surface application of effluent using drip lines has also sometimes been used for decentralized wastewater systems. Some of the same considerations apply, however, as with spray irrigation systems, including the potential for surface runoff of effluent during wet weather periods, providing storage capacity for rainy periods, and impacts to receiving waters. Those are discussed below.

9.1.6 Surface Application of Treated Effluent

Of the final effluent disposition methods discussed in this chapter, surface application is the only one for which pretreatment to acceptable secondary or advanced levels is absolutely essential on behalf of public and environmental health for small- to midsized decentralized systems. An exception might be where the effluent dispersal system can be located sufficiently distant from public contact, and where no surface runoff of effluent would occur. While subsurface drip dispersal of septic tank effluent is not recommended here due to maintenance considerations and buildup of biomat clogging layers at the dripper/soil interface, it is an option that can be employed without public or environmental health impacts in many cases. Surface application, however, particularly for smaller-scale decentralized systems in close proximity to occupants of the facility(ies) served, presumes at least a certain degree of exposure to the effluent. Where public exposure may occur, effluent must be disinfected following treatment. Treatment process considerations and challenges associated with effectively and reliably disinfecting wastewater effluent for decentralized systems were discussed in Chap. 6.

In a few U.S. states surface application of effluent appears to be viewed by regulatory authorities as a relatively inexpensive and acceptable method of effluent dispersal for onsite wastewater systems as compared with other options for certain difficult site conditions. Those conditions may include shallow depth to rock, or very shallow groundwater tables, such as areas around Houston, Texas. In other states, however, there are sufficient concerns about the reliability of and variations in effluent quality produced from predispersal treatment methods that greater caution is employed with the use of surface application, particularly for small-scale decentralized wastewater systems. As discussed previously, flow equalization and attenuation either designed into or that naturally occur with larger-scale systems tends to result in more reliable treatment and better overall effluent quality.

Surface application of effluent for decentralized systems is typically done either through spray irrigation (using sprinkler-type spray heads), or with drip lines placed along the surface of the ground. In the United States, purple coloring is used for piping and other non-potable water (wastewater effluent) components used in surface application systems to help alert persons that the water is not for consumption or general use. Pop-up sprinkler heads used for spray application look essentially the same as automatic sprinkler system heads, though they are often taller. Surface application with drip lines laid along the ground surface has been used for some U.S. projects where there are tree stands along very steep slopes, and where it would be difficult to install subsurface trenches. However, those projects are typically located inland, and not along sensitive coastal or inland watersheds where surface-applied effluent may combine with rainfall and runoff from sites.

Local rainfall and evapotranspiration patterns from historical data must be referred to for sound designs of surface application systems. Most of the water may have to be handled through evaporation and vegetative uptake processes if soil infiltration rates are limited due to soil moisture conditions, soil type, or slopes. Engineers/designers should use an industry-accepted engineering hydrologic model for estimating expected infiltration and runoff rates from the site for design-rainfall events, so that any needed or required storage of wastewater effluent can be provided. A water balance should be performed that accounts for all of the anticipated natural and wastewater system water entering and potentially exiting the site (water gains and losses). Some states have developed models for use with surface application of effluent that also include nutrient balances. Sustainable surface application should always assume zero surface discharge of effluent from the site, unless effluent can be reliably treated to levels acceptable for direct stream/surface discharge.

Cold weather climates present further challenges for the design of surface application systems, with the potential for freezing of piping and other effluent delivery components. Adequate burial depths or insulation of components must be provided, along with any storage needed for excessively cold or wet weather periods. Since plant growth and moisture uptake is limited during cold weather periods, designing for effective water management during those periods is critical.

For residential scale spray irrigation systems, control panels with timers should be used to apply effluent at night or other times of the day when human activity around the system is likely to be minimal.

Household pets should be kept away from application areas. In selecting spray application areas, prevailing wind directions and proximity to buildings and human activities should be considered along with compliance with applicable setback requirements. While treatment methods should be selected that best ensure reliable treatment, there may be times when the treatment process is subject to lesser performance particularly for small residential scale projects. Odors would be a concern in that case, and should be considered with respect to wind direction along with spray mist carriage.

To avoid potential odors, and to ensure adequate disinfection of surface applied effluent where there may be public exposure, predispersal treatment methods should be carefully selected. Chapter 6 discussed treatment methods that have been found to be the most effective for producing effluent with very low levels of ammonianitrogen, which is critical for odor prevention as well as chlorine disinfection for decentralized wastewater systems. Fixed film/attached growth treatment processes, and in particular packed media filters have been observed to perform the most reliably and often be the most cost-effective for accomplishing that. Of the treatment methods described in Chap. 6, single pass intermittent sand filters designed and constructed in accordance with methods and materials described in Chap. 7 would be expected to produce the lowest effluent ammonia levels on average, as well as low suspended solids. Low ammonia levels are very important for rapid disinfection using chlorination. Very low levels of suspended solids are important for UV disinfection (to minimize any potential for shadowing effects).

Surface application of effluent is often selected because it is less costly to construct as compared with methods requiring significant excavation and/or materials brought to the site for dispersal trench or bed installation, particularly where it's allowed for single family residences. However, relatively large setback distances needed or required in many regulatory jurisdictions coupled with lower application/ loading rates to the soil can result in significantly larger land area requirements for spray irrigation as compared with other dispersal options. Ongoing operation and maintenance for higher levels of treatment needed for surface application, and those associated costs should also be considered. For sensitive watersheds having intense rainfall events or rainy seasons, surface application of effluent may not be a suitable or sustainable option.

It is difficult for smaller-scale wastewater systems using some available treatment methods to reliably produce consistently low levels of key wastewater pollutants such that receiving waters would not be negatively impacted over time from systems discharging directly to surface waters. Advanced methods of treatment are continuously being developed that may consistently produce very high effluent quality. However, considerations such as energy usage, reliability of power supplies, and how those power supplies are produced weigh heavily on sustainability. As technological improvements continue, the ability to safely surface-apply and/or discharge to waters directly will surely increase.

Table 9.1 summarizes advantages and limitations for options described earlier. Installation/capital cost considerations are not included in the table because those will tend to vary greatly by project location and site conditions. Maintenance and power usage factors affecting long-term costs are, however, included in the table.

9.2 Matching Dispersal Methods with Site Conditions and Levels of Treatment

Selecting the most sustainable and cost-effective decentralized wastewater systems entails utilizing the soil/land's natural ability to attenuate pollutants as much as possible, and supplementing that as needed with added treatment to prevent adverse impacts. This section presents recommendations for matching dispersal methods discussed previously with levels of predispersal treatment for specific soil types and depths.

Programs overseeing onsite/decentralized systems may determine that either higher or lower soil loading rates are best suited to certain conditions in those settings. Suggested soil loading rates and predispersal treatment levels coupled with dispersal methods presented in this chapter are considered to be relatively conservative on behalf of water resource protection and long-term systems usage. Because decentralized systems are considered permanent wastewater service solutions in this context, recommendations are intended to avoid such problems as excessive biomat buildup in dispersal beds or trenches. Suggested criteria presented herein are for general guidance, and are based on research with soil treatment processes for small- to midsized decentralized wastewater systems and observations on their performance.

Soil loading rates are presented only for *subsurface* soil absorption dispersal methods having specific site conditions and predispersal treatment levels. No application rates are included for surface application systems, since those are much more dependent on local rainfall patterns and climate. Climate, and weather conditions including evaporation and transpiration rates, are also factors for subsurface dispersal methods and appropriate loading rates, and should be considered alongside of recommendations presented in this chapter and applicable regulatory requirements.

Tables 9.2, 9.3, and 9.4 provide recommendations for *minimum* depths of soil beneath bottoms of dispersal systems excavations or application zones for each textural soil class, along with minimum recommended levels of predispersal treatment. The three tables apply to three ranges of slopes. An important assumption for the recommended

Dispersal Method	Advantages	Limitations
1. Conventional gravity absorption beds	• Conventional/common design approach • Nonproprietary methods and materials used • Low maintenance requirements • No power requirements • Design, and methods and materials of construction are often well-described and prescribed in local onsite rules, with extensive literature available on design details.	• Requires favorable site conditions, including either relatively flat slopes or terraced areas • Gravity distribution is more likely to be uneven and develop clogging mat along bed/soil interface • Requires use of more materials (drain rock/media) as compared with conventional gravity trenches and LPD • Likely significantly shorter useful service life as compared with LPD pressure dispersal
2. Conventional gravity trenches	• Conventional/common design approach • Nonproprietary methods and materials used • Can be installed on moderate slopes • Low maintenance requirements • No power requirements • Design, and methods and materials of construction are often well-described and prescribed in local onsite rules, with extensive literature available on design details.	• Space requirements are somewhat higher than for absorption beds. • Requires favorable site conditions • Trenches must be placed along slope contours. • Gravity distribution is more likely to be uneven and develop clogging mat along trench/soil interface. • Likely significantly shorter useful service life as compared with LPD pressure dispersal
3. Chambers	• May provide greater storage capacity of effluent in chambers • Eliminates need for gravel, thereby reducing need for dredging or quarrying • Low maintenance requirements • No power requirements (unless pressure distribution is used)	• Proprietary units, requiring shipment to site unless produced locally • Absence of gravel/rock in trench may reduce natural treatment capabilities, and result in reduced infiltration rates over time due to increased formation of biomat clogging layers along bottoms of trenches. • Chambers may reduce natural evapotranspiration rates occurring with rock/gravel trench and bed systems.

288

	Advantages	Limitations
4. Gravelless trench technologies (other than chambers)	• May be well-suited for sites with steep and/or unstable slopes where movement of heavy equipment and rock/gravel may not be feasible or have adverse impacts • Simulates rock/gravel in trenches, with comparable levels of in-trench treatment provided • Low maintenance requirements	• Reductions in ET rates and absence of gravel/rock media with moisture retained by surface tension phenomena may contribute to overly rapid infiltration rates for rapidly draining soils, or ponding for more slowly draining soils. Without sufficient effluent/hydraulic retention time in aerobic soil conditions, lesser and possibly inadequate soil treatment would occur, with accompanying ground water impact concerns. • Depending on product and materials used, performance over long periods of use and service life may be unknown. • Typically proprietary products needing to be shipped to site if not produced locally
5. Dosing siphons	• Enables intermittent field dosing without electromechanical power • Simple operation • Easy to service and maintain • Long useful service life if noncorrosive components	• Requires filtered effluent, and preferably effluent of at least secondary treatment level, to avoid biomat buildup and potential malfunctioning of unit • Must be checked routinely to verify continued proper functioning, and preferably with dosing event counter • Limited to sites with sufficient elevation drop between dosing siphon and dispersal field area • Dosing manifold piping must have very smooth transitions and constant grade between dosing tank and field lines to prevent air from being trapped in line. • Limited to set discharge volumes per dose
6. Flout	• Enables intermittent field dosing without electromechanical power • Can be used on relatively flat terrain • Simple operation	• Proprietary product with less performance history as compared with other passive dosing methods, including dosing siphons and tipping buckets • Limited pressure head associated with discharged effluent, as compared with dosing siphons

TABLE 9.1 Effluent Dispersal Methods Summary of Advantages and Limitations (*Continued*)

Dispersal Method	Advantages	Limitations
6. Flout (*Cont.*)	• Easy to service and maintain • Can be used to retrofit dosing siphons without digging up or replacing the dosing tank • Should have long useful service life (though relatively newer product as compared with others, so less performance history)	
7. Tipping buckets	• Enables dosing of discrete volumes of effluent at a time without electromechanical power • Simple operation • Long useful service life if noncorrosive components • Can be used on sites with relatively flat slopes	• Limited bucket size availability for units constructed of materials suitable for onsite wastewater systems (noncorrosive materials). • Relatively small discharge volumes may be dosed at a time. • No pressure head associated with discharged effluent
8. Evapotranspiration (lined) beds (ET beds)	• May be used where soil and subsurface conditions preclude use of absorption-type dispersal technologies • May not require power, depending on design • Nonproprietary dispersal method • Lower maintenance requirements than LPD (assuming nondosed ET bed), drip or spray dispersal • Relatively straightforward design and construction methods	• Somewhat more complex design than conventional absorption systems, including need to match loading rate with local climate/rainfall rates • Most climates not suitable for ET beds due to seasonal moisture/rainfall conditions • Larger areas needed as compared with methods also relying on infiltration • Salt accumulation may be a problem over time. • Saturation of bed may occur during periods of extended or intense rainfall. • Requires installation of monitoring ports for checking moisture conditions in ET bed • Landscaping maintenance needed • Liner must remain intact for proper functioning of lined ET beds. • Significantly shorter expected useful service life as compared with soil absorption dispersal methods

9. Low-pressure dosing using a pump(s)	• Relatively even distribution of effluent across field(s)	• Engineered design typically needed, especially for systems larger than those serving single family dwellings
	• Dosing/resting cycles found to provide much better soil treatment as compared with gravity flow trenches or beds	• Requires electric power [unless slopes allow dosing siphon(s) to be used]
	• Relatively shallow trenches (16–24 in or 40–60 cm) enable applying effluent in plant root zone, and moisture and nutrient uptake by vegetative cover.	• Requires availability of drain rock/gravel or other type of media for trench construction
	• Can be installed on significant slopes	
	• Allows flexibility with site planning and layout (due to ability to dose areas upslope of buildings served)	
	• Less disturbance of vegetation than for conventional gravity beds or trenches	
	• Narrower trenches require less drain rock/ gravel media than wider conventional gravity trenches or beds.	
	• LPD field area can be used for light recreational activity or parks, yard space, shallow-rooted landscaping or xeriscaping.	
	• Burial depth of lines coupled with appropriate soil loading rates tends to minimize opportunity for saturation or effluent breakout.	
	• Nonproprietary system, often allowing use of local materials and equipment	
	• Design approaches developed over time have proven very successful, and with long useful service lives.	

TABLE 9.1 Effluent Dispersal Methods Summary of Advantages and Limitations (*Continued*)

Dispersal Method	Advantages	Limitations
10. Mound systems	• Can be used on sites with limestone rock outcrop, karst or fractured bedrock, very shallow or unsuitable soils, or high groundwater table • Utilizes a natural means of treatment, with proper methods and materials offering very good effluent quality prior to effluent infiltration into ground • Nonproprietary dispersal method • Lower maintenance requirements than drip or spray dispersal	• Requires local availability of select sand/fill material • Typically requires hauling/trucking in large amounts of sandy/select fill material to the site • Limited use of the area occupied by a mound • Carefully engineered design needed to prevent breakout and failure • Requires electric power • Landscaping maintenance needed • Requires monitoring ports to be installed
11. Drip irrigation system	• Even distribution of effluent across field(s) • Allows effluent application upslope of facilities served • Can provide for better evapotranspiration and nutrient uptake than other dispersal methods with deeper effluent lines • Shallower application depths optimize plant use of moisture and nutrients (agronomic benefits) • Lower application rate can provide for better overall soil treatment • May be used on sites with shallow soils, insufficient depth to restrictive horizon or steep slopes • Typically less site disruption as compared with systems requiring deeper trench or bed excavation	• Engineered design typically needed, especially for systems larger than those serving single family dwellings • Filtration of effluent needed prior to dispersal • Relatively intense operation and maintenance requirements by sufficiently skilled persons/service providers • Shallow burial depths of emitter lines increase risk of effluent "breakout" (if drip lines are damaged or degrade over time), and surface saturation with effluent particularly during wet weather periods (depending on application rate). • Maintenance contract often required • Uses proprietary components • Requires power consumption • Expected to use more power than LPD distribution (due to frequent back-flushing requirements) • Lower application rates as compared with LPD dispersal, resulting in use of significantly larger dispersal areas for same design flows • Shorter expected useful service life as compared with LPD distribution • Best suited for turf grass, and much less suitable for deeper rooted and natural vegetative covers • Measures must be taken or materials used to prevent root and rodent intrusion into drip lines that may result in clogging or breakout.

12. Surface application/ spray irrigation systems	• Maximizes evaporation of effluent • May be an option where soil and site conditions are such that other dispersal methods are not feasible • Typically causes less site disruption than other dispersal methods requiring trench or bed excavation • Increases availability of moisture and nutrients to surface vegetation as compared with other dispersal methods • Areas upslope of facilities served by system may be used for effluent application (since pumps are used).	• Relatively intense maintenance needed, with maintenance contracts typically required for pre-dispersal treatment, due to public/environmental health risks • Engineered design needed • Larger land areas usually needed for both spray application and necessary setbacks • At least secondary treatment and disinfection of effluent are needed (where there's potential public exposure to the effluent) • Requires electric power • Periodic inspection and effluent testing is needed. • Runoff protection and tail water control typically necessary or required • Greater potential for adverse public health and/or environmental and watershed impacts as compared with most other dispersal methods • Areas used for surface application should not be used for human activity, or for pets on a regular basis, particularly for smaller/residential scale systems being monitored and checked less frequently for effluent quality. • Odors are a concern for systems during periods when treatment process may not be functioning adequately.

Table 9.1 Effluent Dispersal Methods Summary of Advantages and Limitations (*Continued*)

293

Depth of Soil above Rock or Groundwater[a,b,c]	Class Ia Soils		Class Ib, II, or III Soils		Class IV Soils	
	Minimum Recommended Treatment Level for Dispersal Method	Candidate Dispersal Methods Recommended for Consideration	Minimum Recommended Treatment Level for Dispersal Method	Candidate Dispersal Methods Recommended for Consideration	Minimum Recommended Treatment Level for Dispersal Method	Candidate Dispersal Methods Recommended for Consideration
6–12 in (0.15–0.3 m)	• Septic tank pretreatment for Lined ET • Advanced treatment with total nitrogen removal for mounds • Advanced treatment with total nitrogen removal and disinfection for subsurface drip, LPD, and surface application	1. LPD 2. Mound 3. Subsurface drip 4. Lined ET 5. Surface application	• Septic tank pretreatment for Lined ET • Advanced treatment with total nitrogen removal for mounds • Advanced treatment with total nitrogen removal and disinfection for subsurface drip, LPD, and surface application	1. LPD 2. Mound 3. Subsurface drip 4. Lined ET 5. Surface application	• Septic tank pretreatment for Lined ET and mounds • Secondary/advanced treatment for LPD • Secondary treatment w/filter for subsurface drip[d] • At least advanced/secondary treatment and disinfection for surface application	1. LPD 2. Mound 3. Subsurface drip[d] 4. Lined ET 5. Surface application
12–24 in (0.3–0.6 m)	• Septic tank pretreatment for Lined ET[a] • Advanced treatment with total nitrogen removal for mounds • Advanced treatment with total nitrogen removal and disinfection for subsurface drip, LPD and surface application	1. LPD 2. Mound 3. Subsurface drip 4. Lined ET 5. Surface application	• Septic tank pretreatment for LPD • Secondary treatment with filter for subsurface drip[d]	1. LPD 2. Subsurface drip[d]	• Septic tank pretreatment for LPD • Secondary treatment with filter for subsurface drip[d] • At least advanced/secondary treatment and disinfection for surface application	1. LPD 2. Subsurface drip[d] 3. Surface application
24–36 in (0.6–1 m)	• Septic tank pretreatment for Lined ET[a] • Advanced treatment with total nitrogen removal for mounds	1. LPD 2. Mound 3. Subsurface drip 4. Lined ET	• Septic tank pretreatment for LPD • Secondary treatment with filter for subsurface drip dispersal[d]	1. LPD 2. Subsurface drip[d] 3. Conventional gravity trench[e]	• Septic tank pretreatment for LPD • Secondary treatment with filter for subsurface drip dispersal[d]	1. LPD 2. Subsurface drip[d] 3. Surface application

Depth			
	• Advanced treatment with total nitrogen removal and disinfection for subsurface drip, LPD, and surface application 5. Surface application	• Advanced treatment with total nitrogen removal and disinfection for conventional gravity trench	• At least advanced/secondary treatment and disinfection for surface application
>36 in (>1 m)	• Septic tank pretreatment for Lined ET[d] • Advanced treatment with total nitrogen removal for mounds • Advanced treatment with total nitrogen removal and disinfection for subsurface drip, LPD, and surface application 1. LPD 2. Mound 3. Subsurface drip 4. Lined ET 5. Surface application	• Septic tank pretreatment for conventional gravity trenches/beds or LPD • Secondary treatment with filtration for subsurface drip dispersal[d] 1. LPD 2. Subsurface drip[d] 3. Conventional gravity trench or bed[e]	• Septic tank pretreatment for LPD • Secondary treatment with filtration for subsurface drip dispersal[d] • At least advanced/secondary treatment and disinfection for surface application • Secondary treatment for conventional gravity[f] 1. LPD 2. Subsurface drip[d] 3. Conventional gravity trench[e,f] 4. Surface application

[a]Minimum depth to groundwater should be the higher end of depth ranges shown. Where soil depth above rock or groundwater is less than 6 in (15 cm), advanced treatment with total nitrogen removal to acceptable levels for groundwater or receiving watershed, and disinfection should be provided for all soil dispersal methods allowing effluent infiltration/absorption, except mounds, for which disinfection may not be needed.

[b]Depth of undisturbed soil above rock or groundwater is measured from the bottom/base of the excavation for LPD and conventional gravity bed/trenches, or from distribution lines for subsurface drip dispersal.

[c]For mound treatment/dispersal systems, the depth shown above rock or groundwater is from the base of the mound.

[d]For drip dispersal, secondary treatment is the minimum recommended level of predispersal treatment, along with some method of filtration.

[e]Gravelless drain field products might be used instead of rock/gravel media for either conventional gravity flow beds or trenches, or for LPD trenches.

[f]For conventional gravity trenches installed in class IV soils, due to less uniform distribution and tendency for biomat formation, some method of reliably providing at least secondary treatment is recommended.

TABLE 9.2 Minimum Recommended Vertical Separation Distances and Treatment Levels for Different Textural Soil Classes and Dispersal Methods for Sites with *Slopes Ranging from 0 to 15 Percent*

Depth of Soil above Rock or Groundwater[a,b]	Class Ia Soils		Class Ib, II, or III Soils		Class IV Soils	
	Minimum Recommended Treatment Level for Dispersal Method	Candidate Dispersal Methods Recommended for Consideration	Minimum Recommended Treatment Level for Dispersal Method	Candidate Dispersal Methods Recommended for Consideration	Minimum Recommended Treatment Level for Dispersal Method	Candidate Dispersal Methods Recommended for Consideration
6–12 in (0.15–0.3 m)	• Advanced treatment with total nitrogen removal and disinfection for subsurface drip and LPD	1. LPD[c] 2. Subsurface drip	• Advanced treatment with total nitrogen removal and disinfection for subsurface drip and LPD	1. LPD[c] 2. Subsurface drip	• Secondary/advanced treatment for LPD • Secondary treatment w/filter for subsurface drip[d]	1. LPD[c] 2. Subsurface drip[d]
12–24 in (0.3–0.6 m)	• Advanced treatment with total nitrogen removal and disinfection for subsurface drip and LPD	1. LPD[c] 2. Subsurface drip	• Septic tank pre-treatment for LPD • Secondary treatment with filter for subsurface drip[d]	1. LPD[c] 2. Subsurface drip[d]	• Septic tank pre-treatment for LPD • Secondary treatment with filter for subsurface drip[d]	1. LPD[c] 2. Subsurface drip[d]
24–36 in (0.6–1 m)	• Advanced treatment with total nitrogen removal and disinfection for subsurface drip and LPD	1. LPD[c] 2. Subsurface drip	• Septic tank pre-treatment for LPD • Secondary treatment with filter for subsurface drip dispersal[d] • Advanced treatment with total nitrogen removal and disinfection for conventional gravity trenches	1. LPD[c] 2. Subsurface drip[d] 3 Conventional gravity trench[e]	• Septic tank pretreatment for LPD • Secondary treatment with filter for subsurface drip dispersal[d]	1. LPD[c] 2. Subsurface drip[d]

>36 in (>1 m)	• Advanced treatment with total nitrogen removal and disinfection for subsurface drip and LPD 1. LPD[c] 2. Subsurface drip	• Septic tank pre-treatment for LPD • Secondary treatment with filtration for subsurface drip dispersal[d] 1. LPD[c] 2. Subsurface drip[d] 3. Conventional gravity trench[c,e]	• Septic tank pretreatment for LPD • Secondary treatment with filtration for subsurface drip dispersal[d] • Secondary treatment for conventional gravity trenches[e] 1. LPD[c] 2. Subsurface drip[d] 3. Conventional gravity trench[c,e,f]

[a] Minimum depth to groundwater should be the higher end of depth ranges shown. Where soil depth above rock or groundwater is less than 6 in (15 cm), advanced treatment with total nitrogen removal to acceptable levels for groundwater or receiving watershed, and disinfection should be provided for all soil dispersal methods allowing effluent infiltration/absorption, except mounds, for which disinfection may not be needed.

[b] Depth of undisturbed soil above rock or groundwater is measured from the bottom/base of the excavation for LPD, or from distribution lines for subsurface drip dispersal.

[c] Gravelless drain field products might be used instead of rock/gravel media for conventional gravity and LPD trenches.

[d] For drip dispersal, secondary treatment is the minimum recommended level of predispersal treatment, along with some method of filtration.

[e] Great care must be taken with the design/installation of conventional gravity trenches placed along steeper slopes to maintain proper grades. Greater spacing between trenches along contours should be used.

[f] For conventional gravity trenches installed in class IV soils, due to less uniform distribution and tendency for biomat formation, some method of reliably providing at least secondary treatment is recommended.

TABLE 9.3 Minimum Recommended Vertical Separation Distances and Treatment Levels for Different Textural Soil Classes and Dispersal Methods for Sites with *Slopes Ranging from 15 to 30 Percent*

Depth of Soil above Rock or Groundwater[a,b]	Class Ia Soils		Class Ib, II, or III Soils		Class IV Soils	
	Minimum Recommended Treatment Level for Dispersal Method	Candidate Dispersal Methods Recommended for Consideration	Minimum Recommended Treatment Level for Dispersal Method	Candidate Dispersal Methods Recommended for Consideration	Minimum Recommended Treatment Level for Dispersal Method	Candidate Dispersal Methods Recommended for Consideration
6–12 in (0.15–0.3 m)	• Advanced treatment with total nitrogen removal and disinfection for subsurface drip and LPD	1. LPD[c,f] 2. Subsurface drip	• Advanced treatment with total nitrogen removal and disinfection for subsurface drip and LPD	1. LPD[c,f] 2. Subsurface drip	• Secondary/advanced treatment for LPD • Secondary treatment w/filter for subsurface drip[d]	1. LPD[c,f] 2. Subsurface drip[d]
12–24 in (0.3–0.6 m)	• Advanced treatment with total nitrogen removal and disinfection for subsurface drip and LPD	1. LPD[c,f] 2. Subsurface drip	• Advanced treatment with total nitrogen removal for subsurface drip and LPD	1. LPD[c,f] 2. Subsurface drip	• Septic tank pre-treatment for LPD • Secondary treatment with filter for subsurface drip[d]	1. LPD[c,f] 2. Subsurface drip[d]
24–36 in (0.6–1 m)	• Advanced treatment with total nitrogen removal and disinfection for subsurface drip and LPD	1. LPD[c,f] 2. Subsurface drip	• Septic tank pre-treatment for LPD • Secondary treatment with filter for subsurface drip dispersal[d]	1. LPD[c,f] 2. Subsurface drip[d] 3 Conventional gravity trench[c,e,f]	• Septic tank pre-treatment for LPD • Secondary treatment with filter for subsurface drip dispersal[d]	1. LPD[c,f] 2. Subsurface drip[d]

>36 in (>1 m)	• Advanced treatment with total nitrogen removal and disinfection for subsurface drip and LPD 1. LPD[c,f] 2. Subsurface drip	• Septic tank pretreatment for LPD • Secondary treatment with filtration for subsurface drip dispersal[d] • Advanced treatment with total nitrogen removal and disinfection for conventional gravity trenches 1. LPD[c,f] 2. Subsurface drip[d] 3. Conventional gravity trench[c,e,f]	• Septic tank pretreatment for LPD • Secondary treatment with filtration for subsurface drip dispersal[d] • Advanced treatment with total nitrogen removal and disinfection for conventional gravity trenches 1. LPD[c,f] 2. Subsurface drip[d] 3. Conventional gravity trench[c,e,f]	• Septic tank pretreatment for LPD • Secondary treatment with filtration for subsurface drip dispersal[d] • Advanced treatment with total nitrogen removal and disinfection for conventional gravity trenches 1. LPD[c,f] 2. Subsurface drip[d] 3 Conventional gravity trench[c,e,f]

[a] Minimum depth to groundwater should be the higher end of depth ranges shown. Where soil depth above rock or groundwater is less than 6 in (15 cm), advanced treatment with total nitrogen removal to acceptable levels for groundwater or receiving watershed, and disinfection should be provided for all soil dispersal methods allowing effluent infiltration/absorption, except mounds, for which disinfection may not be needed.

[b] Depth of undisturbed soil above rock or groundwater is measured from the bottom/base of the excavation for LPD, or from distribution lines for subsurface drip dispersal.

[c] Gravelless drain field products might be used instead of rock/gravel media for conventional gravity and LPD trenches.

[d] For drip dispersal, secondary treatment is the minimum recommended level of predispersal treatment, along with some method of filtration.

[e] Great care must be taken with the design/installation of conventional gravity trenches placed along steeper slopes to maintain proper grades. Greater spacing between trenches along contours should be used.

[f] LPD and gravity trenches installed along contours of steep slopes may need to be dug manually to prevent excessive disturbance or dangerous conditions for earth-moving equipment. Greater distances between LPD trenches should be used.

TABLE 9.4 Minimum Recommended Vertical Separation Distances and Treatment Levels for Different Textural Soil Classes and Dispersal Methods for Sites with *Slopes Greater Than 30 Percent*

levels of treatment and soil depths is that the soil does not have a structure such as columnar or prismatic that would offer highly preferential flow patterns, and significant channeling of effluent. Soil types and structures allowing excessively rapid infiltration cannot be expected to offer significant natural treatment capabilities for key wastewater pollutants due to reduced hydraulic residence times for treatment processes and/or lower soil organic content.[15,16]

Recommendations in Tables 9.2 through 9.4 are based on a wide range of technical literature, field data, and empirical observations on system performance.[17] The recommendations are not intended to in any way minimize the importance of informed engineering/scientific judgments on suitable and sustainable decentralized systems designs, including necessary levels of treatment and appropriate methods of final effluent dispersal/disposition. They are, however, intended to offer basic guidance information where either no local design guidelines exist, or where applicable guidelines may not seem protective enough of public health or environmental resources. The recommended design criteria in the tables can hopefully help lead site planners and wastewater systems engineers/designers to a "middle ground," for which systems are not designed so conservatively as to unnecessarily drive up costs, or insufficiently protective such that local resources are adversely impacted.

Recommendations provided here should be applicable for the majority of smaller- to midsized projects having conditions consistent with the given soil type and depth, and slope category. They may also be used as general guidelines for designs in locations where there is no local regulatory guidance on practices found over time to be suitable for those settings, and which protect local public health and environmental resources. It must be emphasized that the recommended criteria in Tables 9.2 through 9.4 may not necessarily be the most suitable or sustainable for certain sensitive conditions. Each system should be evaluated independently, based on all of the relevant conditions and project scope, to determine the most appropriate and sound approach. The recommended soil depths, loading rates, and associated treatment levels might need to be adjusted for certain geographic areas such as those in extreme cold weather. Deeper soil depths and/or higher levels of treatment may be needed for those conditions. For sites located in regions with prolonged and excessive rainfall, and in particular regions where slope stability may be a concern (e.g., seismically active regions), the appropriateness of specific treatment and dispersal methods for a site should be examined based on year-round local conditions. More detailed geotechnical investigations may be needed, along with further evaluation of soil geochemistry and nutrient balances, particularly for larger projects.

The leftmost column of Tables 9.2 to 9.4 shows the depth range above rock or groundwater (whether seasonal, "perched," or relatively static) for the bottoms of excavations for trenches or beds.[17] The high end of those depth ranges should be used for depth to groundwater, due to the potential for more direct communication between effluent and groundwater. In the case of drip lines placed or plowed into the soil without media in trenches, the distance is below the soil/drip line interface. For mounds, the depth indicated is measured from the base of the mound.

Dispersal methods listed for slope ranges and textural soil classes are generally prioritized from top to bottom in the order recommended for overall sustainability for those soil and slope conditions, and most projects and sites. Several key sustainability factors considered in that ordering include

- Ability to meet watershed and water quality goals for key wastewater pollutants

- Potential public and/or environmental health risks associated with use of that method

- Power usage for achieving those water quality goals

- Level of ongoing maintenance needed for continued proper functioning

- Ability to use nonproprietary and/or local materials, labor, and equipment for implementing the dispersal method

- Expected useful service life of the dispersal method

Where the use of low-pressure-dosing (LPD) effluent dispersal is indicated, it is assumed that trenches and lines are placed as shallow as conditions will reasonably allow, and that the system is constructed in accordance with methods and materials detailed in Chap. 10 of this book. Installation costs have not been included in prioritizing recommendations due to the great variation in construction costs for each project depending on location and availability of materials/resources needed for each dispersal method. However, that is clearly an important consideration along with all other sustainability factors.

The depths shown are the minimum depths considered sustainable for most sites and geographic conditions, and may need to be increased depending on the specific site conditions. Depths shown assume that the entire soil profile between the bottom of the dispersal trench or bed excavation, or zone of application, is comprised of that type of soil. For soil profiles having soil horizons of multiple textural categories (which is the likely the majority of cases), the soil type resulting in the more conservative and protective design should be used. For very sensitive watersheds needing increased protections from nutrients and/or other pollutant loadings, increased soil depths beneath some types of soil absorption systems may be needed, or added predispersal treatment provided. That may be particularly so for class Ia, Ib, and II soils having higher infiltration rates. Dispersal methods used with each loading rate are critical. Recommendations for soil application rates for dispersal methods relying on absorption/infiltration are presented in Tables 9.5 to 9.7.

(a)

Soil Textural Class[†] (USDA Textural Class)	Corresponding Percolation Test Rate (If Percolation Testing Is Performed)		Application Rate[†] R_a (gal/day/ft²)
	in/h	min/in	
Class Ia[§]	>12	<5	0.5
Class Ib	10–12	5–6	0.4
Class II	4–10	6–15	0.25
Class III	1.3–4	15–46	0.20
Class IV	<1.3	>46	0.10

(b)

Soil Textural Class[†] (USDA Textural Class)	Corresponding Percolation Test Rate (If Percolation Testing Is Performed)		Application Rate[†] R_a (cm/day)
	cm/h	min/cm	
Class Ia[§]	>30	<2	2.0
Class Ib	25–30	2–2.4	1.6
Class II	10–25	2.4–6	1.0
Class III	3.3–10	6–18	0.8
Class IV	<3.3	>18	0.4

*The application rates shown above are the maximum recommended ones for drip dispersal, regardless of the level of treatment provided. The minimum recommended level of treatment for drip is secondary treatment using or followed by a method of filtration removing particles larger than 100 μm.
†For these soil classes and recommended soil loading rates, it is assumed that the soils have "blocky," granular or other soil structure that does not pose significant concerns with preferential flow patterns.
‡Minimum levels of treatment and soil depths are as recommended in Tables 9.2, 9.3, and 9.4, depending on slope and soil type.
§The minimum recommended level of treatment for class Ia soils is advanced treatment with total nitrogen reduction to acceptable levels, and disinfection.

TABLE **9.5**   Recommended Effluent Loading Rates for Subsurface Effluent Dispersal*

There are sites and conditions for which other dispersal methods may need to be considered. For example, while conventional gravity dispersal has not been listed as a recommended method in many cases due to reasons discussed earlier in the chapter, it may need to be considered due to the absence of available power or elevation drop

Soil Type[†] (USDA Textural Class)	Application Rate[†] R_a (gal/day/ft²)	Application Rate[†] R_a (cm/day)
Class Ia	0.5[§]	2.0
Class Ib	0.6	2.4
Class II	0.5	2.0
Class III	0.4	1.6
Class IV	0.15	0.6

---

*Assumes intermittent sand filter is designed and constructed in a manner consistent with the methods and materials described in Chap. 7.

†Minimum soil depths presented in Tables 9.2 to 9.4 apply for these soil types and loading rates.

‡These application rates do not apply to subsurface drip dispersal. The maximum recommended rates for drip are those in Table 9.5.

§Application of effluent to class Ia soils requires a higher level of treatment than can typically be provided by intermittent sand filters. See footnote "§" of Table 9.5.

TABLE 9.6 Recommended Effluent Loading Rates for Subsurface Dispersal of Treated Effluent Where Intermittent Sand Filtration Is Used*

across sites for dosed methods of effluent distribution. Additional predispersal treatment might be used in that case to help counteract issues with gravity dispersal, including uneven effluent distribution and lesser expected service life.

Where gravity dispersal is used on steeper slopes, added care must be taken with design and installation methods and materials to prevent breakout of effluent. Greater distances between trenches should be used for both LPD and gravity trenches as slope steepness increases. Care should also be taken to ensure rock/gravel media is free of fines, since that has been found to contribute to reduced infiltration rates. Avoiding overcompaction will also help significantly in that respect, and is recommended for all site areas used for effluent dispersal/application. Tracked excavating and material-moving equipment, or very low tire pressure vehicles should be used over dispersal field areas before or during installation. Those types of considerations will be discussed further in Chap. 10. The use of gravelless trench products may also be used to reduce compaction in excavations.

Surface application has not been recommended in the tables for steeper slopes, due to surface runoff considerations and potential watershed impacts. Where adequate runoff controls and storage can be provided, and depending on climate and watershed sensitivity, there may be cases where it's reasonable to consider surface application. That's particularly so for larger and/or more remote projects, for which application areas can be located away from human activities and sensitive watersheds. And as mentioned throughout this book, applicable regulatory requirements may preclude the use of

Soil Type[†] (USDA Textural Class)	Application Rate[§] R_a (gal/day/ft²)	Application Rate[§] R_a (cm/day)
Class Ia	0.5	2.0
Class Ib	0.8	3.3
Class II	0.6	2.4
Class III	0.4	1.6
Class IV	0.15	0.6

*Recirculating packed media treatment processes would include recirculating sand/gravel systems and packed textile media filters. For recirculating media filters using more coarse media that may allow greater pass-through of suspended solids or lower nitrification/denitrification rates, some type of filtration process may be needed to ensure that clogging or excessive biomat formation doesn't occur at the soil's infiltrative surface. Information presented in Chap. 6 may be helpful for selecting approaches resulting in adequate overall treatment.

†It is assumed that these loading rates only be used in conjunction with predispersal treatment processes capable of reliably providing for at least 50-percent total nitrogen reduction or 20 mg/L, whichever is less. For larger systems and for sites located in sensitive watersheds, higher levels of nutrient removal may be needed along with removal of other wastewater pollutants.

‡Minimum soil depths in Tables 9.2 to 9.4 apply.

§These recommended application rates do not apply to subsurface drip dispersal, regardless of level of pretreatment provided. Those are as shown in Table 9.5.

TABLE 9.7   Recommended Effluent Loading Rates for Subsurface Dispersal where Recirculating Packed Media Treatment Processes* with Total Nitrogen Reduction[†] Are Used

surface application or other methods entirely, as well as prescribe design details and setback requirements for some or all dispersal methods that may be used.

In some cases surface irrigation *could* be used for deeper soil and milder slope conditions where it was not included in the candidate list of systems recommended for consideration. Where there is sufficient depth of soil and site conditions that lend themselves well to optimizing the use of natural soil treatment processes, the level of treatment needed, long-term maintenance and monitoring requirements, energy consumption, and overall "carbon footprint" associated with surface application do not in many cases support its sustainability for small- to midsized decentralized wastewater systems. An important exception to this, however, is in geographic areas and climates where surface reuse of effluent is a vital element of satisfying water supply needs and long-term water resources management.

Mounds and ET beds have been included in the list of recommended methods to consider only for sites with lesser slopes, due to construction issues for installing those systems along steeper slopes. ET beds would be expected to have seasonal water balance and eventual salt accumulation problems, and accompanying shorter service life than other options. Along with slope considerations it has therefore only been included as an option to consider where the use of other dispersal methods tends to be very limited. Mounds have been given a slightly higher ranking than subsurface drip dispersal for sites having relatively mild slopes, due to the lesser long-term maintenance and predispersal treatment needs for mounds. However, that may not be the most sustainable or cost-effective approach if there aren't sources of suitable sand/soil for constructing mounds within a reasonable transport distance from the site.

Tables 9.5 to 9.7 are to be used in conjunction with Tables 9.2 to 9.4. Effluent disinfection, for example, is not mentioned in Tables 9.5 to 9.7, but may be needed depending on soil type, depth, and/or slopes. Table 9.5 provides recommended soil application rates ("loading rates") for each major textural soil class. Again, the recommendations assume that problematic soil structures such as columnar or prismatic are not present, and for which added treatment and/or lower application rates may be needed. Applicable regulatory requirements may call for lower application rates, or may allow higher rates.

Most soils are of course neither ideal nor homogeneous, and some may not fall neatly into one of the textural categories. Preferential flow patterns will reduce a soil's ability to effectively attenuate pollutants. Judgments will need to be made in each case what level of treatment and hydraulic loading rate together best match the specific site conditions. Soils determined to be within a particular textural class, such as class III soils, may have widely varying infiltration and long-term effluent acceptance rates, and treatment capabilities. Therefore, corresponding percolation rates are provided for each textural class. Where percolation testing is performed to better characterize soils, Table 9.5*a* and *b* provides percolation test result ranges that may correlate to each textural class, and application rates, in U.S. Customary and SI units, respectively. Those respective percolation test rates may also be used for the same soil classes in Tables 9.6 and 9.7.

Tables 9.6 and 9.7 provide recommendations for increased soil loading rates with added types and levels of treatment for subsurface effluent absorption methods other than drip dispersal. Because drip systems were developed and best used for applying effluent at lower agronomic rates, and due to potential problems with overloading and saturating fields with shallow emitter burial depths for drip systems, only the rates in Table 9.5 are recommended. As stated in Tables 9.2 to 9.4, the minimum level of treatment recommended here for drip is secondary treatment followed by filtration to remove particles larger than about 100 μm in diameter, or drip dispersal following treatment with an intermittent sand filter designed and built consistently with the methods/materials in Chap. 7. Applying effluent of a quality that doesn't contribute to excessive clogging or pollutant buildup in lines and/or soil/emitter interfaces is needed on behalf of reasonable maintenance levels and useful service lives.

While an increased loading rate is shown here for class Ib soils, the high percolation rates and rapid nitrification potentials associated with these soils can contribute to excessive nitrogen (and other pollutants) loading to groundwater or surface water supplies. Such factors as local ET rates and overall climate should be considered for determining suitable effluent application rates, and levels of predispersal treatment needed. The amount of organic content in the soil and any tendencies for preferential flow patterns will each affect natural treatment capabilities, with the former favorably and latter adversely.

Low-pressure-dosing dispersal is only recommended in class Ia soils if advanced treatment with nutrient removal is provided. Trenches should be excavated as shallow as possible to maximize ET processes and nutrient uptake. The use of subsurface drip dispersal may offer significant advantages in that regard due to the shallow burial depth of lines, and better drainage in class Ia soils that would tend to offset potential problems with biomat buildup around emitter lines in tighter soils with higher organic content. Use of gravity dispersal in soils with rapid infiltration rates may result in adverse groundwater or surface water impacts, unless relatively high levels of predispersal treatment are provided. A careful evaluation of the site and surrounding conditions is needed to ensure that adequate predispersal treatment is provided for the specific soil and subsurface conditions.

The loading rates recommended in Table 9.6 recognize the reliably low BOD, TSS, and pathogen levels expected on average for intermittent sand filters (ISFs) designed and built like the one detailed in Chap. 7. ISFs with that loading rate, sand gradation and depth, and using timed dosing would also be expected to remove about 25 to 30 percent total nitrogen on average. Increased loading rates following ISF treatment have been successfully used in a number of U.S. states as allowed by their onsite wastewater programs. Research on the long-term acceptance rates (LTARs) of ISF effluent applied to soils has found the acceptance rate to be as much as 8 to 10 times higher for some soils than for septic tank effluent.[18–20] An exception was clay soils, for which much lower increases in acceptance rates were observed. That amount of increase in loading rate (8 to 10 times) would well-exceed a soil's capacity to treat pollutants of concern such as nitrogen, pathogens, PPCPs, and other household chemicals, and might not be sustained as the LTAR (hydraulically) over time. It does, however, demonstrate the effective use of LDP and sustainable design principles as related to water loading and predispersal treatment levels.

Only a very modest increase in the soil application rate for class IV soils is recommended in Tables 9.6 and 9.7 with higher levels of treatment because of the low water infiltration rates for that textural soil category. Due to the rapid infiltration rate and typically lower organic content associated with class 1a soils, and low hydraulic retention time for critical natural treatment processes, those soils were not recommended for increased loading rates in Tables 9.6 or 9.7. Lower application rates using methods that uniformly distribute effluent across the soil, and small dosing volumes spread out over the day will tend to increase the more limited natural treatment capabilities of class 1a soils.

Loading rate recommendations in Table 9.7 recognize the benefits of significant predispersal total nitrogen reduction for employing increased application rates with certain soil types, and those relationships in LDP principles.

Matching a soil's natural ability to treat pollutants and water infiltration rate with levels of pre-soil-dispersal treatment needed to protect the public and environment are essential and central to sustainable land treatment system designs. Failure to recognize and use natural treatment capabilities in designs tends to result in excessive land usage and/or monetary systems costs. Not accounting for limitations in a soil's natural treatment capacity in designs can result in costs to water quality and public health by not providing for sound overall treatment. As with predispersal treatment systems, sustainable dispersal methods must not only be compatible with the geophysical setting, but also with the level of management accompanying their use in the long term.

Due to the high variability of subsurface conditions across even single sites, and variations of climate and watershed conditions, it is difficult to assign loading rates that will be with absolute certainty the most suitable or sustainable in each environmental setting. Soil loading rates presented in this chapter are intended to not exceed the land's natural capacity to assimilate and attenuate pollutants from domestic small- to mid-sized wastewater system *in most geophysical settings*. They will therefore in some cases appear to err on the conservative and more lightly loaded side of decentralized systems planning.

The use of conservatively sized soil dispersal systems is in many ways compatible with low impact development concepts as related to watershed protection. Because impervious land coverages aren't placed over effluent application areas, subsurface dispersal fields can serve multiple functions in water quality management by offering soil and vegetative covers for nutrient uptake from rainfall and site runoff. They also offer aesthetic benefits in the way of green spaces, with many larger dispersal field areas used for parks and light recreational purposes. All of these elements in the aggregate contribute to more sustainable development and land use.

References

1. E. S. Amerson, E. J. Tyler, and J. C. Converse, "Infiltration as Affected by Compaction, Fines and Contact Area of Gravel," *Proceedings of the 6th Symposium on Individual and Small Community Sewage Systems*, St. Joseph, MI, ASAE, 1991, pp. 243–247.
2. D. E. Radcliffe, L. T. West, and J. Singer, "A Model Comparison of Chamber and Conventional On-Site Systems," *Proceedings of the 10th National Symposium on Individual and Small Community Sewage Systems*, St. Joseph, MI, ASAE, 2004, pp. 256–262.
3. B. Grimes, S. Steinbeck, and A. Amoozegar, "Analysis of Tire Chips as a Substitute for Stone Aggregate in Nitrification Trenches of Onsite Septic Systems," Small Flows Quarterly, Fall 2003, Vol. 4 No. 4 pp. 18-23, NESC, West Virginia University, Morgantown, WV.
4. S. Van Cuyk and R. L. Siegrist, "Fate of Viruses in the Infiltrative Surface Zone of Systems that Rely on Soil Treatment for Wastewater Renovation," *Proceedings of the 10th National Symposium on Individual and Small Community Sewage Systems*, St. Joseph, MI, ASAE, 2004.
5. P. R. Owens, E. M. Rutledge, M. A. Gross, S. C. Osier, and R. W. McNew, "The Response of Effluent Absorption Rates to Resting Trenches of a Serially Loaded Septic Filter Field," *Proceedings of the 10th National Symposium on Individual and Small Community Sewage Systems*, St. Joseph, MI, ASAE, 2004.
6. R. J. Miles, R. Rubin and L. T. West, "Fecal Coliform Distribution under Pressure Dosed Onsite Wastewater Systems," *Proceedings of the 11th National Symposium on Individual and Small Community Sewage Systems*, St. Joseph, MI, ASAE, 2007.
7. C. G. Cogger, B. L. Carlile, D. Osborne, and E. Holland, "Design and Installation of Low-Pressure Pipe Waste Treatment Systems," UNC Sea Grant College Publication UNC-SG-82-03, Raleigh, NC, May 1982.
8. Phase 1 Monitoring Report, Alternative Wastewater Management Project, Prepared by Community Environmental Services, Inc. for the City of Austin, TX, Water & Wastewater Utility, 2000.
9. S. M. Parten and H. M. Liljestrand, "Evaluation of Wastewater Treatment Capabilities of Caliche-Type Soils," Department of Civil Engineering, University of Texas at Austin, 1995.
10. R. Siegrist and P. Jenssen, "Nitrogen Removal during Wastewater Infiltration as Affected by Design and Environmental Factors," Proceedings *of the 6th Northwest On-Site Wastewater Treatment Short Course*, Seattle, Washington, September 1989.
11. D. Weymann, A. Amoozegar, and M. T. Hoover, "Performance of an On-Site Wastewater Disposal System in a Slowly Permeable Soil," *Proceedings of the 8th National Symposium on Individual and Small Community Sewage Systems*, ASAE, St. Joseph, MI, 1998.
12. Phase II/III Study Report, Alternative Wastewater Management Study, prepared for the City of Austin, TX, Water & Wastewater Utility by Community Environmental Services, Inc., 2005.
13. M. Rowan, K. Mancl, and O. H. Tuovinen, "Clogging Incidence of Drip Irrigation Emitters Distributing Effluents of Differing Levels of Treatment," *Proceedings of the 10th National Symposium on Individual and Small Community Sewage Systems*, ASAE, St. Joseph, MI, 2004.
14. S. M. Parten, Interviews with U.S. state regulators as reported in *Analysis of Existing Community-Sized Decentralized Wastewater Treatment Systems*, Water Environment Research Foundation, Alexandria, VA, July 2008.
15. M. McLeod, L. A. Schipper, and M. D. Taylor, "Preferential Flow in a Well Drained and Poorly Drained Soil under Different Overhead Irrigation Regimes," *Soil Use Manage* 14:96–100, 1998.
16. L. Barton, L. A. Schipper, G. F. Barkle, M. McLeod, T. W. Speir, M. D. Taylor, A. C. McGill, et al., "Land Application of Domestic Effluent onto Four Soil Types: Plant Uptake and Nutrient Leaching," *Journal of Environmental Quality* 34:635–643, 2005.

17. "Vertical Separation: A Review of Available Scientific Literature and a Listing from Fifteen Other States," Office of Community Environmental Health, Washington State Department of Health, Environmental Health Programs, Olympia, WA, October 1990.

18. E. J. Tyler and J. C. Converse, "Soil Acceptance of Onsite Wastewater as Affected by Soil Morphology and Wastewater Quality," *Proceedings of the 7th National Symposium on Individual and Small Community Sewage Systems*, ASAE, St. Joseph, MI, 1994.

19. D. Sievers, "Pressurized Intermittent Sand Filter with Shallow Disposal Field for a Single Residence in Boone County, Missouri," *Proceedings of the 8th National Symposium on Individual and Small Community Sewage Systems*, ASAE, St. Joseph, MI, 1998.

20. T. L. Loudon, G. S. Salthouse, and D. L. Mokma, "Wastewater Quality and Trench System Design Effects on Soil Acceptance Rates," *Proceedings of the 8th National Symposium on Individual and Small Community Sewage Systems*, ASAE, St. Joseph, MI, 1998.

CHAPTER 10

Low-Pressure-Dosing Effluent Distribution

Subsurface low-pressure dosing (LPD) is considered one of the most sustainable methods of effluent dispersal for applying the limiting design parameter (LDP) principles. The relatively wide range of effluent application rates to soils possible for LPD distribution enables its use in conjunction with appropriate levels of pre-dispersal treatment to match the land's natural treatment capabilities. Factors contributing to its long-term sustainability as a dispersal method include

- Relatively uniform distribution of effluent across the dispersal field.

- Provides aerobic trench conditions with high volume of pore space in gravel/rock (or other) media, offering added fixed film/attached growth treatment in media.

- Trench storage capacity helps prevent saturated soil conditions at soil/trench interface, which is particularly important during wet weather conditions (as long as proper grading is used around trenches).

- Trench depth is within plant evapotranspiration intervals yet deep enough for protection from cold weather and digging rodents, which can be factors for drip dispersal.

- Design and loading rate flexibility.

- Well suited to a very wide range of site conditions, including slopes and soil types.

- Low maintenance requirements relative to other pressure distribution methods such as drip or surface application (with drip needing routine back-flushing of emitter lines and prefiltration to prevent emitter clogging, and surface application needing monitoring of effluent quality).

- Uses nonproprietary system components that can usually be purchased locally or at least regionally throughout the world.
- Very long useful service life when properly designed and installed.

The basic elements of an LPD system are

1. Effluent pump tank

2. Pump(s) (high- or low-head effluent pumps, depending on flow and head requirements)

3. Effluent filter or screen in pump vault (recommended unless very high effluent quality is reliably entering the LPD field dosing tank)

4. Pump controls, including floats, timers, control panel, etc.

5. Discharge piping and valving

6. Manifold piping leading to LPD field(s) (typically Schedule 40 PVC)

7. Valves en route to LPD field(s)

8. LPD trenches, with drain media (rock/gravel or synthetic media) and geotextile fabric

9. LPD distribution lateral lines (perforated Schedule 40 PVC pipe)

Additional components may be needed, depending on the system configuration.

10.1   Low-Pressure Distribution Design Approaches

Several good reference documents have been developed since the early 1980s on the design of LPD, or low-pressure pipe (LPP) distribution systems. An excellent resource mentioned in Chap. 9 is the "Design and Installation of Low-Pressure Pipe Waste Treatment Systems," a 31-page publication by the University of North Carolina Sea Grant College (UNC-SG-82-03). That guide, while somewhat dated relative to pumps, controls, and other materials and equipment used for LPD systems, presents a relatively easy-to-follow simplified approach to designing small-/residential-scale LPD systems. Since the 1980s there have been many improvements with components used for effluent dosing and controls, much of which is discussed in this chapter. One potential problem with use of the UNC guide is that the friction loss calculations are not made directly using engineering hydraulic equations that can be tailored to the specific system configuration, but rather from preassumptions made about (1) coefficient of friction to be used for piping and (2) amount of minor friction losses due to pipe fittings and valves. The guide is therefore considered best suited to small projects because of the lesser potential error from those assumptions, as compared with larger systems having longer runs of pipe and greater numbers of valves and appurtenant components. However, the basic design approach presented is sound.

Another very good and much more current design resource for LPD dispersal is "Recommended Standards and Guidance for Performance, Application, Design, and Operation & Maintenance: Pressure Distribution Systems," published by the Washington State Department of Health, Division of Environmental Health, in July 2007. Appendices in that document present basic hydraulic equations used for designing piping networks and sizing pumps, along with detailed explanations for important parts of designs and comparisons of different approaches. The document also includes some good information on dosing siphons, which may be used on sites with enough elevation change, and where there may not be a reliable source of power for pumps or sufficient power supply. Although the context of the guidance document is compliance with rules and design practices for onsite wastewater systems in the State of Washington, it provides valuable information about methods and materials currently used for LPD systems. The link for downloading that document can be found on this book's website under Chap. 10 of Resources and Helpful Links.

10.2   LPD Design Procedures and Calculations

An LPD design approach is presented here with explanations of key steps in the process. The approach has been found to result in successfully functioning systems in a wide range of geographic terrains. Basic steps include

1. Have a detailed topographic survey performed for LPD system areas.

2. Lay out (on paper or computer drawings) LPD lateral lines in dispersal field area(s), manifold pipe routing, and any valves needed between the pump tank(s) and field area(s).

3. Calculate flows from lateral lines and dosed field zones based on layout.

4. Calculate total dynamic head requirements for pump(s) based on the system layout and flows.

5. Select type and model of pump(s) based on flow, head, and other system requirements.

6. Determine control settings for pump(s), including whether demand or timed dosing will be used, relative float setting elevations (relative to top of pump tank or some other reasonable permanent elevation on the pump tank).

Each of these steps in the process is discussed below.

10.2.1 Topographic Survey Used for the Design

An accurate topographic survey of the entire LPD system area should be obtained and used for the design, including the pump tank(s) that will dose the field area(s), the route(s) to the field(s), and the field areas. The surveyed area should include enough added distance around the perimeter of the field for planning, with at least approximately 25 to 50 ft (7 to 15 m) upslope and laterally, and about 50 to 100 ft (15 to 30 m) downslope from the field area(s). Contour intervals of no more than 1 ft (0.3 m) should be used for the field area(s) survey. For steeper sites, it may be acceptable to use larger contour intervals for long runs of gravity or pressure pipe leading to and from the dosing tank and LPD field, but usually no more than 2 ft (0.6 m) elevation difference between contour lines is recommended unless slopes are very steep. Judgments may need to be made in that regard relative to survey costs, character of terrain (including difficulty for survey crews to get detailed topography), and distances between components for which grades are critical.

10.2.2 LPD Field, Manifold Piping, and Valve Layout

The total length of distribution piping needed depends on the design flow, the soil loading rate to be used, and an assumption made as to the absorptive area served by each linear foot of distribution line. That will depend on trench width, and tends to vary by soil type, with greater lateral/trapezoidal movement to be expected with increased clay and fines content. Recommendations for soil loading rates were discussed in Chap. 9. Most permitting authorities regulating decentralized/onsite wastewater systems specify allowable soil loading rates for specific soils and levels of treatment. They also may specify allowable assumptions for absorptive area per length of trench. That assumption is based on a determination of the soil treatment area encountered by downwardly migrating effluent, typically trapezoidal in shape (except for very coarse or sandy soils). For class II, III, and IV soils, and distribution trench widths ranging from 6 to 12 in (0.15 to 0.3 m), an assumption of 3 ft^2 of absorptive area per linear foot of trench (0.91 m^2 per every linear meter of trench) is reasonable. With sands, for which there's a significantly greater downward component of infiltration, the assumed area might need to be reduced by about one-third per length of distribution trench (2 ft^2 of absorptive area per linear foot of trench, or 0.6 m^2 per every linear meter of trench).

Several basic rules of thumb can be helpful when laying out LPD field lines to achieve sound designs. Those include

- Lay out all dosing/distribution lines along contours.
- Limit distribution line lengths between manifolds delivering effluent to lateral lines and ends of lateral lines to no more than about 70 ft (21 m).

- Distances between effluent dispersal/lateral lines should be no less than 3 ft (1 m) unless soils are very sandy, and typically 4 to 5 ft (1.2 to 1.5 m) to allow construction equipment to straddle trenches without driving on them, depending on the particular excavating equipment and trench width used. This will be discussed in detail later in this chapter. Greater distances between lines and trenches may be needed for steeper slopes or where distribution lines need to avoid trees or other features.

- The elevation difference between the lowest and uppermost distribution lines in any one zone should not be more than 2 to 3 ft (0.61 to 0.91 m). For sites with steeper slopes, greater numbers of effluent delivery manifolds with appropriate valving may be needed.

- Limit total length of distribution/lateral lines in each zone to (1) what pumps of the size and type to be used can handle and (2) flows that can be accommodated by the pump tank size and configuration and dosing volumes to be used.

There are a multitude of variations on system configurations that could potentially be used for LPD fields and pumping systems. Experience with laying out LPD fields tends to increase the efficiency of that process, and reduce numbers of iterations needed to produce layouts that will work well for sites.

An example of a layout for a medium-sized residential system is shown in Fig. 10.1. The design flow for the residence served by that LPD field is 420 gal/day (1590 L/day). The soils were found to be class III (a clay loam), and a soil loading rate of 0.2 gal/day · ft² (0.81 cm/day) was to be used. It was determined that primary septic tank treatment would be sufficient prior to subsurface effluent dispersal, based on the depth and type of soil. Dividing the flow by the soil loading rate, a total soil absorption area of 2100 ft² (195 m²) was needed. Assuming 3 ft² per linear foot of distribution trench (0.91 m² per linear meter), a minimum of 700 linear feet (213.4 m) of distribution trench and lateral line piping were needed.

For easier and more cost-effective construction, a center-to-center separation distance of 4 ft (1.2 m) between adjacent lateral lines was used. The overall field size used was therefore 25 percent greater than the minimum required. For this particular site, that was not a problem. For some sites it may be necessary due to space constraints, or for avoiding tree/vegetation removal and/or site disruption, to use the minimum area needed or required. In those cases the engineer/designer of the system should coordinate as needed with the installer to ensure that equipment and construction sequencing adequately allow for protection of adjacent trenches. Where regulatory inspections are needed, it may be necessary to call for additional inspections so that portions of the LPD field can be completed and lightweight track hoes or other very low-pressure equipment can work over the

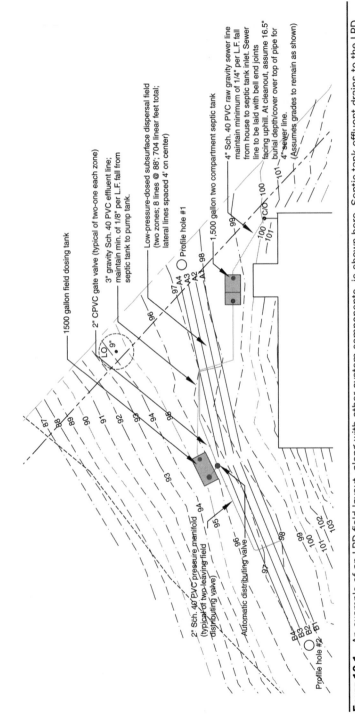

1500 gallon field dosing tank

2" CPVC gate valve (typical of two-one each zone)

3" gravity Sch. 40 PVC effluent line; maintain min. of 1/8" per L.F. fall from septic tank to pump tank.

Low-pressure-dosed subsurface dispersal field (two zones; 8 lines @ 88'; 704 linear feet total; lateral lines spaced 4' on center)

Profile hole #1

1,500 gallon two compartment septic tank

4" Sch. 40 PVC raw gravity sewer line maintain minimum of 1/4" per L.F. fall from house to septic tank inlet. Sewer line to be laid with bell end joints facing uphill. At cleanout, assume 16.5" burial depth/cover over top of pipe for 4" sewer line. (Assumes grades to remain as shown)

2" Sch. 40 PVC pressure manifold (typical of two-leaving field distributing valve)

Automatic distributing valve

Profile hole #2

FIGURE 10.1 An example of an LPD field layout, along with other system components, is shown here. Septic tank effluent drains to the LPD field dosing tank, which is located just downslope of the field. Positive hydraulic pressure head between the pump and the field avoids the need for an antisiphon device, and helps produce more stable hydraulics to the field (U.S. Customary units shown).

completed field areas. Each site and project tends to be a little different in this respect, so it's important for the designer to be cognizant of those considerations when laying out the field.

Two alternating field zones are used for this system to allow dosing and resting cycles for each zone. Based on the width along the contours of a naturally terraced area clear of trees, it was determined that four field lines could be placed each 4 ft (1.2 m) apart. Dividing 700 linear feet (213.4 linear meters) by 2 (for two field zones), and that number divided by 4 (number of distribution/lateral lines), gives 87.5 linear feet (26.7 m) per lateral line, for each of the eight lines in the two zones. Rounding up, each line will be 88 ft long (no rounding needed for SI units here). For the two alternating field zones used for this design, a hydromechanically actuated (nonelectric) valve was used to switch fields dosed each time the pump turns on, and using a "demand" dosing approach. Demand-dosing relies on float settings in the pump tank to actuate the pump, with on-off float elevations set to dose a certain volume from the tank (based on elevation difference in the tank for that volume) at a time. This will be discussed later in this chapter along with timed dosing.

Note on the layout in Fig. 10.1 that the profile holes are located outside of the dispersal field area, though close enough to have provided representative subsurface site information. The lateral/distribution lines are laid along contours on the hillside as much as possible. They were located where there was a slightly terraced area, and where no trees would need to be removed. While they are shown exactly parallel in the drawing, small-diameter PVC lines can assume a certain amount of curvature, so the LPD trenches can actually be excavated so as to align more closely to the contours. The septic tank is located relatively near the house, further up the hill, to prevent long runs of gravity sewer lines carrying raw sewage and potential disturbance to quiescent conditions needed for primary settling, and to allow better access by a pump truck for periodic septage removal. The field dosing tank typically needs pumping less frequently. Both the septic tank and the field dosing tank are also oriented along the contours, since they must be installed level, and this alignment minimizes cut/fill and disturbance. The LPD distribution lines are "center-fed" in this layout. Each of the four lines in the two dosing zones is too long for manifolds to feed lines from one end. Due to friction losses and pressure differences along distribution lines with increasing distance from delivery manifolds, center-feeding is recommended unless lines are less than about 50 ft (15 m) long.

10.2.3   Calculating Flows for Distribution Lines and Zones

To size pumps, flows must be calculated for each of the field zones served. This is calculated for each hole in each of the lateral lines, and then summed for the whole distribution zone. Using the same discharge orifice size for each of the distribution lines, spacing of orifices is adjusted for elevation changes with different lines. In the alternative, orifice sizing could be adjusted. Table 10.1 shows an example of

Line #	Elevation (ft)	Pressure Head (ft)	Line Length from Center to End (ft)	GPM per Orifice	Calculated # of Orifices	Rounded # of Orifices	Orif. Spacing (ft)	Orif. Spacing (in)	GPM/ Line	GPM per Full Trench Length (Column J × 2)
A1	98.00	2.0	44	0.407	11	11	4.10	49.2	4.48	8.96
A2	97.75	2.3	44	0.432	10.37	10	4.56	54.7	4.32	8.64
A3	97.35	2.7	44	0.468	9.56	10	4.56	54.7	4.68	9.36
A4	97.10	2.9	44	0.490	9.14	9	5.13	61.5	4.41	8.82
								Total flow for zone A:		35.78

Line #	Elevation (ft)	Pressure Head (ft)	Line Length (ft)	GPM per Orifice	Calculated # of Orifices	Rounded # of Orifices	Orif. Spacing (ft)	Orif. Spacing (in)	GPM/ Line
A1	98.00	2.0	88	0.407	22.5	23	3.91	46.9	9.36
A2	97.75	2.3	88	0.432	21.68	22	4.10	49.1	9.50
A3	97.35	2.7	88	0.468	19.68	20	4.53	54.3	9.37
A4	97.10	2.9	88	0.490	19.10	19	4.78	57.3	9.31
							Total flow for zone A:		37.54

Line #	Elevation (ft)	Pressure Head (ft)	Line Length from Center to End (ft)	GPM per Orifice	Calculated # of Orifices	Rounded # of Orifices	Orif. Spacing (ft)	Orif. Spacing (in)	GPM/ Line	GPM per Full Trench Length (Column J × 2)
B1	98.00	2.0	44	0.407	11	11	4.10	49.2	4.48	8.96
B2	97.80	2.2	44	0.427	10.49	10	4.56	54.7	4.27	8.54
B3	97.30	2.7	44	0.473	9.47	9	5.13	61.5	4.26	8.52
B4	96.80	3.2	44	0.515	8.70	9	5.13	61.5	4.63	9.26
								Total flow for zone B:		35.28

Based on Fig. 10.1 layout; English units.

TABLE 10.1 Example of Lateral Line Orifice Spacing Calculations

these calculations done for the field layout shown in Fig. 10.1, using the same sized discharge orifice for each of the eight field lines. Torricelli's equation, derived from Bernoulli's equation, is used for these calculations, with

$$Q = C_d A_o (2gh)^{0.5}$$

where Q = discharge from the orifice, ft^3/s or m^3/s
  C_d = unitless orifice coefficient (assumed here to be 0.6)
  A_o = cross-sectional area of the orifice, in ft^2 or m^2
  g = acceleration due to gravity = 32.2 ft/s^2 in U.S. Customary units, or 9.81 m/s^2 in SI units
  h = residual water pressure head at the orifice, in ft or m

Consistent units need to be used for the above equation, whether U.S. Customary or SI, with conversions made as needed. For Fig. 10.1 which is in U.S. Customary units, Table 10.1 gives the discharge rates from holes for the lines calculated at each of the four elevations, with h being the only factor that varies.

Orifice discharge coefficients for pipe tend to vary based on orifice size, pressure heads, and flow rates. A coefficient (C_d) of 0.60 is assumed here for the range of residual pressure heads to be used. The orifice size is assumed to be 5/32 in (4 mm). Effluent screens commonly used for decentralized systems typically pass particles up to 1/8 in in diameter, so using a slightly larger orifice size will tend to let those particles pass and not clog the field lines. For effluent expected to be very reliably free of suspended solids and organic matter, orifice sizing of 1/8 in (3.2 mm) or possibly as small as 3/32 in (2.4 mm) might be used, particularly if it helps with the flows/hydraulics and system configuration for pump sizing.

Field pressures are set so that the top line in each zone has a residual pressure head (h) of 2 ft (0.6 m) above the lateral line as measured at the distal end. For center-fed manifolds, the first orifices next to the manifold are best placed about 1.5 to 2 ft (0.46 to 0.6 m) away for each half of the line, and 6 to 12 in (15 to 30 cm) away from the ends.

The calculations to determine the number and spacing of orifices for each line can be done either by (1) treating each half of the lines separately or (2) calculating orifice spacing across the entire line. The former allows placement of the manifold exactly in the middle of each line, but only allows numbers of orifices to change from line to line by multiples of two, with less balancing of the orifice numbers and flows between lines for sloped sites to achieve more uniform distribution. The latter spacing calculation has the disadvantage of requiring that the manifold be offset from the center for lines with odd numbers of orifices, since the middle orifice will land exactly where the manifold would tee into the distribution line. Depending on numbers of lines and the orifice spacing, this can result in the manifold being too close to one or more orifices, and requires more attention to detail by the installer of the system. The example below presents results for both approaches, as summarized in Table 10.1.

First calculating orifice spacing for each half of each distribution lines at each contour elevation, the number of holes for each half is determined to be

$$[(L_d/2 - D_{oc} - D_{oe})/D_o] + 1 = N_o$$

where L_d = full length of the lateral/distribution line at that contour elevation

D_{oc} = distance selected for use between the center-feed manifold and adjacent orifices

D_{oe} = distance between the last orifice on the line and the end of the distribution line

D_o = spacing between orifices

N_o = number of orifices

For this example, the following assumptions are made initially:

L_d = 88 ft (26.8 m)

D_{oc} = 2 ft (0.61 m)

D_{oe} = 1 ft (0.3 m)

D_o = 4 ft (1.22 m)

N_o is first calculated with the distance between holes *initially assumed* to be 4 ft (1.22 m) for the highest line in the field, and with that distance increasing for lines further downhill. That result gives 11.25 orifices. Rounding to the nearest whole number gives 11 holes. Recalculating, and this time solving for D_o gives an exact distance between orifices for the uppermost line of 4.1 ft (1.25 m) between each hole for the top lateral line.

Table 10.1 shows the basic calculations carried forward. For lines located further down the hill, the residual pressure head (h) will increase. Torricelli's equation is used to calculate the discharge from the orifices at each elevation (fifth column from the left). To maintain approximately the same effluent distribution, the spacing between holes must increase in a linear fashion with decreased elevation. For uniform distribution across the field zone, the combination of orifice spacing and orifice discharge calculated for that line elevation should balance so that the total discharge from each field line is approximately equal (with less than 10 percent variation). The number of holes for lines further down the hill can therefore be calculated using the total flow calculated for the uppermost line, and dividing that by the flow from a single orifice for the lower distribution lines. Those results are then rounded to the closest whole number, and the exact spacing then calculated. After calculating the subtotaled flow for that half-line using that number of orifices, it is doubled to determine the flow for the full length of line.

The rightmost column in Table 10.1 shows that flows for each line in zone A are well within 10 percent of each other with this layout. For field zones that are not center-fed, the orifice spacing for each entire length of line can be calculated in a similar fashion for the full length of line. The length of each line used in this example would exceed the recommended length for end-fed lines, but the method for calculating orifice spacing is essentially the same.

For calculating the flows for entire lengths of lines, D_{oc} is eliminated (since the exact orifice spacing across the whole line including the middle section will be calculated), D_{oe} is multiplied by 2, and L_d is used instead of $L_d/2$. The equation then looks like this:

$$[(L_d - 2D_{oe})/D_o] + 1 = N_o$$

(orifice spacing calculation used for entire length of line)

The difference in flows for the two scenarios is due to (1) the slight variation in spacing at the middle of the line and (2) rounding up from 22.5 to 23. If 22.5 orifices were instead rounded down to 22, the total calculated flow from zone A would be 35.73 gpm, almost exactly what was calculated for the half-lines.

The results of these calculations for the field zones can then be specified on the plans for the system installer, and orifice holes marked and drilled in the proper locations. Figure 10.2 shows how trenches and distribution lines must be slightly offset for this example, when doing the calculations for whole distribution lines. The manifold is not exactly at the center of some of the distribution lines. For these types of reasons, experienced installers often drill lateral lines and lay them out with the manifold prior to completing trench excavation, to avoid having to put additional fittings into manifolds. The more fittings needed, the higher the system friction/energy losses and materials costs.

Pumps are manufactured with certain flow and pressure head ranges, so the total flow per field zone needs to be in a range that falls within relatively efficient operating ranges for pumps to be used. A variety of other factors must be considered also, including minimum pump run times, minimum recommended field dosing volumes (to be discussed later in this chapter), float settings, and pump tank

Figure 10.2 Offset LPD distribution lines to prevent conflicts with orifice positions.

depth and dosing volume per unit depth, and the desired number of effluent doses applied to the field daily. Changing one or more of these factors can significantly change other factors with potentially unwanted results. Balancing these various elements of systems designs is learned through experience with a variety of system configurations. While some software has been developed that allows input of some or perhaps even all of these factors, care must be taken to ensure that assumptions made by the software model for those calculations are applicable to and appropriate for the specific system.

10.2.4 Calculating Pump Pressure Head Requirements

The flow requirements for the pump(s) dosing the field(s) are calculated using the results of the previous step (sum of last column in Table 10.1), and the configuration of the piping, valve(s) and appurtenances between the pump and the distribution orifices. Based on that flow requirement, elevation differences for the total system, and "major" and "minor" friction losses, the pressure head needed for delivering that flow to the LPD field can be determined through system head calculations. That total head should include at least 2 ft (0.61 m) of residual head at the uppermost distribution field lines.

"Static head" requirements are calculated as the difference between the water surface in the pump tank and the elevation of the residual head for the uppermost field line in the zones dosed by that pump. Since the water surface elevation in the pump tank changes, the lowest possible water surface elevation for any operating conditions should be used. Table 10.2 shows calculations for system pressure head requirements for the example shown in Fig. 10.1 and Table 10.1.

The basic hydraulic equation used for this is the Hazen-Williams formula, which is an empirical formula developed for water flow in pipes flowing full with a moderate velocity range (<3 m/s or 10 ft/s).[1] In U.S. Customary units the equation is

$$V = 1.318 C_{HW} R_H^{0.63} S^{0.54}$$

where V = velocity of flow through the pipe = Q/A, where Q is the flow rate and A is the pipe's cross-sectional area

S = slope of the energy gradient line (EGL) = head loss per unit length of pipe = h_f/L

R_H = hydraulic radius = the water flow cross-sectional area A divided by the wetted perimeter P; for circular pipe, $A/P = (\pi D^2/4)/\pi D = D/4$

C_{HW} is the Hazen-Williams coefficient for friction losses through pipe, and ranges from about 150 for very smooth PVC or polyethylene pipe down to about 80 or 90 for very old rough and deteriorating

Q_M(gpm)	Q_M(ft³/s)	D_M(ft)	A_M(ft²)	V_M(ft/s)	H_s^*(ft)	L_M(ft)	C(Haz.)	Sum of K_M	H_f(ft)	H_m(ft)	$H_f+H_m+H_s$	
15	0.033	0.167	0.022	1.53	18.6	65	130	31.5	0.43	1.15	20.18	
20	0.045	0.167	0.022	2.04	18.6	65	130	31.5	0.74	2.04	21.38	
25	0.056	0.167	0.022	2.55	18.6	65	130	31.5	1.11	3.19	22.90	
30	0.067	0.167	0.022	3.06	18.6	65	130	31.5	1.56	4.59	24.75	
35	**0.078**	**0.167**	**0.022**	**3.57**	**18.6**	**65**	**130**	**31.5**	**2.07**	**6.25**	**26.92**	**26.92**
40	0.089	0.167	0.022	4.08	18.6	65	130	31.5	2.65	8.16	29.42	
45	0.100	0.167	0.022	4.60	18.6	65	130	31.5	3.30	10.33	32.33	

Q_D(gpm)	Q_D(ft³/s)	D_D(ft)	A_D(ft²)	V_D(ft/s)	H_s(ft)	L_D(ft)	C(Haz.)	Sum of K_D	H_f(ft)	H_m(ft)	$H_f+H_m+H_s$	
3	0.007	0.104	0.009	0.78	0	44	130	2	0.15	0.02	0.17	
4	0.009	0.104	0.009	1.05	0	44	130	2	0.25	0.03	0.28	
5	**0.011**	**0.104**	**0.009**	**1.31**	**0**	**44**	**130**	**2**	**0.38**	**0.05**	**0.43**	**3.45**
6	0.013	0.104	0.009	1.57	0	44	130	2	0.53	0.08	0.61	
7.5	0.017	0.104	0.009	1.96	0	44	130	2	0.80	0.12	0.92	
10	0.022	0.104	0.009	2.61	0	44	130	2	1.36	0.21	1.58	
										Total head required		**30.37**

* Includes 2 ft of head above the uppermost distribution pipe elevation in the field, and losses for distributing value and discharge assembly at given flow.

Notes: $H_{f(pipe)}$ = headloss from pipe flow (pipe friction losses $[H_{f(pipe)}]$ are calculated using Hazen-Williams formula.)
$H_{m(minor)}$ = headlosses from fittings and valves
$H_{s(elev)}$ = elevation head
C = coefficient used for PVC pipe is 130; Q_M = flow through the manifold from the pump tank to the distribution lines
Q_D = flow through each leg of the distribution lines (for this example, there are eight legs for each zone havng four lines.)

To simplify calculations, it is assumed here that the pressure head at each of the distribution lines is equal. That will not be the case for multiple distribution lines placed along hillsides. However, the error introduced by that assumption is very minor.

Minor losses associated with manifold piping to field zones includes the following: pump discharge assembly (counted in static head), including check valve and ball valve; 2 to 1-1/2 in reducer; 90° elbow; distributing valve (counted in static head); swing check valve; 1-1/2 to 2 in expansion fitting; gate valve (assumed 3/4 in closed here); 90° elbow; 3-tees, straight through; 1-tee, and 90° turn and 1/2 size reduction (actually 2 to 1-1/4 in). The distribution lateral line losses include a reducing tee.

TABLE 10.2 Pump Pressure Head Calculations for LPD Pump Sizing

324

pipe; C_{HW} is a dimensionless coefficient. Roughness values of 120 to 140 are typically used for designs with PVC pipe, depending on assumptions about any buildup inside the pipe. In SI units, the Hazen-Williams formula is: $V = 0.85 C_{HW} R_H^{0.63} S^{0.54}$

Nomographs (or nomograms) have been developed that solve the Hazen-Williams formula, and are often used for calculating friction losses h_f. Solving the equation for S gives friction losses per length of pipe from $S = h_f/L$, in either SI or U.S. Customary units using the correct conversion factor k:

$$h_f = L[(V/kC)(4/D)^{0.63}]^{1.852}$$

where h_f = major friction losses in feet or meters
L = length of pipe run in feet or meters
V = flow velocity in either ft/s or m/s = Q/A
Q = flow in either ft^3/s or m^3/s
A = cross-sectional area of pipe in either ft^2 or m^2
k = 1.318 for U.S. Customary units and 0.85 for SI units
C = Hazen-Williams roughness coefficient
D = pipe diameter in either feet or meters

"Minor" friction losses from valves and fittings in piping systems can be calculated from

$$h_m = K(V^2/2g)$$

where h_m = minor friction losses in feet or meters
K = sum of minor friction coefficients
V = pipe flow velocity in ft/s or m/s
g = acceleration due to gravity = 32.174 ft/s^2 in U.S. Customary units, or 9.8066 m/s^2 in SI units

Values for K have been developed for a wide range of pipe bends, fittings, and valves. Some of the common K factors used are listed below, along with their values used for the example above.[2] For a particular LPD system layout, these are summed together, and multiplied by $V^2/2g$ to calculate the friction/head losses due to fittings.

Some manufacturers of decentralized systems components provide head loss curves for pumping system equipment they sell (e.g., Orenco), including piping discharge assemblies for pump tanks and hydromechanically actuated distributing valves (like the one indicated on Fig. 10.1 that alternates dosing between field zones). For the calculated flow for the system, manufacturers' information can be referred to for determining the amount of head loss from those components, and simply added to the static head for the manifold piping.

Fitting or Valve Description	K Factor (or Coefficient), Unitless*
90° elbow or bend, standard 2 in	0.6
45° elbow or bend	0.3
22-1/2° elbow or bend	0.2
Tee (straight through)	0.6
Tee, reduced by 1/2 with a 90° turn	2.0
1-1/2 to 2 in (40 to 50 mm) expansion ($d/D = 0.75$)	0.2
2 to 1-1/2 in (50 to 40 mm) reducer ($d/D = 0.75$)	0.2
2 to 3 in (50 to 80 mm) expansion	0.4
Swing check valve	2.1
Gate valve, 1/4 open	24.0
Gate valve, 1/2 open	5.6
Gate valve, full open	0.19

*Values cited vary by source, and will change with pipe size and radius of curvature. Values shown are approximate for pipe sizes used for the system in Fig. 10.1. Hydraulic literature should be referred to for selecting K factors appropriate to each system.

As a rule of thumb, pipe sizes should be selected so as to maintain "scouring velocity," which is taken to be between about 2 and 7 ft/s (0.6 to 2.1 m/s). Using the middle to lower end of that range will help prevent scum or particulate buildup in lines while not generating excessive friction energy/head losses.

10.2.5 Pump Selection

Based on the flow, static head requirements and head loss calculations, the pump(s) needed to serve the system can be selected. For this LPD system example, a pump capable of delivering at least 36 gal/min (2.3 L/s) at a pressure head of 35.5 ft (1.06 Pa, or N/m²) is needed. Figure 10.3 shows the pump curve for the vertical turbine 1/2-hp (0.37-kW) pump used for this system. The pump is capable of passing up to 1/8-in (3.2-mm) solids, which would be needed for a pump dosing septic tank effluent from a screened pump vault passing solids up to that size. The design operating point for the LPD system is plotted under the curve, using the calculated flow and pressure head requirements from Tables 10.1 and 10.2. Operating points in the middle of the curve tend to fall into the most efficient operating

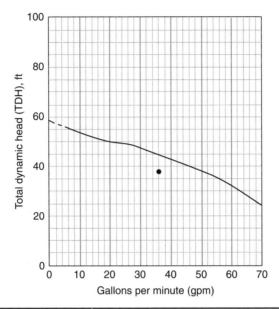

FIGURE 10.3 This manufacturer's pump curve is for a 1/2 hp (0.37 kW) high head effluent pump.

conditions, and since the point is near but still below the curve, this particular pump seems to be a good fit for this system.

Two types of effluent pumps commonly used for small- to mid-sized decentralized wastewater systems are lower-head centrifugal pumps and higher-head vertical turbine pumps. Most manufacturers of low head effluent pumps offer 1-year warranties on those pumps. Vertical turbine effluent pumps tend to have significantly longer service lives, with certain manufacturers (e.g., Orenco) offering 5-year prorated warranties for their high head effluent pumps. An advantage of the Orenco high head pumps is that they will pass solids up to 1/8 in (3.2 mm) in diameter, whereas high head pumps currently sold by most if not all other manufacturers will only pass solids up to 1/16 in (1.6 mm). That will either significantly increase servicing needs for the pump or reduce its service life for systems dosing septic tank effluent or those relying on effluent filters or screens, since effluent filters/screens are typically manufactured to pass solids up to 1/8 in (3.2 mm) across. If the intake screen on the pump cannot pass the solids, it may blind and overheat the pump.

10.2.6 Pump Controls and Settings

One of two basic approaches may be taken to control field dosing: demand dosing or timed dosing. Demand dosing applies effluent to the field intermittently based on the rate that effluent enters the dosing tank or chamber. Timed dosing operates the pump(s) on specified "on" and "off" intervals. Demand dosing may therefore result in fewer alarm conditions over time given the same quality of equipment, though timed dosing enables use of more system-tailored controls. Where it can be used, timed dosing enables the use of greater numbers of "microdoses" per day, which offers better treatment performance for soil dispersal systems or for dosing predispersal treatment media. Demand-dosing float controls have simpler and typically less expensive panels, but have the disadvantage of having to dose only in volume increments possible for the given tank dimensions and float setting capabilities, and tend to be significantly less precise than is possible with timed dosing.

Mercury level floats are the most common type of float used for small- to midsized decentralized systems, and are used for both timed and demand dosing. Smaller low profile tanks commonly used for residential systems do not have much vertical flexibility for spacing floats on a float "tree." Where floats have to be set relatively close together, there is the risk of pump cycling. Some control panel manufacturers address that by including minimum pump-run-time capabilities in control panels, with that interval set based on the specific float configurations, pump flow rate and amount of draw-down in the tank when the pump operates for a certain amount of time. Figure 10.4 shows a pump vault using four floats, including high water alarm, "on" and "off" floats, and redundant low-level "off" float. The function of the "redundant off" float is to prevent the pump from being damaged by running dry in the event the "off" float fails to turn the pump off.

Notice in Fig. 10.4 the difference in reserve storage capacity above the high water alarm when the entire pump tank is watertight. If tank

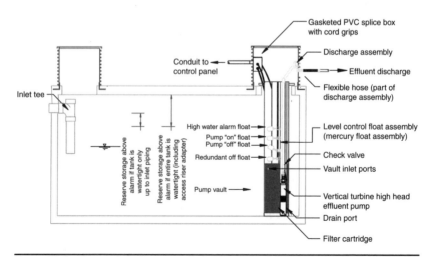

FIGURE 10.4   Demand dosing pump float control configuration.

inlet fittings are not cast so the tank is watertight up to at least the under-side of the top surface/lid of the tank, the reserve storage capacity in the event of a power outage or pump failure is significantly reduced.

The discharge assembly shown in Fig. 10.4 includes a flexible hose assembly. If flexible hose is used instead of Schedule 40 or Schedule 80 PVC, it is recommended that only relatively high-pressure thicker-walled hose is used. It should be rated for say 200 psi (1400 kN/m²) or higher to withstand stresses associated with temperature and pressure changes, including proximity to tank lids in warmer climates that can become quite hot.

Timed dosing has the advantage of enabling delivery of smaller volumes of liquid at a time. For LPD dispersal and sand filter treatment systems, spreading dosing events out through the day has been found to contribute to better overall treatment. Configuring panel settings to use timed doses that won't unnecessarily trigger alarm conditions can, however, be challenging for systems serving residences or commercial facilities having less predictable usage and flow patterns. Controls for timed dosing frequently include timer-override floats and timer settings, for occasions when system flows exceed normal design flows, and water levels in the pump tank rise at a rate exceeding what the normal timer settings can handle. Override settings can be used to reduce the "off" time by some factor, and increase pumping rates until normal operating conditions are restored.

Figure 10.5 shows a float configuration using timed dosing, and with duplex effluent pumps that alternate each time a pump is activated by the timer. For this float setup, the same float turns the pump "on" and "off." If the liquid level is above the "on" position for the float, the timer will activate the pump through its dosing period. It's

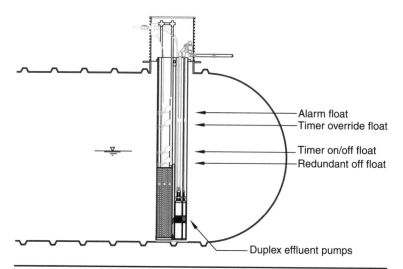

FIGURE **10.5**   Float-tree configuration for timed dosing.

important to make sure that there is sufficient liquid depth between the pump "on"/"off" float and the redundant "off" float that the latter isn't unnecessarily activated when the liquid level is just at or above the "on" position, and the pump draws down the liquid to below the "off" position. This float arrangement also shows a timer-override function. Panels can be built so that some type of cautionary alarm light or signal is activated when the pump cycle goes into an override condition.

When mercury-type floats become damaged or require replacement, they should be carefully handled and properly disposed of, to prevent mercury from entering the environment or water supplies.

The system engineer/designer should carefully evaluate usage patterns for the facilities served, including special events/occasions when flows may be significantly higher, to incorporate enough reserve capacity into the pump tank and to establish float and control settings that won't trigger alarm conditions unnecessarily. Power outages should be anticipated and planned for, in terms of reserve storage capacity in both treatment and dispersal tanks and with controls. The system designer should make sure that if the controls revert to default settings following a power outage, they are appropriate for the specific design, or make sure that someone having experience with the system controls or with detailed enough instructions is available to reset the timer to the correct settings.

It's important to make sure that the specified pump run time or demand-dosed volume doesn't excessively cycle the pump motor, using too short operating times. For dosing pumps housed in pump vaults using effluent filters/screens, the vault inlet hole elevations should be set to allow enough fluid depth above the pump intake to avoid vortexing and lowering of liquid to levels that may damage and/or reduce the service life of the pump.

There are two design considerations when configuring pumping systems for LPD distribution systems that are sometimes found to be at cross-purposes: (1) achieving sufficient pipe "flushing" volumes, and minimum dosing volumes to achieve uniform flow distribution, and (2) using greater numbers of dosing events with smaller dosing volumes to accomplish higher levels of soil treatment. This is particularly so for LPD systems dosing septic tank effluent, for which minimum flushing volumes are needed to prevent excessive buildup of biomat or solids in lines with higher BOD and TSS levels, and where uniform distribution is important for soil treatment processes. Minimum dosing volume is especially important to achieving uniform flow distribution where orifices are directed downward, and draining and refilling of pipes is occurring.

A rule of thumb for dosing septic tank effluent in LPD systems is that the minimum dosing volume is equal to the volume of the delivery pressure manifold plus 5 times the volume of the LPD distribution lateral lines. For the example illustrated in Fig. 10.2, the delivery manifold is relatively short (65 ft or 19.8 m) since it's located adjacent to the field lines. With 88 ft (26.8 m) lateral lines, and for a 2 in (50 mm) pressure manifold and 1-1/4 in (32 mm) lateral lines, the minimum dosing volume would be 122.8 gal (465 L). With a longer manifold, this would be even greater. For the design flow of 420 gal/day (1590 L/day) in that example, there will be on average between three and four doses per day using float settings that will deliver the minimum volume.

A key element of decentralized wastewater systems that contributes to their cost-effectiveness and sustainability relative to the level of overall treatment possible is their optimal use of natural soil treatment processes. A greater number of doses of smaller volumes of effluent takes much better advantage of that potential than smaller numbers of larger doses for domestic wastewater pollutants relying on residence time in soils for treatment. To achieve as much soil treatment as possible, a minimum of four to six dosing events per day is recommended for LPD dispersal systems, regardless of the level of predispersal treatment provided. For soils at each extreme of the textural classification range (very fine or coarse) a greater number of dosing events per day is recommended. For class IV soils that will allow more time for infiltration between dosing events and prevent saturation. Septic tank pretreatment is oftentimes all that may be needed for LPD dispersal in soils with higher clay content, as long as dosing/resting cycles are employed. Therefore, float level or timer settings should be used that deliver minimum dosing volumes the maximum number of times daily. These considerations tend to favor the use of timed dosing where its use is feasible due to its ability to deliver greater numbers of smaller volumes/doses of liquid over a 24-hour period.

It may not be possible to use a preferred higher number of daily doses when applying septic tank effluent, where it's important to flush sufficient volumes of effluent through lines to prevent buildup during each dose. For each project, it's necessary to strike a balance between sometimes competing factors such as these.

For coarse soils that rapidly infiltrate, a higher number of smaller doses will allow more attached growth treatment on the surface of sandy or gravelly soils. Using the minimum recommended treatment levels for class Ia soils in Chap. 9, there is less concern with minimum flushing volumes since a higher effluent quality would tend to contribute much less to buildup and orifice clogging over time. Where higher levels of predispersal treatment must be provided due to marginal natural soil treatment capabilities, and since minimum dosing volumes are not so much of an issue relative to flushing and potential line clogging, a greater number of smaller effluent doses (6 to 10 per day) should be used if at all possible. In that circumstance, there will likely be more frequent routine maintenance visits to the system during which field lines can be checked for any buildup, as compared with systems using only septic tank pretreatment requiring less frequent checks and servicing. That offers an added safeguard and opportunity to do field checks of the more complex control panels and settings often used for timed dosing.

The method of dosing control used in combination with the number of effluent doses to be used per day based on the design flows should match the specific system and conditions, along with any applicable regulatory requirements. Systems engineers/designers should consult with pump and control manufacturers as needed to ensure that the system dosing configuration used is suitable both for the equipment and the project needs.

10.3   LPD Construction Methods and Materials

Most of the equipment and materials of construction used for the installation of LPD effluent dispersal systems are among those commonly used for site development and infrastructure construction activities worldwide. There are some specialized components discussed in this chapter used to enhance reliable and long-term proper functioning of LPD systems, but they are not essential when unavailable in remote locations. Most of those components are relatively lightweight, however, and can be shipped to most locations around the world.

LPD systems discussed in this section use relatively shallow trenches to better facilitate vegetative uptake of nutrients, and evaporation and transpiration processes. Relatively shallow trench depths along with alternate dosing and resting cycles for effluent application have been found to accomplish much better soil treatment with LPD dispersal. Narrower trenches will tend to better facilitate the use of trench sidewalls for effluent absorption and nutrient uptake. Shallower and narrower trenches cause less site disturbance and excavated material that may need to be hauled from the site. Projects having LPD trenches with widths ranging from 6 to 12 in (15 to 30 cm) are described here. Other important aspects of LPD dispersal systems have been discussed in previous chapters, including watertight pump/dosing tanks, and watertight riser adapters, risers, and lids.

As mentioned previously, when planning an LPD system it's important to consider the equipment to be used for excavating trenches, to ensure that there is adequate separation distance between field lines, and that access to each trench for placing aggregate or final cover with equipment is made possible either through the layout, or with the sequencing of construction activities. LPD construction using both a 12-in-wide (30-cm) digging bucket and a 6- to 8-in wheel width (15- to 20-cm) rock trencher are described here.

A potential advantage of using trenchers is that their digging depth is set on the machine, and if they can be aligned to drive along contours, this can help maintain level trench bottoms. However, alignments have to be carefully set to grade and with relatively smooth surface conditions (since bumps will raise the height of the equipment and decrease trench depth). A disadvantage of trenchers in some soils, and especially those with higher clay content, is that sidewalls and bottoms of trenches may become very smeared and sealed, reducing their infiltration potential. In rocky conditions, however, rock trenchers may be needed. Where they can be used, digging buckets will not tend to seal trenches as much because they don't use a rotary action.

Also, trenchers capable of excavating trenches 8 to 12 in (20 to 30 cm) or wider tend to be very expensive to own or rent and less available in many locations, as compared with mini-excavators or backhoes with digging buckets. Trenchers typically use more fuel per length of excavated trench as compared with mini-excavators, particularly if conditions are somewhat rocky. Trenchers are, however, able to cut a much "cleaner" and even sidewall and trench bottom as compared with excavator-mounted digging buckets in fairly rocky areas. If a trencher is to be used, before laying out the field the distance between its wheels should be measured or otherwise determined, to avoid driving over adjacent trenches once they're excavated.

Figures 10.6 and 10.7 below show trenches cut by a digging bucket and trencher, respectively, on the same site having very rocky soils. The digging bucket has dragged larger rocks out of the side and bottom of the trench (Fig. 10.6), and in spots a backhoe-mounted rock hammer was needed to break up rock that was too large (these trenches

Figure 10.6   This trench was cut using a digging bucket, and requires much more bedding sand and backfill material than a trench cut with a trencher (see Fig. 10.7).

Figure 10.7   This photo shows a trench cut with a rock trencher for piping between wastewater system components. Much less bedding sand and backfill material are needed as compared with the trench in Fig. 10.6, although trenchers typically use significantly more fuel per linear foot of trench cut than excavator-mounted digging buckets.

were not used for LPD distribution lines, but for conduit and effluent piping leading to the field area). Rock-trencher-excavated trenches (Fig. 10.7) need much less cushion sand below and above the piping before backfilling, as compared with the digging-bucket-excavated trenches.

Figures 10.6 and 10.7 illustrate some of the trade-offs of using trenchers versus excavators. Operation of trenching equipment requires greater attention to alignments, since there is much less space in the trench for laying piping and fittings. Smaller-diameter Schedule 40 PVC pipe can conform to the small amount of bend in this trench. Fines at the base of narrow trenched excavations can be more difficult and time consuming to remove, while much greater amounts of sand bedding and dispersal field aggregate must be used for wider trenches. Where there are lesser supplies of aggregate, trenched lines may be more feasible. The trenched fines in Fig. 10.7 may be used for backfilling this sewer line trench after placing cushion sand above and below the line.

When trenching with a rock-trencher in very rocky soils, a relatively large number of cutting teeth can be lost from the cutting wheel in the process. Extra teeth should be kept on hand whether the equipment is owned or leased for projects. Cutting trenches with significant numbers of teeth missing results in much less efficient excavation and excessive fuel usage.

As seen in Figs. 10.7 and 10.10, rock trenching generates considerably more dust on sites, which needs to be considered in sensitive watersheds, in areas with steeper slopes where surface runoff may carry sediments, and elsewhere dust may be a problem. Whether trenching or using an excavator, adequate erosion and sedimentation controls should be used. One of the project examples described in Chap. 11 illustrates the use of LPD in a highly sensitive tropical marine watershed where trenching would generate far too much dust and fines, and where trenches must be excavated with a digging bucket, and in some cases hand-excavated, along steeper slopes.

10.3.1 LPD Field Layout and Construction

Before laying out field lines for excavation, access routes for equipment needed for all portions of the project should be identified. If for example, a concrete septic tank or field dosing tank is to be used, and a large concrete tank delivery truck must place the tank(s) near the LPD trenches, ample space must be provided so that the truck will not need to drive over or too close to the field lines. Overcompaction of dispersal field soils should be avoided with heavy equipment, along with any damage to excavated or completed dispersal field trenches or components. Lightweight, tracked, or very low ground pressure (not more than about 5 psi or 35 kN/m^2) wheeled equipment should be used on and around dispersal field areas during construction. It is common for contractors to complain of being "boxed in" by a designer's layouts, while attempting to comply with property owners' wishes to protect trees, vegetation, and other site features.

After determining the type and size of equipment to be used for excavating trench lines and for constructing other portions of the system, the LPD field lines can be laid out. Figure 10.8 shows ground marking around the perimeter of this LPD field area. These LPD trenches will be rock-trenched using a trencher that cuts 6 in (15 cm) wide trenches, and will be placed 54 in (137 cm) on center so that the trencher used will not drive over adjacent trenches. Trenches at least slightly wider (8 in or 20 cm) are preferred due to the difficulty of cleaning out trenched material from the bottom of the trenches. Trenched fines should be removed from the trenches following excavation to prevent sealing or reduced infiltration rates along the trench bottom.

LPD field lines are oriented as closely as possible along surveyed contour lines, typically using a construction laser level. The bottoms of LPD trenches must each be level along their length, with design

Figure 10.8 Perimeter of LPD field laid out and marked for trenches to be oriented along contours.

specifications commonly calling for ±1/2 in (±13 mm). Figures 10.9 and 10.10, respectively, show LPD trenches cut with a digging bucket and a rock trencher.

Standing in narrow trenches to remove trenched material can be difficult. Special narrow ditch-cleaning shovels are needed for clearing fine trenched material out of the excavations before placing drain rock. Trenchers vary in the way that trenched material exits from one or both sides (as in Fig. 10.7). For trenchers sending material out of both sides of the unit and where LPD lines are relatively close together, it may be possible and useful to block off the side adjacent to already excavated trenches, to prevent that material from adding to what must already be cleared from the trench. Sheets of plywood can also be placed over excavated trenches and moved along beside the trencher, with excavated material dumped to one side of the previously excavated trench.

The soils shown in Fig. 10.10 have relatively high clay content and consisted largely of weathered bedded limestone with pronounced lateral bedding planes. Effluent applied to that type of soil would tend to have significant lateral movement, and so the use of narrower trenches is better supported in those soils as compared with those having a much more vertical/downward direction for effluent migration (such as sands).

Trenching/excavation for manifold piping, valves, and all other components in the dispersal field area, which if done later would disrupt or damage the field area, should be completed before cleaning out the

FIGURE 10.9   Trenches excavated with a 12-in (0.3-m) digging-bucket are shown here, placed 4 ft (1.2 m) on center parallel along contour lines on this relatively flat site. These soils were not nearly as rocky as the soils in Fig. 10.10.

FIGURE 10.10   A 6-in rock trencher (15-cm-wide cutting wheel) cut these trenches. This trencher is capable of cutting to a maximum depth of 24 in (0.6 m), which the deepest that LPD trenches should ideally be excavated. A long tape was pulled and staked here to help align the trencher while cutting.

LPD trenches for placement of drain rock media. After leveling and cleaning out bottoms of trenches, rock/gravel media is then placed. The preferred media size range is 3/4 to 2 in (2 to 5 cm), and must be washed free of fines. It should be a hard rock that will not tend to either generate significant fines when being placed in the trenches, or dissolve relatively easily when exposed to water/moisture continuously. Media of that hardness would have a Mohs hardness rating of at least about 4, and cannot be scratched with a fingernail or copper penny. Dolomite rock is about the softest rock that should be used for dispersal field media (Mohs hardness of 3.5 to 4), and doesn't rapidly dissolve or effervesce with dilute hydrochloric acid unless it's scratched or powdered. Figure 10.11 shows washed river rock that is limestone-based but sufficiently siliceous (calcite dissolves and is replaced by silica over time) to be plenty hard enough for use as dispersal field drain rock.

Referring to the LPD trench detail in Chap. 9 (Fig. 9.4), 6 in (15 cm) of the drain rock are placed into the bottom of the trench. Depending on trench widths, numbers of adjacent trenches, available equipment and

FIGURE 10.11   Limestone-based river rock that's sufficiently siliceous for hardness as drain rock.

FIGURE 10.12   A Bobcat-mounted shaker-spreader is used here to place gravel in the dispersal field trenches. This avoids either overcompaction around and between trenches from driving heavier bucket loaders on or around the field, or having to place the gravel by hand with shovels and wheel barrows.

reach of that equipment, various approaches may be used. Figure 10.12 shows drain media being placed by a Bobcat-mounted shaker-spreader, with a close-up view of that equipment in Fig. 10.13. That equipment enables straddling the LPD trenches with the Bobcat, and placement of the media underlying the distribution pipe in this field of about 490 linear feet of trench (150 m) in only 1 to 2 hours.

Schedule 40 PVC distribution/lateral line piping is drilled to the specified orifice hole size and spacing. Orifice holes on lateral lines may be directed upward (12 o'clock) or downward (6 o'clock), depending on the specific system needs and details. Orifice shields (Fig. 7.25) or semicircular pipe sections as shown in Fig. 10.14 can be used to redirect effluent spray onto the drain media if the 12 o'clock configuration is used. An advantage of orifices located on the top/crown of distribution lines is that the piping will remain full of effluent, or "primed," between doses, and achieve steady and more uniform flow much more quickly. For projects in colder climates, the burial depth of

FIGURE 10.13 Close-up view of a shaker-spreader, also used for placing concrete.

FIGURE 10.14 Semicircular pipe sections used as one way of redirecting dosed effluent downward.

lines must be considered relative to frost lines especially for lines remaining full. Another factor is the potential for buildup of solids or biomat in distribution lines, which may occur more for top-oriented orifices especially where septic tank effluent is being dosed, possibly requiring more frequent line-flushing. That is less of a concern with well-filtered or advanced treated wastewater such as sand filter effluent, though lateral lines using either approach should be checked at least annually to see if suspended solids or biomat buildup is occurring and flushing is needed. Where it's desirable to use the 12 o'clock orientation and drain distribution lines between doses, one out of every 4 to 5 orifice holes can be directed downward (6 o'clock).

Orifice holes should be drilled clean and free of shavings, and if possible using a drill press. PVC shavings and any dust or debris should be cleared from the lines prior to installing them in trenches. Figure 10.15 shows distribution lines, each with equal hole spacing for a very flat site, marked, drilled, and ready for installation after placement of gravel in the trenches. Figure 10.16 shows LPD piping being assembled, with the manifold being connected to the four distribution lines before placement of the next 6 in (15 cm) of drain media over the distribution lines. The orifices are facing in the downward (6 o'clock) position in Fig. 10.16.

Drain media is then placed over the LPD piping, as shown in Fig. 10.17 for another project using a loader bucket and shovels to

Figure 10.15 LPD effluent distribution lines with hole spacing marked and drilled clean. It's best to use a drill press to achieve clean orifice holes that will help avoid clogging and buildup around the hole over time.

FIGURE 10.16 LPD lateral/distribution lines and manifold pipe being solvent-welded using PVC pipe primer/cleaner and PVC cement.

FIGURE 10.17 A bucket-loader is used here to place gravel in LPD trenches. These effluent distribution trenches had to be excavated and completed one at a time to avoid driving the loader over adjacent trenches for placing the gravel.

FIGURE 10.18   Large LPD field ready for placement of final soil cover.

smooth/level the media. After placement of the upper layer of drain rock media, geotextile fabric is placed and anchored over the rock media before placement of the final soil cover. The same type geotextile fabric is used for trenches as was shown for sand filters in Chap. 7 (Fig. 7.28), although it can be purchased in narrower rolls and/or cut to fit the width of the trench.

Figure 10.18 shows a large LPD field ready for placement of final cover in a wooded area in a state park where trees are left interspersed between field zones.

Notice in the photos that ends of distribution lines have turn-ups with threaded/screw-on capped ends. The turn-ups and capped ends allow for (1) temporarily connecting a pipe extension (using a length of pipe solvent-welded to a female threaded adapter) for testing pressure head in the field, and (2) flushing field lines if/as needed. It's best to place covered valve boxes over the end of at least one line in each field zone, or metal locating tape or wire in trenches before final cover so that ends of lines can be located later.

One of multiple methods can be used to alternate dosing zones, including manual valves that are periodically switched (e.g., Jandy valves), hydromechanically operated nonelectric valves (e.g., Hydrotek®), and motor-actuated valves. Manually switched valves dose a single field area for some period of time, and are then switched, and so are more limited in the dosing/resting benefits. Valves that cycle between fields are preferred, because they help avoid over applying to a field area while continuing to dose each area and maintain bacterial populations needed for natural treatment processes. Hydroteks have the advantage of not needing

Figure 10.19   A three-port Hydrotek distributing valve.

electric power, and have been found to perform very reliably for properly designed systems, and when properly installed.

Figures 10.19 to 10.21 show 3-port, 4-port, and 6-port Hydrotek distributing valves, respectively, for dosing larger systems having more dosed areas. Some Hydrotek valves are sold without the inlet/outlet

Figure 10.20   Two 4-port Hydrotek distributing valves leading to eight effluent distribution zones.

Figure 10.21   A six-port Hydrotek distributing valve.

pipe fittings attached to the valve. In that case, much greater care needs to be taken with field installations to ensure, for example, proper valve orientation, and that PVC cement doesn't improperly run into parts of the valve and cause malfunctioning. The Hydroteks shown in Figs. 10.19 to 10.21 have inlet and outlet pipe fittings already in place, along with unions and clear pipe sections to visually check operation and cycling of valve ports.

Figure 10.22 shows a 2-port distributing valve enclosure followed by bronze swing check valves, and CPVC gate valves for setting pressure heads in the field zones. Bronze gate valves are also commonly used, but are subject to corrosion and best used for lines conveying secondary or advanced treated effluent. Gate valves are adjusted so that there is the right amount of residual water pressure head above the uppermost distribution line in each of the two zones (typically about 2 ft or 0.6 m). The number of turns from fully closed position is recorded for each valve and kept for future reference. It's best to mark the number of turns on the pipe beside the valve also with a permanent marker. Before final cover is placed, covered valve boxes will be placed over the gate valves. These may have locked covers where there are concerns about tampering (e.g., public park areas), or handles may be removed from the valves and stored.

Hydrotek valves function well as long as the effluent flowing through them is relatively free of grit or other particulate matter that can cause the stem and disk assembly inside the valve to not reset properly as it rotates between outlets to different field zones. The outlet arrangement varies by numbers of ports, so when laying out the

Figure 10.22 A two-port Hydrotek distributing valve used for a small 2-zone dispersal field.

dispersal field and valve piping it's best to check the valve port configuration for the particular number of field zones served, and orient the valve so that fewer bends and fittings are used that would increase head losses and costs. Smooth runs of piping into and out of the distributing valves are helpful, to avoid turbulence and possible air pockets. It may be best to leave about 4 to 5 ft (1.3 to 1.6 m) separation distance between pipe bends or valves and the inlet bend into the Hydrotek valve, with at least about 2 to 3 ft (0.6 to 1.0 m) of pipe on the discharge side before any other valves. If there are high points in the pressure manifold leading to field areas, an air relief valve should be installed regardless of whether a switching valve of some type is used or not.

These automatic effluent distribution valves and their appurtenant field components need to be located so as to avoid air pockets or trapped air in and around the valve, and prevent effluent from draining back through the valve. Depending on the elevation of the dosing tank relative to the dosed field and the orientation of orifice holes in field lines (6 or 12 o'clock), siphoning of effluent into the field also needs to be avoided. Maintaining fully charged/filled lines leading to these valves will help avoid air from being trapped in lines and causing the valve to malfunction. In this case, check valves should be used following Hydroteks if the Hydroteks are more than about two feet (0.6m) below the effluent discharge point at the field, and should be placed before field pressure-setting valves (typically gate valves).

In cold climates it may be necessary to drain lines leading to and away from the Hydroteks between doses to prevent freezing of effluent lines. In that case the effluent distributing valve should be located at the high point in the line, with some means of air release at or just ahead of the valve (e.g., air relief valve), and all or some field line orifices oriented downward (6 o'clock). Some of these valves are sold having an air relief component integral to the valve to avoid problems with trapped air. However that may also result in small releases of effluent out of the valve and onto the ground inside the valve enclosure. While this is probably a very small volume, that element has been removed from some manufacturers' models.[3] The specific valve model manufacturer's literature should be referred to in all cases to ensure proper use, design details and installation procedures.

For setting residual field pressures for each zone, a vertical pipe extension of the same diameter pipe is screwed onto one of the turned-up ends of the distribution line for the uppermost field line. It is easiest to use clear piping, as shown in Fig. 10.23, marking the desired elevation above the lateral line on the tube, and then activating the pump. Where timed dosing is to be used, the time needed to achieve relatively steady flow after the pump is turned on should be recorded, and considered when setting "on" time intervals in the control panel.

After setting field zone pressures, final cover is placed and graded so that surface water drains around and away from the field area. For dispersal fields along slopes, a low profile berm can be constructed at a short distance upslope of the field areas to prevent surface water run-on to the dispersal field. Vegetative cover is then established over the field as soon as possible to prevent erosion and downstream sedimentation, and to provide the necessary plant growth for enhancing treatment processes and nutrient uptake. Only very lightweight equipment should be used for placing final cover over dispersal fields, such as lighter tracked equipment and small skid steers. Soil cover used over dispersal fields should be relatively free of clays, with no finer textural character than a class Ib or II soil. The dispersal field needs to remain as aerobic as possible to facilitate bacterial treatment processes and evaporation of moisture.

Figure 10.24 shows part of a shallow LPD dispersal field located along the slopes of a small garden area. The trenches had to be hand-dug, and needed to be located as close together as possible due to the shortage of space (about 3 ft or 1 m on center). The system was installed to replace an existing system inadequate for remodeling of a home located along a rocky bluff over a sensitive river watershed. Silt fencing and seeded erosion control blankets were used to maintain the final soil cover and quickly reestablish vegetation over the area with the help of a temporary irrigation/sprinkler system placed over that area.

For larger LPD fields, several pieces of equipment that can be useful for raking, seeding, and incorporating the seeded field are shown in Figs. 10.25 to 10.27.

FIGURE 10.23   This clear length of pipe is fitted with a female adapter to screw onto the turned-up section of PVC pipe at the end of the highest distribution line. A mark was made at a level 2 ft (0.6 m) above the distribution lateral, and gate valve opened from a full-shut position, with the number of turns needed to achieve the marked level (arrow) recorded.

Figures 10.28 and 10.29 show well-established grassy areas where LPD field zones are located that serve residential cluster systems. Several park areas like this one are interspersed around the development serving a number of cluster systems. Prior to subsurface soil dispersal, wastewater receives only septic tank/primary treatment. These systems have been in place for about 20 years, and according to regulatory inspectors have been very trouble free.

As discussed in Chap. 7, control panels and breaker boxes should be located in protected areas, such as shaded and covered sides of buildings, or housed in cabinets to protect them from extreme weather conditions. Exposure to very hot temperatures, electrical storms, and

FIGURE 10.24 Hand-trenched LPD field area, with grass seed blanket placed, and temporary irrigation system used to quickly establish the final vegetative cover.

FIGURE 10.25 Grader rake.

moisture can significantly shorten panel service lives and cause malfunctioning of components. Applicable national and local electric codes should always be followed for installing electrical components, and panels and wiring installed by licensed electricians as required and/or appropriate. Pump and panel manufacturers' detailed specifications and instructions should of course always be followed

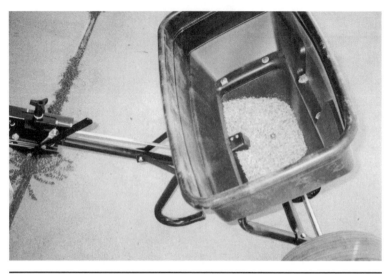

FIGURE **10.26** Seed spreader.

FIGURE **10.27** Seed incorporator (small acreage flexible harrow), pulled behind a Skid Steer.

with installations. These should be left with the system owner or secured in the control panel box, along with all settings used for floats (elevations referenced for example to top of tank or bottom of riser) and timers recorded in writing. Figure 10.30 shows a system control panel and separate breaker box for each of three circuits (one for each of two pumps operating independently and one for the controls) mounted on cross-ties of two cedar posts located adjacent to the

FIGURE 10.28   LPD field serving a clustered development.

FIGURE 10.29   Other LPD field areas of the same cluster system development as Fig. 10.28.

pump tank about 100 ft (30 m) away from the buildings served. A protective enclosure similar to the one shown in Fig. 7.33 will be built around the two boxes. Figure 10.31 shows the breaker box and controls for a much larger LPD system serving a public park, with the panel located in a shaded/protected location on a building. The control panel has a caged high water alarm mounted on top.

FIGURE **10.30** Breaker box and control panel before constructing protective housing around the boxes.

FIGURE **10.31** Breaker and Panel boxes mounted on exterior wall in shaded/ covered spot for a large LPD system serving a public park.

For owners wishing to operate their system using solar-electric-powered pumps, or to have that option in the future, the system design engineer and an electrician knowledgeable on photovoltaic power installations should discuss the system's operation and configuration as needed to ensure that the pump(s) and all of the electric components are compatible.

High water alarms should always be placed on separate electrical circuits from pumps. Since many electricians may not be familiar with panels used specifically for decentralized wastewater systems intended to operate in certain ways, engineers and/or installers for systems should be present when controls are installed and made operational. A separate external breaker for each circuit is safest in case internal breakers in the panel fail to trip. External breakers should be housed in a box either adjacent to the system's control panel, also in a protected location, or within view of the pump station and controls, and in keeping with applicable electric requirements.

Electric splice boxes needed for pumps and floats in pump stations should be gasketed and sealed, and sufficiently resistant to moisture. Appropriate materials of construction will vary by climate, conditions, and applicable electric codes. For splice boxes located outside of pump tanks, conduit should be used for wiring between tanks and the splice box. Depending on the system configuration and venting, and any applicable electric code requirements for commercial and residential systems, wastewater pump tanks may be considered to have ignitable concentrations of flammable gases and subject to methods and materials suitable for those environments. In the United States, the National Fire Protection Association (NFPA) classifies such conditions under their NFPA 820 standard, and whether, for example, certain conditions would be subject to Class 1 Division 1 standards under U.S. electric codes. This is much more likely to be an issue for raw wastewater or septic tank effluent with those associated gases, than for secondary or advanced treated wastewater. Proper safety precautions along with installation approaches ensuring long useful service lives should be used in any case.

Where approved under applicable electrical standards, gasketed and sealed PVC splice boxes may be mounted on sturdy pump tank risers, as shown in Fig. 10.32 for a duplex pump station (indicated by arrow) and in Figs. 10.33 and 10.34. Notice in Fig. 10.32 that all of the wiring, once installed, is neatly looped out of the way of the piping and discharge assembly, and secured with a plastic electric tie next to the splice box. The splice box is secured to the PVC riser with conduit installed through the riser in a grommeted hole (cut with a hole-saw either on site by the installer, or at the engineer-specified height by the manufacturer) as shown in Fig. 10.33. The black electric cord grips are tightened around and secure the wire in the splice box for each of the floats and pump. Notice also the gray PVC unions on the discharge piping in Fig. 10.32 that enable pulling pumps if/as needed.

Wiring should always be placed in conduit of the right type for the specific location, and buried-conduit properly bedded and covered with cushion sand prior to backfilling. Conduit and pull boxes should be carefully located on the as-built plans, with distances referenced to permanent structures or features.

FIGURE 10.32   Duplex effluent pump station with screened pump vault.

FIGURE 10.33   PVC splice box mounted on interior of PVC access riser of pump tank. The black cord grip nuts are tightened around electric cords running to the box from pump(s) and floats in the pump tank.

Figure 10.34 After wire is pulled from the control panel (as shown) to the splice box, and wiring connected for the pump(s) and floats, the splice box will be sealed using a gasketed PVC cover, secured with screws inserted into holes as shown.

10.4 Sand-Lined Trenches and Bottomless Sand Filters

A modification to LPD dispersal commonly used in some U.S. states is the lining of trenches with sand before placement of drain rock media and distribution piping. This approach is intended to provide added treatment for soils having inadequate treatment capacities, such as class Ia soils. About 18 to 24 in (0.46 to 0.6 m) of sand is placed above the natural soil in the trench or bed. This approach is similar to a bottomless sand filter. Monitoring data collected for this type of dispersal method raises concerns about the use of this approach with respect to total nitrogen removal, as well as other wastewater constituents of concern in rapidly infiltrating soils with lower organic content, (e.g., class Ia soils). It does, however, appear effective for providing pathogen reduction to levels comparable to many sand filters in service today.

Figure 10.35 shows this type of dispersal system being constructed following a residential subsurface flow wetland treatment system (described in Chap. 8), and monitored over a period of about 2 years. Approximately 18 in (0.46 m) of concrete sand (ASTM C-33) were placed in the bottoms of LPD trenches, along with two monitoring gravity lysimeters installed at the base of the lowest trench.

FIGURE **10.35**   Sand being placed to depth of 18 in (0.46 m) in trenches.

Lysimeters were built as shown in Fig. 10.36, with samples collected using a Lexan bailer from the sump of the lysimeter (indicated by the arrow). The lysimeters were cleaned and rinsed with distilled and deionized water obtained from a local utility lab prior to placing them. The augered annular space around the monitoring sump in Fig. 10.37, indicated by the arrow, was sealed with bentonite pellets to prevent short-circuiting of the infiltrating effluent into the rock below. Clean/washed pea gravel was placed around the French-drain style perforated lysimeter piping into which effluent drained for sample collection. After installation of the lysimeters, sand was placed in the trench and it was finished off in the same way as the trenches to the left in Fig. 10.38. Semicircular pipe was used over these lateral lines having orifices in the 12 o'clock position (see Fig. 10.14).

Final cover was placed over the field with wheel barrows and shovels, due to the very shallow burial depth of field lines and pipe sections covering the lateral lines. The area was seeded and grass cover established, and the system placed into service. Figure 10.39 shows the LPD field after the grass cover was established.

Monitoring of effluent from each of the unit processes (septic tank effluent, subsurface flow wetland, and field monitoring lysimeters) was then carried out during the next 2 years. Background samples were taken from each of the two field lysimeters before the system was

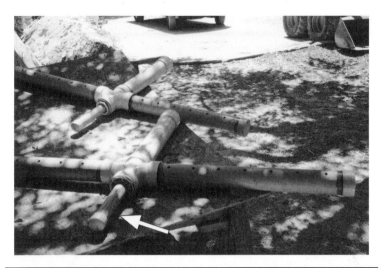

FIGURE 10.36   PVC gravity lysimeters installed for field monitoring (arrow points to sump into which Lexan bailer is dipped for sample collection).

FIGURE 10.37   Installed monitoring lysimeter; arrow points to the sump location, where the annular space around the lysimeter sump is sealed with bentonite up to the bottom of the excavation.

Figure 10.38 Monitoring lysimeters installed in trenches located at most downhill elevation, ready for placement of sand, drain rock, and pipe.

Figure 10.39 Sand-lined trench (pressure dosed) dispersal field with grass cover established.

placed into service. Table 10.3 shows the wetland effluent monitoring results again along with the sand-lined LPD trench monitoring results. There were a number of occasions when there was not sufficient effluent collected in the field lysimeters for analyses, so the

Measured Parameter	Wetland Effluent		Trench Lysimeters Monitoring Results	
	Average	Range	Average	Range
BOD_5 (mg/L)	28.3 (19)	12–64	9.3 (7)	4.2–18
COD (mg/L)	116.4 (23)	46–202	34.3 (15)	8.3–57.9
TOC (mg/L)	34.2 (17)	13–145	32.9 (10)	5.8–160
TSS (mg/L)	17.2 (7)	4–62	ND*	ND
NH_3-N (mg/L)	48.0 (22)	34–74.1	0.73 (13)	0.05–3.01
TKN (mg/L)	45.6 (23)	19.1–66.3	2.3 (14)	0.76–4.37
$(NO_3 + NO_2)$ – N (mg/L)	0.2 (23)	0.02–1.67	40.9 (14)	19.5–111
Fecal coliform (col./100 mL)	2.1×10^4 (15)	733– 1.5×10^5	339 (10)	3–1,000[†]
Fecal streptococcus (col./100 mL)	3.0×10^4 (4)	1.2×10^3– 8.6×10^4	200 (3)	200
Chloride (mg/L)	99.4 (20)	70.2–200	85.5 (10)	55–120
Alkalinity (mg/L)	591 (4)	492–735	ND	ND

Numbers in parentheses for averages indicate numbers of samples collected and analyzed for this parameter.
*TSS was not considered a meaningful parameter here, and was therefore not measured.
[†]One outlier for fecal coliform measured at 11,200 colonies/100 mL was not included in this data summary because there was approximately 1.5 in (4 cm) of rainfall during the 24-hour period preceding collection of that sample that would be expected to contribute surface fecal sources to the lysimeters.

TABLE 10.3 Sand-Lined Trench Monitoring Results[4]

wetland data tends to have significantly higher numbers of sample events.

Background levels (results of samples collected following rainfall events, and prior to commencing effluent dosing to the field) were analyzed for total nitrogen and fecal coliform, in addition to some other parameters. Total nitrogen and fecal coliform were 1.62 mg/L and 11 colonies/100 mL, respectively, for one of the lysimeters, and 0.58 mg/L and 20 colonies/100 mL for the other lysimeter.

Due to the larger volume of sample needed to conduct reliable laboratory tests for such parameters as BOD and fecal coliform, there were only a limited number of analytical results for those. The monitoring results here are not particularly surprising, since they are very similar to what would be expected from a dosed sand filter. The sand treatment media in the trenches provides for very good nitrification, but very little denitrification and vegetative uptake of nitrogen due to the rapid infiltration rates and very low organic content of the sand.

Based on the TKN levels in the wetland effluent, and the levels of TKN and total nitrogen for the lysimeters, it appears that most of the effluent nitrogen from the wetlands is exiting the bottoms of trenches in the nitrate form.

The performance of the sand-lined trenches was found to be much poorer with respect to total nitrogen removal than other traditional LPD systems monitored in similarly rocky limestone conditions. That is likely due to the much longer soil retention times possible in shallower trenches constructed in natural conditions. The added 18 in (0.46 m) of sand caused the depth of the sand-lined trenches to be fairly deep as compared with LPD trenches constructed as described earlier in this chapter. Shallower trench depths allow application of the effluent well within plant root zones, and into soils having higher organic content and therefore larger bacterial populations responsible for treatment processes, including denitrification. A longer residence time in those conditions will also be a key factor for providing much higher levels of treatment for a wide array of other wastewater pollutants.

Based on the above and the fairly large body of performance data available for sand filter treatment units, a better approach for systems to be constructed in very coarse soils having fairly high infiltration rates (e.g., class Ia) might be to use some method of advanced treatment capable of significant nitrogen removal (following septic tank pretreatment), followed by a traditional shallow LPD dispersal field. Sand-lined trenches may, however, be effectively used in clay soils to better nitrify, offer pathogen reduction, and reduce biomat formation along the soil/trench interface by providing better treatment with the sand media for applied effluent. Due to the higher nitrate levels for effluent migrating from the trenches, there would, however, need to be sufficient depth of slowly permeable soils, and lateral distance to any nearby drainage-ways, to prevent adverse impacts to groundwaters or surface waters.

References

1. Ned H. C. Hwang, *Fundamentals of Hydraulic Engineering Systems*, Prentice-Hall, Inc., Englewood Cliffs, N.J., 1981.
2. Chicago Pump Bulletin 9900, "Hydraulics and Useful Information," FMC Corporation 1981; References to various technical sources and universities for different valves and fittings.
3. Telephone communications with Orenco Systems, Inc. technical sales personnel, Sutherlin, Oregon.
4. "Alternative Wastewater Management Project, Phase II," Final Report, prepared for the City of Austin, TX, Water Utility by Community Environmental Services, Inc., August 2005.

CHAPTER 11

Project Examples

11.1 Guiding Principles for Sustainable Systems Planning

Some of the most important elements of implementing sustainable projects must occur in the very earliest stages of planning. These concepts have been discussed in previous chapters. Below is a short list of things that often distinguish the most sustainable projects from others:

1. Get together (whether in person or remotely) with other key project team members very early in the planning process, even before conceptual plans and layouts are developed. Don't leave critical elements like wastewater planning until the end, when structures/roads and other utilities have already been located even preliminarily. It adds unnecessarily to project costs and project team frustrations to have to try to relocate critical project elements. It's best not to wait until team members become attached to their layouts. Project team "politics" may become the driver more than an unbiased evaluation of the most sustainable and cost-effective overall site plan and design. If a decentralized wastewater system is to be used, the ability to evaluate soils/conditions throughout the site for their acceptability for final disposition of effluent is critical to selecting the most sustainable system configurations. Sites are best evaluated from a relative "blank slate," locating all infrastructure elements so they can fit together in an integrated way.

2. The more sustainable systems tend to be those that most effectively use natural processes to accomplish the specific goal. Consuming energy and resources, and generating excess heat and pollutants (added "carbon footprint") to accomplish what naturally available mechanisms can do, is much less sustainable. The most effective and efficient use of natural land/soil treatment processes, and designs that serve to

enhance those capabilities, will result in the most sustainable decentralized wastewater systems. The billions of bacteria naturally resident in just a small cube of most soils are far more efficient with the decomposition of most potential pollutants in domestic waste streams than energy-consuming electromechanical processes. They must simply be given suitable amounts of time and other necessary conditions to do that work. Evaluation and selection of site areas that offer the most effective use of natural processes will contribute to the most sustainable wastewater systems. This relates back to the first item above.

3. The best practitioners are those who understand the limits of their knowledge. Find help and ask questions when an area of the planning exceeds your area(s) of true expertise and understanding. This item relates to both of the above guiding principles for sustainable planning.

11.2   Project Examples

Examples of several decentralized wastewater systems projects are described in this chapter that are located in a variety of settings and land types. While some of these projects are located in relatively pristine and sensitive settings, no judgments are made here relative to the overall sustainability of any of the projects based on their geographic locations. The reality is that development continues to occur in vulnerable settings throughout the world. The goal here is to present ways in which the wastewater service elements of those projects can be accomplished in a sound and sustainable manner, and integrated with the project as a whole. As low impact development (LID) practices are increasingly developed and shown to be effective for other aspects of projects in preventing degradation of natural resources, those too should be implemented alongside of sound wastewater systems.

The project examples are generally presented in order of progressively challenging site conditions, project goals, and availability of material resources. Those elements are described for each. All five of the systems discussed are designed to handle domestic wastewater flows. For each example, the wastewater service needs, owner priorities as discussed with the engineer, and reasons for selecting a particular wastewater service approach are presented. Different combinations of sustainability factors discussed in Chap. 2 are the focus of each project based on communications between the engineer and project owners. Those are outlined for each of the project examples.

11.2.1 Example 1: Rural Youth Camp

Facilities Served and Design Flow

Facilities served by the system include

- Eleven cabins, with nine serving 12 to 16 persons each, and two serving two to three persons each
- Administrative offices
- A dining hall for preparing and serving three meals per day for up to 150 persons
- An infirmary/medical care facility with staff residential quarters
- An indoor recreation facility
- Laundry facilities

The design flow for this system was determined to be 4500 gal/day (17,000 L/day) during peak season (summer use). The camp serves attending youths, staff, and visiting family during summer months, with cabins rented to the public during nonsummer months.

Geophysical Setting

The camp is situated in a rocky and hilly area of the southwestern United States. The average year-round temperature is about 70°F (21°C), ranging from freezing or below-freezing temperatures in the winter months to over 100°F (38°C), in the summer. Average annual rainfall is about 35 in (89 cm)/year. Intense thunderstorms are common to the area, with occasional extended periods of rain, particularly in late spring and early fall months.

Owners' Expressed Priorities

Based on discussions with the camp's owner, the following were identified as priorities for the design of the wastewater system:

- Low operation and maintenance requirements (as simple a system as possible), particularly given the seasonal use and rural setting of the facilities
- The most cost-effective system (on a long-term cost basis) that will meet the project's needs and applicable permitting requirements
- Long useful service life for the system
- Ability to power any electromechanical components associated with the wastewater system with solar/photovoltaic power (panels to be mounted on the large dining hall roof and indoor recreational facility)

- Prevent water quality degradation to a river bounding one side of the property (campers and visitors hike and play in that area)
- Aesthetic compatibility of the system with the camp and surroundings
- Desire to not take land areas out of use for the purposes of wastewater service, such as with surface irrigation of effluent, along with concerns about potential health risks associated with children venturing into those areas

Local Regulatory Authority Concerns and/or Focus

The only requirement here was that the system comply with applicable state and local rules, and design and construction criteria.

Physical Site Evaluation

A conceptual site layout was developed by the owner's architect, showing tentative locations of cabins, roads, and other facilities. Upon visiting the site, several areas relatively near the facilities served were selected for excavating profiles holes, and holes were dug for soils evaluations. These areas were located along the hillsides near where the cabins, dining hall, and other buildings were situated to minimize piping or conduit and pumping distances. Very shallow depths of class III soil overlying very rocky weathered limestone and shallow limestone rock were encountered in all of the profile holes dug in areas near the facilities.

The engineer discussed these findings with the owner, and asked if there might be other areas available for final soil treatment/dispersal. The owner suggested checking an area across a creek from the buildings, and in a pasture used for skeet shooting by campers in the summer. Profile holes were dug in that area, with significantly deeper soils encountered in each of several profile holes. Soils were again found to be class III (sandy clay loam with pockets of powdery white caliche), but to an average depth of about 32 in (0.8 m), overlying weathered bedded limestone with pockets of rock and gravel. Harder fractured limestone rock was encountered at a depth of about 4 ft (1.2 m) on average. No signs of mottling or groundwater were encountered.

This potential dispersal field area was very large, with very flat slopes (less than 2 percent), with more than adequate setback distances available from all property lines, drainage-ways, and site improvements. The area was found to be well drained, with no signs of recent or historical flooding.

Recommended System Selection

Based on the findings for the profile hole examinations and overall site assessment, it was determined that the skeet range would be very well suited for a subsurface effluent dispersal field. Based on the depth of class III soils, a relatively low maintenance low-pressure-dosing (LPD) dispersal system could be installed with trench depths of 16 to 18 in (0.41 to 0.46 m), leaving an average of 14 to 16 in (0.36 to 0.41 m) of undisturbed soil for natural soil treatment beneath the trenches for primary treated effluent. State and local rules required only 12 in (0.3 m) of suitable soil beneath LPD trenches, with this system able to exceed those requirements by several inches of soil depth in the proposed dispersal field area. This approach meets the recommendations in Chap. 9 for soil textural categories, vertical separation distances, predispersal treatment levels and methods of effluent dispersal, with a subsurface effluent application rate of 0.2 gal/day · ft^2 (0.8 cm/day) to be used.

It was determined that, particularly considering rocky conditions in the areas where the buildings were to be built, an effluent collection system should be used, with septic tanks serving each of the facilities. Primary treatment would be provided near each of the buildings, with small diameter effluent lines leading to a common effluent lift station that would then pump effluent up a small hill and across a bridge to the subsurface effluent dispersal field area. Based on a topographic survey of the entire site, a collection system layout was developed allowing almost all of the cabins and facilities to be served by gravity effluent collection lines, with only one building needing an effluent pumped connection leading to one of the main effluent gravity collectors. The collection system would therefore be a hybrid STEP/STEG configuration, with a common effluent lift station transferring all effluent to the field dosing station to be located near the skeet range.

Due to the potentially high levels of oil and grease generated by the dining hall on a seasonal basis, and to avoid excessive excavation into rock for exterior grease separation capacity, it was determined that an interior under-the-sink grease separation unit would be installed in the commercial kitchen (similar to unit in Fig. 5.6). Owners and management staff were informed by the engineer that they would need to make sure that kitchen staff were advised of the need to empty the unit on a daily basis as a routine kitchen cleaning activity, and to ensure that its capacity would not be exceeded during the course of a day's cooking and cleaning activities. In conjunction with the interior grease separator, the septic tank serving the kitchen would be oversized to provide additional oil and grease separation.

A 7000-gal (26.5-m^3) two-compartment fiberglass (FRP) tank was selected having a relatively large length to width ratio (almost 4:1), with a height of 6 ft (1.83 m) for serving the dining hall kitchen. Based on the projected daily flows for the kitchen, the tank would provide a hydraulic retention time of about 3.1 times the daily flow, with the relatively long tank length intended to help grease flotation and some added distance for cooling as wastewater enters the tank. Wastewater from the dining hall restrooms was to be handled with two smaller single compartment tanks in series on the other side of the building, due to grades and the location of the restrooms relative to the kitchen.

While FRP tanks would likely offer significantly longer service lives than concrete tanks due to their corrosion resistance for serving cabins and other facilities with smaller flows, high-quality 1000 to 1500 gal (3800 to 5700 L) FRP tanks were not available locally, and would be much more expensive to transport to the site. FRP tanks would also cost more per unit volume of capacity due to the low local costs and ready availability of concrete products. The engineer located a manufacturer of high-quality and thicker-walled precast concrete tanks within approximately 45 minutes trucking from the site. The owner was informed of the options, and decided that concrete tanks should be used (on behalf of lower capital costs) for all septic tanks serving the cabins and other facilities except for the large FRP dining hall kitchen tank.

The proposed system would have minimal power requirements, and could easily be powered with a photovoltaic source of electricity, particularly with the large roof areas available for installing solar panels, and ample battery storage space also available. There would be a total of three effluent lift/pumping stations, including (1) single STEP connection, (2) effluent transfer station across a bridge from the dispersal field, and (3) dispersal field dosing pumps.

Implemented System

Below are a series of photos illustrating key elements of the system.

Figure 11.1 shows bolt-down, gasketed, and watertight fiberglass access riser lids over septic tanks serving cabins along one of the roads through the camp. Figure 11.2 shows the large FRP tank serving the dining hall being filled with water, with piping temporarily run from the building, to test for watertightness.

Figure 11.3 shows a vented riser lid with a carbon filter pack used for tanks located at certain points along the gravity collection line. This will ensure that the effluent lines are adequately vented and no odors generated. The carbon filter packs in these lids must be periodically replaced. These were used in conjunction with normal roof-mounted plumbing vents on each of the buildings served. Figure 11.4 shows a heavier duty riser enclosure cast into a concrete collar over a tank located where there were grazing livestock. The bolt-down gasketed fiberglass lid offers watertightness, with the cast iron collar and lid providing protection against potential damage from surface activities.

Figure 11.5 shows the LPD trenches being excavated, and drain rock, piping, and geotextile fabric placed and covered one trench at a time. Since trenches were cut with a trencher, this prevented soil from filling adjacent cut trenches prior to placing rock, pipe, and fabric during trenching of adjacent LPD trenches. Work proceeded in one direction, avoiding the need for excavation or loader equipment from

FIGURE 11.1 Primary treatment/septic tanks are installed here with bolt-down watertight lids, serving each of the buildings with the effluent collection system and LPD subsurface dispersal system.

FIGURE 11.2 The large FRP tank serving the dining hall kitchen is being filled with water and tested for watertightness here prior to final backfilling to finished grade.

Figure 11.3 Vented bolt-down riser lid with carbon filter (top of photo). Nonvented lid in foreground.

Figure 11.4 Heavy-duty access riser lid enclosure for areas with livestock, or locations with heavier surface loads.

having to drive over excavated or completed trenches. Figure 11.6 shows two of the four 3-port Hydrotek distributing valves leading to six of the twelve LPD field zones. Valve boxes housing check valves and pressure-setting gate valves following these two Hydroteks are also shown here prior to placing final soil cover/backfill. Each of the Hydroteks (and other valves) is bedded in pea gravel and drain rock, with risers and lids brought up to final grade for access and servicing as needed. Figure 11.7 shows the LPD trenches completed and ready for placement of final soil cover and seeding to establish native grasses.

FIGURE 11.5   LPD Trench excavated in one of six field-dosing zones.

FIGURE 11.6   Two 3-port hydromechanically actuated distributing valves serving each of six field-dosing zones.

Figure 11.7 LPD field zones after placement of final soil cover, and before vegetative cover is established.

Routine operation and maintenance of the system entails checking effluent filters for need of cleaning at least once annually, at which time sludge and scum levels in tanks are checked to determine if pumping is needed. Other system checks associated with septic tanks, grease traps and low-pressure-dosing dispersal systems listed in the operation and maintenance guidelines posted on this book's website are to be done at least once annually. For this system, those checks and maintenance activities should be done at the end of the busy summer season.

11.2.2   Example 2: Single Family Residence

A single family residence was to be remodeled and enlarged, precipitating the need for an upgraded onsite wastewater system. Based on an inspection of the existing system and applicable state and local rules, it was determined that a new replacement system would be needed.

Facilities Served and Design Flow

The main house would be the only facilities served by this system. Based on the square footage and numbers of bedrooms, and the intentions of the owner to use only water-conserving fixtures, the design flows were determined to be 360 gal/day (1363 L/day).

Geophysical Setting

The residence was located along a steep bluff overlooking a recreational lake in the southwestern United States. The average year-round temperature for the area is about 68°F (20°C), ranging from freezing or below-freezing temperatures in the winter months to over 100°F (38°C), in the summer. Average annual rainfall is about 32 in (81 cm) per year. Intense thunderstorms are relatively common to the area, with occasional extended periods of rain, particularly in May-June and September-October.

Owners' Expressed Priorities

The owners expressed the following priorities for the new wastewater system:

- Energy efficiency (they wished to power the system with a new photovoltaic power system to be installed for the rebuilt home).

- Low operation and maintenance requirements; the owners expressed particular concerns about what they had heard from other nearby property owners regarding the use of and relatively high service needs of aerated tank package treatment units.

- Protection of local water quality and the environment.

- The use of xeriscaped vegetation as much as possible over treatment and dispersal field areas.

- Long useful service life for the system.

- The most cost-effective system that will meet the above and applicable requirements.

Local Regulatory Authority Concerns and/or Focus

- Protection of water quality for the lake adjacent to the property, especially with regard to pathogens from human sources around the lake

- Compliance with applicable regulatory/permitting requirements

Physical Site Evaluation

The lot's available area (outside of the footprint of the remodeled home, and considering required setbacks from roads, structure's property lines, significant grade breaks, and other features) was determined to be sufficient for a fairly wide array of system options as long as space was used efficiently. Except for a small garden area to potentially use for final effluent disposition with a slope of approximately 23 percent, all of the lot area available for the wastewater system had very flat slopes.

A total of four soil profile holes were dug around two areas that appeared to be the most suitable for locating an effluent dispersal field. The results of those investigations showed class III soils to a depth of at least 18 in (0.46 m) in the flatter area, underlain by approximately 12 in (0.3 m) of weathered bedded limestone and caliche intermixed with some pockets of rock and gravel. Fractured limestone rock was encountered at a depth of 30 in (0.76 m). In the more steeply sloped small garden area, class III soils intermixed with pockets of caliche were found to a depth of 22 in (0.61 m), underlain by at least 16 in (0.41 m) of weathered bedded limestone. Fractured rock was encountered at a depth of about 38 in (0.97 m). All of the lot was well-drained, with no signs of groundwater or soil mottling in the profile holes. The site was high above the lake and out of the potential floodway.

Recommended System Selection
Due to the proximity to, and drainage of surface waters from the entire site toward the lake, it was determined that surface application of treated effluent would not be a feasible option. Further, based on local rainfall and evaporation rates, the land area needed for surface application would be significantly greater than that needed for subsurface application in class III soils.

LPD trenches with a maximum depth of 18 in (0.46 m) could be excavated in the flatter portions of the site for part of the subsurface dispersal area, leaving 12 in (0.3 m) of weathered bedded limestone above fractured rock. In the more steeply sloped garden area, LPD trenches could be hand-excavated to a depth of 12 in (0.3 m), leaving a depth of 10 in (0.25 m) of class III soils over another 16 in (0.41 cm) of weathered limestone beneath the trench bottoms for further treatment.

Monitoring data gathered for low-pressure-dosed systems in these types of soils (as discussed in Chap. 9) shows that while high levels of nitrogen removal can be accomplished with that dispersal method, pathogen reduction may be somewhat limited, and vary greatly by location even within a single site. Based on the unreliability of these soils for offering adequate pathogen reduction, it was determined that treatment for that limiting design parameter (LDP) should be provided.

Figure 11.8 shows a bluff line very near this site with essentially the same subsurface conditions as those found in the profile holes excavated for this project. A to-scale shallow LPD trench cross-sectional view is superimposed on the soil profile view to illustrate the use of that method of dispersal in these conditions. LPD trenches would, of course, stop well short of reaching grade breaks like this, in keeping with necessary and required horizontal setback distances from critical features. The black horizontal line just above the layer of fractured limestone rock represents the vertical limits where significant natural treatment would be expected to occur. Figure 11.9 shows a close-up view of that same section of soil with the trench cross section. Notice that the darker soil having higher organic content extends

Figure 11.8   Class III soils overlying weathered bedded limestone ("caliche") soils. Layer of hard fractured limestone located below the black horizontal line.

Figure 11.9   Closer view of shallow LPD cross section, with monitoring data and expected performance of LPD systems in these conditions as discussed in Chap. 9.

several inches below the trench bottom, and is still well within the plant root zone, offering moisture and nutrient uptake along with further bacterial treatment processes. With the use of relatively narrow 6 to 8 in (15 to 20 cm) wide trenches, the sidewalls of trenches are more exposed to effluent percolating over the drain rock media, and offer better conditions for moisture and nutrient uptake by surface vegetation.

Since neither UV disinfection nor chlorination can be used effectively for septic tank effluent, at least secondary treatment would need to be provided for pathogen reduction. Drip dispersal was considered due to the ability to bury lines in trenches excavated to somewhat shallower depths than LPD trenches. However, if drip dispersal were used there would be added routine operation and maintenance requirements along with the need for at least secondary treatment, as compared with using LPD distribution.

Based on all of the above considerations, treatment using an intermittent sand filter (ISF), and followed by shallow LPD subsurface dispersal was recommended. The following considerations provided the basis for that recommendation:

- Predispersal treatment with an ISF designed and built in accordance with material presented in Chap. 7 has been shown to reliably produce effluent with average fecal coliform levels of less than 200 colonies/100 mL.

- An ISF would reliably reduce total nitrogen levels by 25 to 30 percent.

- A raised ISF could be constructed on this site and be used as a xeriscaped raised garden, as long as surface vegetation was carefully selected and deeper rooted or woody plant species avoided.

- ISFs need only one routine maintenance/service call per year for long-term continued operation, which would also be sufficiently frequent for an LPD dispersal system. A properly designed ISF would be expected to last at least 30 to 40 years.

- An LPD field preceded by ISF treatment would be expected to last for many years, and essentially indefinitely for the field itself. Pumps, controls, and valves would need periodic replacement over time, but those costs are very minor relative to entire system.

- ISFs are one of the most energy efficient methods, if not the most energy-efficient method, of treatment capable of reliably producing effluent with very low effluent levels of BOD (<5 mg/L), TSS (<5 mg/L), and TKN (<2 mg/L), along with very good levels of natural pathogen reduction.

- Pathogen reduction through sand filtration avoids either dosing soils with chlorinated effluent or needing to dechlorinate effluent before dispersal. And it avoids the power usage and frequent maintenance associated with UV disinfection.

A loading rate of 1.0 gal/day · ft² (4 cm/day) was used for the sand filter. A conservative LPD field soil application rate of 0.2 gal/day · ft² (0.8 cm/day) was used to help ensure no adverse impacts from nitrogen or pathogen levels to the adjacent lake. The local regulatory authority felt that no increased loading rates following intermittent sand filtration were justified for this system due to its location in a sensitive watershed. The owners felt that the recommended system met all of their objectives, and would offer the best combination of treatment and dispersal options.

Implemented System
The septic tank, pump tank, and sand filter treatment units were installed in keeping with methods and materials discussed in Chaps. 5 and 7, respectively. A fiberglass tank was used for the septic tank, due to its much greater corrosion resistance and much longer service life as compared with precast concrete. A precast concrete tank was used as the LPD field dosing tank for the ISF effluent. A third compartment of the FRP tank was used to dose the sand filter, and the pump vault was equipped with an effluent filter. No effluent filter was used for the LPD dosing pump, due to the very high quality of effluent entering that tank.

To avoid excessive excavation for the ISF, a raised bed was used, and is shown during construction in Figs. 7.29 and 7.30 in Chap. 7.

Shallow 6- to 8-in (0.15- to 0.2-m) wide LPD trench lines were installed in a flat area beside the sand filter, and trenches manually dug in the smaller garden area to a depth of 12 in (0.30 m). Figure 11.10 shows the trenches dug, rock media placed, and lateral lines drilled and placed alongside the trenches in the steeper garden area.

Sandy loam top soil was used to cover both the ISF and field areas, with the flatter dispersal field area and the ISF xeriscaped with shallow-rooted native species. Due to the steeper slopes, and more shallow line burial depth for the small garden area, grass cover was established in this area to avoid erosion and have more solid vegetative cover, and to prevent root penetration into LPD lines. Grasses also tend to provide better nutrient uptake than many other ground cover species.

This system requires one full system check and servicing per year, performing the activities listed for septic and pump tanks, pump stations and controls, effluent filters and low-pressure-dosed dispersal on the book's website under operation and maintenance guidelines.

Following installation, sand filter effluent quality was monitored for this system during a period of over 2 years to verify adequate

Figure 11.10 Hand-excavated LPD trenches in a smaller and more steeply sloped garden area. Rock media is in place, and trenches ready for placement of drilled LPD lines (shown beside trenches), more rock media, geotextile fabric, and the sandy loam final soil cover.

levels of pathogen reduction. After the first few months of system start-up, and during the following 2-year monitoring period, fecal coliform levels averaged 87 and ranged from 2 to 293 colonies/100 mL. All samples were collected and analyzed in accordance with the then-current edition of *Standard Methods for the Examination of Water and Wastewater*, with analyses performed by a municipal wastewater utility laboratory.

11.2.3 Example 3: Safari Tent Camp

This project demonstrates the use of natural processes and recycled materials to provide a relatively high level of treatment in a remote location with very limited availability of power supplies or material resources commonly used elsewhere for wastewater treatment systems.

A luxury style safari tent camp is to be located in southeastern Africa in a wildlife reserve. The focus of the project is an eco-camp using a resource conservation approach, and which also offers essentially all the comforts of other luxury safari lodging facilities. Guest facilities will include toilets, showers, and dining for morning and evening meals (lunches will be prepared for guests and taken on daily outings). The project is intended to integrate comforts preferred by many tourists, with sustainable building practices and resource conservation. The region is very dependent on tourism for continued support of wildlife preservation and local economies. Visitors to the camp will be provided with educational information about the wastewater system, to encourage the use of more sustainable approaches and reuse of waste materials elsewhere.

Each tent and support facility contributes to the camp's water supply with gutters draining to common cisterns. Due to sometimes prolonged drought conditions in the region, a large amount of cistern capacity will be included in the camp facilities. Solar hot water heaters and pumps are to be used, along with photovoltaic lighting and electric supply for guest and support facilities. A limited amount of PV battery supply is to be provided for use during wet and/or cloudy periods.

Facilities Served and Design Flow
Facilities to be served by the system include

- Twelve luxury tent sleeping accommodations (two persons each), each with toilet, sink, and shower
- Staff sleeping tents, with shared showers and toilet facilities
- Kitchen and dining facilities

All of the plumbing fixtures will be very low flow and water conserving. Rainwater collection and cisterns will be used for the water supply. Peak visitation design flows were estimated to be 1200 gal/day (4542 L/day). That estimate was based on relatively conservative (high) daily usage assumptions, since visitors to the camp would have variable water usage habits and preferences.

Geophysical Setting
The camp is to be located in a relatively dry plains region with gently rolling hills, near a pond that serves as a critical drinking water supply for local wildlife. Proximity to the equator gives very modest variations in temperature year round, with averages ranging from about 80 to 90°F (27 to 33°C). Annual rainfall averages about 10 to 15 in (25 to 40 cm), with two "wet" seasons. Figures 11.11 and 11.12 show the setting for this project.

Owners' Expressed Priorities
The following were identified as priorities for the design of the system:

- Maximum use of local natural resources and recycled materials
- Use of methods not requiring electric power supply; PV supplies and battery storage capacity would be sufficiently limited that a strong preference was expressed for the wastewater system to not rely on that power supply.

FIGURE 11.11 This plains area of southern Africa is relatively dry, with grasslands and often only sparse stands of trees. (*Photo courtesy of Heidi Fischer.*)

FIGURE 11.12 Wildlife depend on water from the pond located downhill from the proposed safari camp. (*Photo courtesy of Heidi Fischer.*)

- Reliable wastewater treatment method capable of protecting local water supplies (nearby pond and groundwater supplies), and periodically recharging the pond during prolonged dry periods

- Approach that would not result in any objectionable odors or unsightly conditions for guests

- Very low ongoing maintenance needs; no need for specially skilled operators or service providers

- Long useful service life

- Prevention of any adverse impacts to local wildlife

Local Concerns and/or Focus

- Wildlife/environmental and public health protection

- Demonstration of sound methods of wastewater treatment

Physical Site Evaluation

Soils on the site where the camp was to be located were determined to consist of one of the following: a sandy clay loam, sandy clay, or clay. Plains grasses were the predominant ground vegetation, with greater numbers of trees located near the pond. A well-drained area along a gently sloping hillside was identified as the most suitable location for the camp and associated facilities. No presence of groundwater or soil mottling was found in the profile holes dug in the proposed dispersal field area.

Recommended System Selection

Any wastewater effluent used to supply the pond with water during dry periods would need to be treated to at least a secondary or advanced level, with some method of pathogen reduction. Sand/media filtration offers the best natural approach for this, along with good nitrification that would prevent odors (associated with ammonia otherwise present). Due to the presence of clay soils and the need to prevent buildup of biomat clogging layers in a final effluent dispersal field, some method of secondary or advanced treatment would also be beneficial in that respect. This was considered particularly important given (1) the relatively large field area needed for the system based on flows, and (2) the inability to use traditional low-pressure dosing to achieve more uniform effluent distribution.

Due to the distance from the camp to regional population centers for sources of precast concrete products and aggregate supplies, it was recommended that fiberglass tanks be used for primary treatment. These tanks could be trucked to the site along with other building supplies and materials, and placed with relative ease as compared with much heavier concrete products.

Two methods of secondary/advanced treatment in combination, following primary treatment, were proposed that could reliably and consistently achieve a high quality of effluent without needing a source of electric power. A subsurface flow wetland followed by an intermittently dosed sand filter would provide for an estimated 25- to 30-percent total nitrogen reduction along with pathogen reduction by predispersal treatment processes. Natural soil treatment processes in these relatively tight soils with high organic content would provide enough nutrient removal for protection of groundwater and surface water supplies. Moisture retention in the upper soil horizons would contribute to nitrification-denitrification processes. The evapotranspiration potential in this region is very high, which would contribute to nutrient uptake by vegetation.

The two potential disadvantages noted in Table 6.3 associated with both of these natural treatment methods—land area requirements and climate effects—would not be a factor for either system in this geophysical setting. A tire chipping facility and a waste glass crushing operation were located that could offer sources of media that could be used for those two treatment processes. The proposed system would consist of

- Primary treatment in two 1500 gal (5678 L) single compartment fiberglass tanks, with the outlet of the second tank fitted with an effluent filter sized for the peak design flow of 1500 gal/day (5678 L/day).

- Gravity drainage from the second septic tank to the head of the subsurface flow wetland.

- The wetland will drain to a 500 to 1000 gal (1900 to 3800 L) level control tank in which a Flout™ effluent dosing device will be installed. The Flout will dose a media filter located downhill from the wetland and dosing tank. The dimensions of the wetland will be approximately 30 × 60 ft (9.1 × 18.2 m). There is more elevation and flow flexibility with a Flout as compared with a dosing siphon. A Flout will also not be subject to potential problems such as losing its prime or "trickling," and should be easier to install in the field as long as the dosing tanks are cast properly.

- The dosed media filter will drain to a second 500 to 1000 gal (1900 to 3800 L) effluent tank fitted with a Flout, which will dose a series of subsurface dispersal trenches located at the base of the hillside. The media filter dimensions will be approximately 30 × 40 ft (9.1 × 12.2 m), with a filtration depth of 3 ft (0.91 m).

- Subsurface dispersal trenches will also use chipped tires for drain media. Trenches with widths of about 1 ft (0.3 m) will be excavated along ground contours to a total depth of approximately 2 ft (61 cm). A fairly flat area was selected at the base of the hill having sandy clay loam soils, and better infiltrative capacity than areas where the wetland and sand filter would be located. The total length of trenches spaced 4 ft apart on center will be 1000 ft (305 m), using a soil application rate of 0.4 gal/day · ft² (1.6 cm/day). Field laterals will be center-fed from the Flout dosing device, and be limited to lengths of about 83 ft (25 m). Three Flouts will each center feed four lateral lines of that length.

The wetland will reduce suspended solids and BOD to a level that should better protect the media filter from clogging and failure over time (10 to 20 mg/L BOD and TSS), as compared with dosing septic tank effluent onto the filter. That will be particularly important given that only a limited amount of pressure head from elevation differences will be available for effluent distribution across the media filter (using the Flout devices). Both the wetland and media filter will be located in relatively flat, terraced areas along the hillside, downhill from the facilities served.

System Implementation

Due to the absence of appreciable rock in the upper soil horizons, construction of the system can be carried out using equipment commonly used for construction, including a backhoe and excavator. Ground glass can be used as a bedding material for the tanks in addition to its use for the media filter. Dispersal field trenches can be installed using a 12-in (0.3-m) digging bucket mounted on the excavator.

To avoid having to synthetically line the wetland or construct a containment structure, it was recommended that the system be configured so that the wetland would be located on an area of the site with extensive clay soils. The excavation will be lined with geotextile fabric and then PVC plastic liner (10 mil is sufficient) to prevent washing of sediments from the bottoms and sides of the wetland excavation into the media. It will not, however, need to be watertight, with the primary function of the PVC plastic to prevent buildup of biomat along the base and sides of the wetland. Rounded river rock will be used to line the bottom and sides of the excavation, and anchor the geofabric and plastic liner with a protective barrier that will prevent puncturing of the fabric with any wires that may protrude from the chipped tires. A total of about 33 yd³ (25 m³) of 3 to 6 in (80 to 150 mm) river rock would be needed for this. Only about two large end-dump trucks (each with about 17 yd³ or 13 M³) of pea gravel would be needed for the wetland by using the chipped tires for the treatment media. This would be placed just above the chipped tires, at the surface of the wetland. A course of 1- to 2-in (25- to 50-mm) rock approximately 2 to 3 in (5 to 8 cm) deep would be placed along the top surface of the tires to keep the pea gravel in place above the tire media, and provide a barrier between any protruding tire wires and pea gravel for workers when vegetation is planted in the wetland. The wetland would be planted with native reed species that are relatively deeply rooting, such that the roots will penetrate to the base of the wetland.

Crushed glass having a gradation comparable to intermittent sand filter media discussed in Chap. 7 will be trucked to the site for use in the media filter. Drain rock and pea gravel will be needed at the base around the filter underdrains, and drain rock placed beneath and above the distribution piping as detailed in Chap. 7. The spacing of distribution piping over the filter will however need to be closer, with 18 in (0.46 m) on center spacing used for both the lateral/distribution line and orifice spacing. The closer spacing will be used because much less pressure head will be available for effluent distribution and line scouring/clearing. Orifices will be directed upward into the filter so that lines will remain fully charged between doses to help with better distribution. Orifices will however be directed downward in the subsurface dispersal field trenches, enabling those distribution lines to fully drain at the end of each dosing cycle. This was expected to help clear lines of any fines exiting the filter.

The excavation for the media filter will be constructed in the same manner as the wetland, in soils consisting of primarily clays. Although some effluent infiltration will occur from the wetland and the glass media filter, both units will drain freely to the two dosing tanks. This, along with the thin PVC liner material at the base of the wetland and media filter, will minimize any opportunity for significant infiltration from the predispersal treatment units.

The Flout dosing devices will need to be shipped from the United States, but these are relatively small and lightweight items. A double Flout will be used to dose the sand filter, and a triple Flout for the subsurface dispersal field.

If a relatively local source of durable and watertight fiberglass tanks can't be found, these can be shipped to the nearest port and transported by truck to the site. The tanks to be fitted with the Flout dosing devices will need to have outlets located near the bottoms of the tank with a fabricated outlet connection cast into the tank, for connection of the Flout discharge piping. That connection will need to be water-tested due to hydrostatic pressure on the joint(s) between dosing events. Schedule 40 PVC piping and other materials needed for construction of the system should be available from the nearest cities.

During periods of drought, final effluent from the treatment system will be used to recharge the pond, with a bypass line connected to the discharge piping outside of the Flout-dosing tank for that purpose. There will be more than ample pressure head to divert treated effluent to the pond by switching valves just beyond the field dosing Flouts. Added media depth was planned for the media filtration unit to help ensure better treatment, including pathogen reduction for potential reuse of the effluent for the pond. Although nitrate levels will likely still be fairly high in the effluent leaving the filter and contributing to algae growth in the pond, the quality of effluent entering the pond will likely be far superior to many of the smaller stagnant ponds used by wildlife in the region during drier periods.

Strong protective barriers and fencing will be placed around all portions of the system having relatively shallow piping or vulnerabilities to damage from wildlife, including the wetland, sand filter and subsurface dispersal field. Tanks will be buried to a depth that allows for 2 ft of final cover for surface load bearing purposes, and watertight access lids will be given added protection using the approach shown in Fig. 11.4 (first project described in this chapter). Essentially all vulnerable elements of the wastewater system or other site infrastructure must be protected from damage by large wildlife including elephants. Vegetation will need to be selected for the subsurface flow wetland that doesn't particularly attract predominant local fauna, especially during dry seasons.

Kitchen cooking and cleaning practices will be used that minimize the entry of oils/fats and greases into the wastewater system. Staff at the camp will be trained on the functioning of the system, and how to perform periodic checks and routine servicing of components.

11.2.4   Example 4: Estate Home Development

This project involves a planned development for a small volcanic island, or cay, located in the northeastern corner of the Caribbean region. Access to the cay is by boat or helicopter only, with the cay located approximately 0.5 mi (805 m) off the coast of the nearest larger and fairly well developed island into which there is international air and boat traffic.

Facilities Served and Design Flow

Facilities served by the system (or systems) include

- Residences including a number of large estate homes and villas
- Staff and visitor housing
- A country club complex, including dining facilities, gathering rooms, a fitness center and a limited amount of overnight guest housing
- A yoga/spa center
- Shops and dock landing facilities, including smaller eateries
- A restaurant
- Nature center
- Two ocean side recreational areas including restrooms and light food service and small beaches

All of the plumbing fixtures in structures will be low flow and water conserving. Rainwater collection and cisterns or reverse osmosis will be used for drinking water on the cay, or a combination of those. Total design flows for all of the facilities at full build-out were determined to be approximately 54,400 gal/day (206 m^3/day).

Geophysical Setting

The eastern Caribbean region is tropical, with year-round temperatures typically ranging from about 70 to 90°F (21 to 32°C). Average annual rainfalls total about 40 in (102 cm). Monthly rainfall rates can vary greatly by season and year, with the most intense and prolonged rainfall events occurring during the hurricane season (July to November). Monthly totals may range from zero (during drier drought periods) up to 20 in (50 cm) or more. Historically most hurricanes have occurred during the months of August and September. Tourism and visitation in the region tend to drop off dramatically during the hurricane season, with the highest tourism tending to be between later November and April.

This cay is volcanic in origin, with land rising steeply to the top ridgeline of the cay from the shorelines around most of its perimeter. The region is characterized by frequent and sometimes strong earthquakes, with a quake of magnitude 6.3 recorded as recently as early fall of 2008. Although a photovoltaic power supply grid will be used throughout the development, the potential for power outages is significant in the region, particularly during periods of extended rainfall and cloudiness, and intense storms.

Certain coral species are present in large numbers along some portions of the cay's shoreline. Water quality protection in general, and in particular protection of sensitive coral species is seen as a key concern to the developers, to the public, and local regulatory authorities and policy makers.

Owners' Expressed Priorities

Based on discussions with the project owners and architects, the following were identified as priorities for the design of the wastewater system:

- Method of wastewater service suitable for a world-class development
- Approaches that would not take significant land out of use, and that would blend well with the aesthetics of the natural environment throughout the cay
- Use of approaches that would be relatively trouble-free, and not result in odors or other objectionable features
- Manageable operation/maintenance and routine service requirements, given the relatively remote setting

- Long useful service life for the system

- Ability to power any electromechanical components associated with the wastewater system(s) with solar/photovoltaic power supplies

- Prevent water quality degradation from the wastewater system to surrounding shorelines and aquatic resources

- Prevent adverse impacts to protected or endangered terrestrial floral and faunal species

- A cost-efficient method of service capable of meeting the above objectives

Based on available soil and general topographic surveys of the cay, it was determined on a preliminary planning basis that several clustered wastewater systems would be the most cost-effective and environmentally appropriate means of serving the development. The majority of the development would be located along the spine (top ridgeline) of the cay, with relatively few shoreline facilities planned. Elevations vary significantly along the cay's top ridgeline, and naturally drain in multiple directions. Serving the cay with multiple cluster systems would facilitate several things favorable to meeting the project objectives and priorities. Those include

- Ability to phase in wastewater system clusters at the same pace as development occurs in each area of the cay.

- Ability to avoid having to finance wastewater infrastructure well before capacity is needed for as yet undeveloped phases of the project.

- Ability to monitor flows and fine-tune treatment capacities relative to design flows and treatment performance, as individual clusters of systems are developed.

- Ability to intersperse park areas around the development that can also be used for final effluent disposition.

- Avoidance of management issues associated with use of many individual onsite systems, particularly in a seasonal and highly variable use setting.

- Avoidance of long runs of piping and larger lift stations needed to convey all flows to a single location, and potentially greater site disruption.

- Ability to maintain wastewater service to other clusters in the event problems occur with one of the systems.

- Avoidance of greater environmental, potential public health and water quality risks, and potential impacts associated with having to convey relatively large volumes of wastewater over longer distances.

- Ability to blend/combine a distributed wastewater service approach with low impact development needs and approaches.

- Avoidance of larger and more visible wastewater infrastructure needed to handle all flows from the development.

These considerations were discussed with the architects and owners, who agreed with the engineer to proceed using this approach.

Essentially the entire perimeter of this cay is characterized by either very steep slopes with thin and unstable soils, or rock bluffs. Figures 11.13 and 11.14 illustrate the two predominant types of terrain on the two sides of the cay's ridgeline. While some soil exists on the steeper slopes illustrated in Fig. 11.13, they are typical of tropical soils on volcanic islands, tending to be relatively high in organic content and unstable along slopes, particularly where natural vegetation is disturbed. It was therefore determined that only flatter areas along the cay's ridgeline should be used for final effluent dispersal (see Fig. 11.15).

Based on the total estimated design flows, several flatter areas along the ridgeline were identified as likely the most suitable locations for final dispersal of treated effluent. Surface application of effluent was ruled out due to the sensitivity of the surrounding shorelines and waters, and topography of the cay. Three areas of sufficient size to accommodate projected flows, stands of native and existing trees, gardens, points of interest, and other aesthetic features were

FIGURE **11.13**   Thin soils overlying rocky, unstable slopes.

FIGURE **11.14** Steep bluffs jutting from the water to the ridgeline along the entire length of the cay on one side.

FIGURE **11.15** Ground surface covered with loose rocks and dead wood under trees and scrub vegetation along the top ridgeline of the cay.

reserved along the top of the cay. Rough estimates of actual dispersal area needs were made using soil loading rates appropriate to soil classes identified for those areas from the soil survey. Land planning of roads and structures was then done around these reserved areas.

Regulatory and Permitting Authorities Concerns and/or Focus

- Protection of endangered and/or protected species (aquatic and terrestrial)
- Protection of water quality resources in general
- Meeting applicable local and national planning and implementation requirements
- Local cultural, socioeconomic, and aesthetic resources

Physical Site Evaluation

A site visit was made to investigate actual soil conditions within each of the three major areas reserved for final subsurface effluent dispersal. Prior to the site visit it was unclear whether soils had been mostly eroded along the cay's ridgeline, leaving only extremely rocky conditions. Wild goats on the cay had eaten essentially all of the grasses on the cay, along with most good ground cover vegetation, leaving mostly trees, vines, cactus, and other scrubland-type vegetation.

Consistent with the general topographic survey, relatively flat conditions were found in the three major dispersal field areas. Figure 11.15 shows surface conditions typical of those areas. The ground surface was found to be almost completely covered with large and small rock fragments, with vegetation having been removed and soils eroded from above the rock. Loose rocks on the surface were moved aside, exposing soil underneath. The rocks covering the surface were determined to be functioning as the primary means of natural erosion control throughout the top portions of the cay.

Since there were no docking facilities yet, and steep hikes to the top of the cay, only very lightweight tools and equipment could be carried and used to dig soil profile holes. A pick and shovel were used to dig holes as deep as possible in each of the three dispersal areas. Class III soils were found in two of the three areas, and class II soils in the third. Profile holes were dug with relative ease to depths of at least 18 in (46 cm), and usually deeper. Figure 11.16 shows the class II soils found in one of the three areas. No hard rock horizons were encountered in any of the profile holes.

Recommended System Selection

Based on the findings from the physical site evaluation, the overall wastewater service approach developed during conceptual/preliminary planning was determined to be the most appropriate.

FIGURE **11.16**   Class II sandy loam soils found in one of the three reserved dispersal field areas.

Due to concerns about potential leaching of nutrients through subsurface rock into shorelines, it was determined that predispersal treatment capable of reliably providing for at least 50-percent total nitrogen reduction along with very low levels of BOD and TSS would be needed. UV disinfection would be provided also to ensure adequate pathogen reduction. Effluent collection was determined to be the most energy efficient and environmentally protective means of sewering the cluster systems, and would also provide primary treatment at each building served. Considerations contributing to that conclusion included

- For buildings located significantly downslope of treatment and dispersal areas, this would allow pumping of only the liquid fraction of the waste streams, and avoid trouble calls, potential overflows and much higher maintenance frequency associated with grinder and/or raw wastewater pump stations.

- The greater tank capacity and reserve storage capacity in effluent collection systems as compared with grinder or raw wastewater lift stations offers much greater water quality protection during power outages.

- Watertight tanks, lids, and related components were considered critical to water quality protection.

- Smaller excavations and retaining walls could be used for primary settling tanks, rather than much larger excavations and greater disturbance associated with larger tanks serving multiple structures.

AdvanTex® recirculating textile media filters were determined to be a suitable method for accomplishing the above treatment goals. Advantages with using that treatment method for this project include

- Reliable performance for properly designed and implemented systems

- Very energy efficient relative to other commonly used processes treating to the same levels

- Effluent TSS levels are sufficiently low for efficient UV disinfection, which will be used in preference to chlorination in this type of sensitive marine setting

- Enclosed, contained unit, offering odor and vector control and protection from intense weather conditions

- Modular units, capable of being phased in with flows as needed

- Low sludge production relative to activated sludge processes

- Easy operation and maintenance, with fewer electromechanical components and lower routine service and operational requirements as compared with most other systems meeting comparable treatment levels

- Corrosion resistant components where components are exposed to the elements or wastewater gases

Effluent loading rates to AdvanTex treatment units would be held to a maximum of about 15 to 20 gal/ft$^2 \cdot$ day (61 to 81 cm/day) to ensure good nitrification levels and low effluent BOD and TSS. That will allow effective UV disinfection to occur and help reduce cleaning frequencies for UV units. The lightweight fiberglass (FRP) AdvanTex unit "pods" could be shipped long distances stacked inside each other in cargo ship containers, and assembled at the project site. Smaller fiberglass tank half-shells could also be shipped in that way, and assembled on-site by someone trained and experienced with those procedures. Larger-sized FRP tanks needing to be assembled at the factory must be shipped on cargo ship racks.

Based on the soils and site investigations, shallow low-pressure dosing was determined to be the most cost-effective and energy-efficient method of subsurface effluent dispersal that would meet the wastewater systems objectives in combination with the predispersal treatment systems. With relatively high evaporation potentials in the region, distribution lines placed at depths of 12 to 16 in (30 to 41 cm) in fields with suitable vegetative covering and proper grading should be able to provide for very good moisture and nutrient uptake. It is expected that nitrogen removal can be provided to essentially background levels, with nitrogen being a pollutant of principal concern for the shorelines surrounding the cay. Soil loading rates of 0.4 and 0.6 gal/ft² · day (1.6 and 2.4 cm/day) would be used for dispersal fields located in class III and class II soils, respectively.

System Implementation

Using the land plan developed by the architects and owners, clusters were laid out with combinations of STEP and STEG effluent collection systems, making use of gravity flow as much as possible. Clusters were configured based on projected timing for phasing in each section of the planned development. AdvanTex treatment units and recirculation tanks were located on site plans in relatively inconspicuous locations at the edges of or across roads from dispersal field areas, to be tucked along hillsides in terraced areas and behind retaining walls. Figure 11.17 illustrates this for a system shown during construction and prior to final backfilling along a relatively steep slope. Treatment units will be located close enough to field dosing tanks that the same control panel location can be used for both. A total of seven cluster systems were planned, ranging in flows from 4000 to 12,000 gal/day (15.1 to 45.4 m³/day).

Figure 11.17   Retaining walls used along steep slopes for locating this AdvanTex treatment system. (*Photo courtesy of Orenco Systems, Inc.*)

Remote telemetric monitoring of the treatment systems will be done along with a full-time operator available to perform routine system checks, maintenance activities and respond to any service calls. Recirculation rates will be adjusted if/as needed by the local systems operator(s) based on seasonal flow conditions and monitored treatment performance. All treatment units and pump stations will require backup power supplies (generators) along with enough reserve storage capacity in STEP and STEG tanks and those at treatment and field dosing stations for at least a certain amount of storage time (at least 1 to 2 days of low to modest wastewater production). Even using a self-contained photovoltaic power grid on the cay, there may be weather or system-related power supply interruptions or occasional shortages. All lift stations used to transfer effluent from STEP or STEG connections uphill to treatment and dispersal stations are to be equipped with backup power (generator) capabilities, due to the limited storage in those smaller effluent lift stations.

For the few facilities located along or near shorelines, extra precautions and fortification will be needed for primary settling tanks, risers, and lids to prevent leakage and overflows. Only duplex and/or redundant pump configurations will be used to avoid pump tank overflows in the event of a pump failure. For facilities located along shorelines, in addition to watertight tanks and lids like those shown for the beach area in Fig. 11.18 (Stinson Beach, CA), protective sea walls will need to be constructed on the water side of wastewater

FIGURE 11.18 Watertight wastewater tank access risers and bolt-down lids shown here at a Stinson Beach, CA public beach, exemplify the types of features needed for shoreline wastewater systems components on this cay project.

components for the eventuality of hurricanes and storm surges. The Stinson Beach, California, area shown in Fig. 11.18 does not, however, tend to experience the same frequency of intense seasonal storms common to the Caribbean region. Antiflotation structures will also be needed for primary settling and pump tanks located in lower lying areas for this project.

Erosion and sedimentation controls during construction are critical to the successful implementation of this wastewater system in a sustainable manner, along with all other construction and long-term management activities associated with this project. Several elements of a sound erosion/sedimentation control approach during construction would include

- Placement and continuous monitoring of appropriate and effective erosion/sedimentation controls before commencement of any construction activities

- Leaving as much existing vegetation and ground cover as possible

- Herding, removal and relocation of the wild goats from the cay

- Removal of ground cover, including both rocks and vegetation only where needed, and only for the specific area to be disturbed during the course of that day's activities

- Encouraging sheet flow of surface run-off from disturbed areas

- Reestablishing vegetation in disturbed areas as quickly as possible. Two grass species were identified from the region that would be suitable for dispersal field vegetation

Rocks cleared from areas can be used for certain erosion/sedimentation control measures downhill from disturbed areas, including those shown in Fig. 11.19a and 11.19b. Lines of bagged aggregate and sand should be placed downhill and adjacent to the type of structure shown in Fig. 11.19a because they will help prevent blow-outs. The shape of the bags will conform to the land surface and prevent channeling and short-circuiting of sediment laden runoff past the sedimentation control structures. Where sheet flow through those types of controls can't be maintained, check dams like those illustrated in Fig. 11.19(b) are used to slow the velocity of flow along drainage routes and prevent further erosion. Check dams will also naturally trap a certain amount of sediments from run-off.

Due to the relative unfamiliarity with this type of treatment system in the region along with the very sensitive geophysical conditions, a construction crew will be brought from the United States to install the systems having substantial experience with the installation of both AdvanTex and low-pressure dosing systems in similar rocky/hilly conditions. Persons formally trained with the assembly and

(a)

(b)

FIGURE 11.19   (*a*) Excavated rock from construction activities used for sedimentation controls, along with silt fencing. Bagged aggregate and sand placed just downhill of the silt fencing provide greater structural stability for steep slopes, and will reduce the potential for blow-outs or short-circuiting of sediment laden runoff waters. (*b*) Check dams like these will slow water velocities and trap sediments for drainage pathways along the steep slopes surrounding this cay.

installation of both AdvanTex units and FRP tanks are needed for the project. Most of the smaller-sized FRP tanks serving individual residences and smaller facilities will be shipped as half-shells and assembled as needed. More experienced members of the construction team will train local workers about the systems' details and operation as they are being installed. Local plumbers and electricians involved

with the installations will be trained in providing routine maintenance activities as well as trouble-shooting.

The owners of the development will assume long-term responsibility for the care and management of these wastewater cluster systems. All lots on the cay will have legally recorded affidavits stating
that the property is a part of a designated cluster system, and the
owner will be a party to a management/ownership agreement associated with that system. This affidavit will be added to the real property deed on which the cluster system is located. No properties will
be transferred to new owners without the new owners being advised
that the property is a part of a cluster system, and that the new owner
must be a party to that agreement. A full-time systems operations
manager working for the owners will oversee the long-term operation and maintenance of the systems.

All primary settling, dosing, and recirculation tanks will be placed
on an inspection schedule to determine pumping needs. When pumping is needed, a locally licensed pump truck will clean tanks and
deliver sludge/septage to a centralized treatment facility located on
the adjacent much larger and more populated island. Tanks will be
sized so as to avoid frequent pumping schedules (on average not
more than every 5 to 7 years), due to the need for pump trucks to
travel by ferry between the two islands.

11.2.5 Example 5: Part-Time Residence and Vacation Rental Home

This project is also located in the eastern Caribbean region, though on
an island with very different geophysical conditions. The property is
located along the coastline of a very flat, rocky island in the British
West Indies. The residence is to be used on a part-time basis by the
owner, and rented to guests when not used by the owner.

Facilities Served and Design Flow

The residence has three levels, with living space on the upper two
floors. Except for a portion of the area used for a concrete cistern, the
bottom level is to be left open to the air and used for parking due to
the likelihood of flooding, as described in the site evaluation information below. There are a total of three bedrooms, with potential additional overnight occupancy in other parts of the living areas. Occupancy is expected to range from two to six persons, and flows from
about 150 to 600 gal/day (568 to 1171 L/day), varying by a factor of
four. Usage will vary seasonally, with much less usage during the
hurricane season.

Geophysical Setting

Physical conditions for this project are very similar to the third example
above, except that the geophysical character of the island is quite

FIGURE **11.20**   Hard rocky coral formations are present along large segments of the island's shorelines.

different. This island was not formed from volcanic activity, and is a very flat island formed of coral and limestone. A quarry on the island produces limestone aggregates, and cement and concrete products are available on-island. Due to the low profile of the island, storm surges from hurricanes and seismic activity, including tsunamis, are an important planning factor here. Figures 11.20 and 11.21 show the terrain along the shore for this lot, which is typical of much of the island's shorelines.

Owners' Expressed Priorities

The owner of this project is a sustainable building designer and planner. Essentially every element of this project was conceived from the point of view of implementing a truly sustainable project. The owner's goal was to demonstrate the use of sustainable methods and materials for a residence located on an exceptionally difficult site that is also very vulnerable to a variety of natural forces, including rising water, hurricane force winds, and earthquakes. Basic criteria for this included

- Use of a very energy efficient treatment and dispersal system (the residence would be completely off-the-grid, and fully powered using a solar photovoltaic system, including battery storage)

Figure 11.21 Other portions of the property have sandy beaches like this one, and dunes next to the water's edge.

- Ability to withstand more intense hurricane and seismic events
- Long useful service life
- Low operation and maintenance requirements
- Prevention of adverse water quality and shoreline impacts from the wastewater system
- The most cost-effective system capable of meeting the above criteria

Regulatory and Permitting Authorities Concerns and/or Focus

- Protection of local water quality and resources
- Use of local labor resources as much as possible for projects
- Use of sound methods and materials
- Need to minimize sludge production due to limited capabilities on-island to deal with wasted/pumped sludge and septage
- Use of energy efficient systems

Physical Site Evaluation

The property consists of just over 1/2 acre (2200 m²) of land along the shoreline, with a small stretch of dunes lining the shore between the

FIGURE 11.22 Mildly sloped hill consisting of very hard coral rising up away from shoreline before dipping back down to about sea level away from the shoreline.

home site and the water's edge. Just beyond the dunes, the terrain turns to very hard, jagged coral as shown in Figs. 11.22 and 11.23. The topography is undulating, dipping back down to low-lying areas away from the shore. Over most of the site, the groundwater table is just below the ground surface, which is essentially sea level.

Recommended System Selection

It was determined that the wastewater system should be located on the back side of the house away from the shoreline for several reasons, including aesthetics (with ocean views central to site planning in the Caribbean), protection of intense wave action and storm surges, and greater water quality protection. Historical storm surges for the region were reviewed, and it appeared likely that at some point in the system's life the shoreline would be likely to experience a storm surge of at least 10 to 12 ft (3 to 3.7 m) around the house. The use of completely watertight predispersal treatment units would be critical to protecting shoreline water quality in the event of tidal or storm surges. It was recommended to the owner that the predispersal wastewater treatment units be located inside of a protective closed retaining wall, or closed sea wall, connected to the back side of the house. A structure adjacent to the house with a height above natural grade of about 14 ft (4.3 m) was planned, with the design of that structure to

FIGURE 11.23 Very hard jagged coral rock covered with cactus and sea cliff plants. The arrow points to some of the coral rock in this photo.

be prepared by the structural engineer on the project. Sufficient wall height would be needed to protect the fiberglass surface of both the AdvanTex unit and tank lids from damaging winds and debris during storms. It would also provide a wind block for someone to open and service the AdvanTex unit, given the very breezy ocean setting.

Septic tank pretreatment followed by treatment with an AdvanTex treatment unit would be used. This was selected for many of the same reasons as for the fourth example above, including enclosed/ protected treatment unit that would help prevent odors or aesthetic issues, and very small treatment system footprint as compared with other packed media processes capable of reliably providing comparable treatment levels. Corrosion resistance would also be critical to this marine setting.

Due to the presence of groundwater at or near the ground surface, final effluent dispersal and further natural treatment processes would be carried out in raised vegetated beds. The bottomless raised dispersal beds will be intermittently dosed on a demand basis to avoid possible alarm conditions when the residence is occupied by guests, and for simpler control settings. Raised beds will be located outside of but adjacent to the protective sea wall, with a total above-grade height of about 4.5 ft (1.4 m). A loading rate of 0.8 gal/day · ft² (3.3 cm/day) based on peak seasonal use of the house will be used,

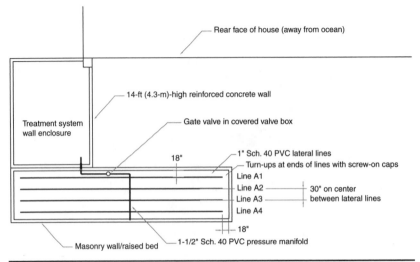

Rear face of house (away from ocean)

14-ft (4.3-m)-high reinforced concrete wall

Treatment system wall enclosure

Gate valve in covered valve box

18"

1" Sch. 40 PVC lateral lines
Turn-ups at ends of lines with screw-on caps
Line A1
Line A2
Line A3
Line A4

30" on center between lateral lines

18"

Masonry wall/raised bed

1-1/2" Sch. 40 PVC pressure manifold

FIGURE 11.24   Site layout of treatment system enclosure and raised dispersal bed adjacent to residence (not to scale).

with soils near the surface of the raised bed to be loamy sand. That should be sufficiently conservative because peak use of the residence will occur during the drier nonhurricane season months when few tropical storms are occurring. Much more moisture and nutrient uptake will be occurring in the raised vegetated bed during those periods.

System Implementation

A schematic of the system's layout as planned next to the residence is shown in Fig. 11.24. A more detailed view of the treatment enclosure is shown in Fig. 11.25. Piping on the system layout has been routed so as not to be exposed outside of either the treatment enclosure or the raised bed. Concrete antiflotation "deadmen" will be used for the tanks and AdvanTex unit. Tanks and the AdvanTex unit will be backfilled as shown in Fig. 11.26, with porous pavers placed along the surface so that service personnel can move around the units for checks and maintenance. An exterior masonry stairwell will lead to an opening in the top of the treatment enclosure next to the house.

The treatment units' structural enclosure/retaining walls will be designed so as to be able to provide external loading support for the FRP tanks buried in saturated soil (to be conservative). Recall from Chap. 5 that FRP tanks are designed and built primarily for external loading. It's therefore necessary to simulate backfilled conditions with the above-ground enclosure.

Sewer stub-outs from the residence will enter the walled enclosure directly from the house walls through sleeves cast into the wall of the structure. The pressure manifold leading from the field dosing basin to the raised final effluent dispersal bed will pass through a

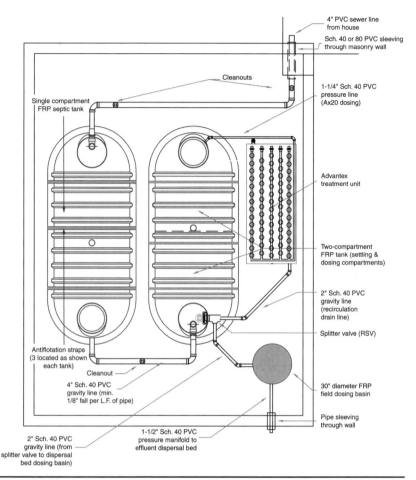

FIGURE 11.25 Plan view of treatment system and field dosing basin (not to scale; U.S. Customary units shown).

pipe sleeve between the treatment enclosure and the raised bed's wall, to avoid exterior exposure. Although due to its height, the raised bed will always be somewhat exposed to higher storm surges, those occasions should be relatively rare. Distribution piping inside the bed will be strapped to concrete anchoring placed at pipe intersections and ends of lines, to secure them in the event of rising water. The bed will have approximately 12 in (0.3 m) of "freeboard" space above the vegetated surface and below the top of the wall around the bed, which will tend to limit the extent that soil or sand would be washed out if water rises to a level below the top of the wall. Rock media placed around the distribution piping will also help anchor the pipe against disturbance. The concrete retaining wall structure around the treatment units will be anchored into the underlying rock to prevent vertical or lateral movement.

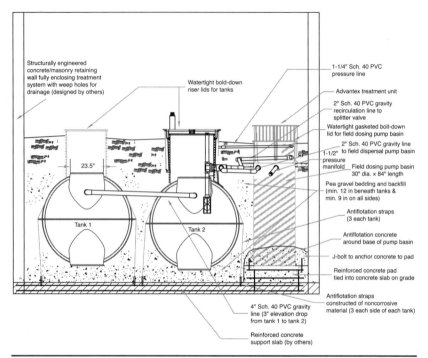

FIGURE 11.26 Cross-sectional profile/side view of treatment system and field dosing basin and piping (not to scale; U.S. Customary units shown).

Effluent distribution lines in the raised beds will be placed in manually dug trenches on 6 in (15 cm) of drain media. Approximately 12 in (0.3 m) of loamy sand will underlie the drain rock, with about 18 in (0.46 m) of coarser sand beneath the loamy sand. Note in the layout that distribution lines in the raised bed are placed at 30 in (0.76 m) instead of the typical minimum distance of 3 ft (0.91 m). The reason for that is the use of loamy sand underlain by sand.

In situations where very serious storms and potentially high waters are expected, the system will be shut down, with any flows to it temporarily stored in the combined reserve capacities of the dosing basin, recirculation tank, and septic tank.

As with the third example, due to local unfamiliarity with the assembly and installation of this type of system, a construction consultant experienced with all aspects of this system's installation, including FRP tank assembly, will oversee the installation. Again, the local plumber and electrician involved with the project will be trained on the operation and long-term care of the system. A local designated long-term service provider will take responsibility for the routine checks and maintenance needed, and respond to any trouble calls.

This system should require a minimum of two and as many as four routine service visits per year with the higher number possibly needed due to highly variable use conditions. Activities performed during those visits would include distribution line flushing and pressure checks, effluent filter cleaning, and other manufacturer-specified service activities for the treatment unit, along with other recommended operation and maintenance activities associated with septic tanks and low-pressure dosing provided on this book's website for this type of system. The system will be connected to remote telemetry and its operation tracked, with any adjustments needed to recirculation rates or other features made by the local service provider given instructions by either the engineer and/or the manufacturer.

For planning projects like this one in remote settings, it is always more cost-effective to try to coordinate multiple projects at once. In some cases, persons might *not* opt to use a particular sound and appropriate treatment system because they may have to individually pay for all or portion of a cargo container to transport the materials and equipment for just that one project. For this project example, because several AdvanTex units can be stacked inside each other, and similarly for smaller FRP tanks (up to 1500 gal or 5677 L), it offers significant cost benefits to coordinate planning and construction for several projects for which those same products are needed. Advance planning and coordination along those lines can sometimes be the difference between affordable and unaffordable ways of implementing the most sustainable wastewater service options for projects in such geographic settings.

Index